AF342607

Trends in Mathematics is a series devoted to the publication of volumes arising from conferences and lecture series focusing on a particular topic from any area of mathematics. Its aim is to make current developments available to the community as rapidly as possible without compromise to quality and to archive these for reference.

Proposals for volumes can be sent to the Mathematics Editor at either

Birkhäuser Verlag
P.O. Box 133
CH-4010 Basel
Switzerland

or

Birkhäuser Boston Inc.
675 Massachusetts Avenue
Cambridge, MA 02139
USA

Material submitted for publication must be screened and prepared as follows:

All contributions should undergo a reviewing process similar to that carried out by journals and be checked for correct use of language which, as a rule, is English. Articles without proofs, or which do not contain any significantly new results, should be rejected. High quality survey papers, however, are welcome.

We expect the organizers to deliver manuscripts in a form that is essentially ready for direct reproduction. Any version of TeX is acceptable, but the entire collection of files must be in one particular dialect of TeX and unified according to simple instructions available from Birkhäuser.

Furthermore, in order to guarantee the timely appearance of the proceedings it is essential that the final version of the entire material be submitted no later than one year after the conference. The total number of pages should not exceed 350. The first-mentioned author of each article will receive 25 free offprints. To the participants of the congress the book will be offered at a special rate.

Algebraic Cycles, Sheaves, Shtukas, and Moduli

Impanga Lecture Notes

Piotr Pragacz
Editor

Birkhäuser
Basel · Boston · Berlin

Editor:
Piotr Pragacz
Institute of Mathematics of the
Polish Academy of Sciences
ul. Sniadeckich 8
P.O. Box 21
00-956 Warszawa
Poland

e-mail: p.pragacz@impan.gov.pl

2000 Mathematical Subject Classification: 14-02

Library of Congress Control Number: 2007939751

Bibliographic information published by Die Deutsche Bibliothek. Die Deutsche Bibliothek lists
this publication in the Deutsche Nationalbibliografie; detailed bibliographic data is available in
the Internet at http://dnb.ddb.de

ISBN 978-3-7643-8536-1 Birkhäuser Verlag AG, Basel - Boston - Berlin

© 2008 Birkhäuser Verlag AG
Basel · Boston · Berlin
P.O. Box 133, CH-4010 Basel, Switzerland
Part of Springer Science+Business Media
Printed on acid-free paper produced from chlorine-free pulp. TCF ∞
Cover Design: Alexander Faust, CH-4051 Basel, Switzerland
Printed in Germany
ISBN 978-3-7643-8536-1   e-ISBN 978-3-7643-8537-8

9 8 7 6 5 4 3 2 1   www.birkhauser.ch

Contents

A tribute to Józef Maria Hoene-Wroński

Preface

The articles in this volume are an outgrowth of seminars and schools of Impanga in the period 2005–2007. Impanga is an algebraic geometry group operating since 2000 at the Institute of Mathematics of Polish Academy of Sciences in Warsaw. The present volume covers, besides seminars, the following schools organized by Impanga at the Banach Center in Warsaw:

- *Moduli spaces*, April 2005,
- *Algebraic cycles and motives*, October 2005,
- *A tribute to Hoene-Wroński*, January 2007.

More information about Impanga, including complete lists of seminars, schools, and sessions, can be found at the web-page:

$$\texttt{http}: //\texttt{www.impan.gov.pl}/ \sim \texttt{pragacz/impanga.htm} \ .$$

Let us describe briefly the contents of the lecture notes in this volume. [1]

Jean-Marc Drézet, in his first article, discusses fine moduli spaces of coherent sheaves, i.e., those endowed, at least locally, with universal sheaves. Whereas the most known fine moduli spaces appear in the theory of (semi)stable sheaves, the author constructs other, the so called "exotic" fine moduli spaces; the corresponding sheaves are sometimes not simple.

The subject of the second article of Jean-Marc Drézet is the study of moduli spaces of coherent sheaves on multiple curves embedded in a smooth projective surface. The author introduces new invariants for such curves: canonical filtrations, generalized rank and degree, and proves a Riemann-Roch theorem. A more detailed study of coherent sheaves on double curves is presented.

Tomas L. Gomez gives an outline of constructions of different moduli spaces. His starting point is the Jacobian of a smooth projective curve, and the final aims are moduli spaces of principal sheaves. A pretty complete account of the theory of principal bundles and sheaves is presented; a special emphasis is put on their stability properties. Orthogonal and symplectic sheaves serve as instructing examples.

---

[1]The lecture notes by J.-M. Drézet, T.L. Gomez, A.H.W. Schmitt, and Ngo Dac Tuan stem from the first school, the article by V. Srinivas from the second school, the opening article of P. Pragacz from the third school, and finally the articles by A. Langer, P. Pragacz, and that by P. Pragacz-A. Weber from the seminars of Impanga.

Adrian Langer gives a comprehensive introduction to torsion free sheaves and the moduli spaces of (semi)stable sheaves in any dimension and arbitrary characteristic. The author discusses carefully the (semi)stability conditions and restriction theorems. One of the main goals is to give the boundedness results, which are crucial to construct moduli spaces using the techniques of the Quot-schemes. Line bundles on the moduli spaces are also described, and generic smoothness of the moduli spaces of sheaves on surfaces is showed.

Piotr Pragacz discusses some topological, algebraic, and geometric properties of the zero schemes of sections of vector bundles, namely the connectedness and the "point" and "diagonal" properties. An overview of recent results by Vasudevan Srinivas, Vishwambhar Pati, and the author on these properties is presented.

Piotr Pragacz and Andrzej Weber generalize Thom polynomials from singularities of maps to invariant cones in representations of products of linear groups. With the help of the Fulton-Lazarsfeld theory of positivity of ample vector bundles, they show that the coefficients of Thom polynomials expanded in the basis of the products of the Schur functions, are nonnegative.

Alexander H.W. Schmitt gives an account of classical and new results in Geometric Invariant Theory (especially the theory relative to a base curve), and present a recent progress in the construction of moduli spaces of vector bundles and principal bundles with extra structure (called augmented or "decorated" vector or principal bundles). The problems of taking various quotients and stability conditions are widely discussed and illustrated by numerous examples.

Vasudevan Srinivas shows some applications of the intersection theory of algebraic cycles to commutative algebra. A special emphasis is put on the study of the groups of zero-dimensional cycles, modulo rational equivalence, on smooth projective or affine varieties (in particular, surfaces). Their applications to embedding and immersion of affine varieties, indecomposable projective modules, and the complete intersection property are given.

Ngo Dac Tuan presents a "friendly" introduction to shtukas, the stacks of shtukas, and their compactifications. The notion of a "shtuka" was first introduced by Drinfeld and used in his proof of the Langlands correspondence for GL_2 over function fields. It recently has been used by Lafforgue in his proof of the Langlands correspondence for higher groups GL_r over function fields.

We dedicate the whole volume to the memory of **Józef Maria Hoene-Wroński** – one of the most original figures in the history of science. The opening article by Piotr Pragacz discusses some aspects of his life and work.

Acknowledgments. The Editor thanks the authors for their scientific contributions, to Adrian Langer and Halszka Gasińska-Tutaj for their help with the school on moduli spaces, and finally to Dr. Thomas Hempfling from Birkhäuser-Verlag for a pleasant editorial cooperation.

Warszawa, July 2007   *The Editor*

Algebraic Cycles, Sheaves, Shtukas, and Moduli
Trends in Mathematics, 1–20
© 2007 Birkhäuser Verlag Basel/Switzerland

Notes on the Life and Work of Józef Maria Hoene-Wroński

Piotr Pragacz

To reach the source, one has to swim against the current.
Stanisław J. Lec

Abstract. This article is about Hoene-Wroński (1776–1853), one of the most original figures in the history of science. It was written on the basis of two talks delivered by the author during the session of Impanga "A tribute to Józef Hoene-Wroński"[1], which took place on January 12 and 13, 2007 in the Institute of Mathematics of the Polish Academy of Sciences in Warsaw.

1. Introduction and a short biography. This article is about Józef Maria Hoene-Wroński. He was – primarily – an uncompromising searcher of truth in science. He was also a very original philosopher. Finally, he was an extremely hard worker.

When reading various texts about his life and work and trying to understand this human being, I couldn't help recalling the following motto:

Learn from great people great things which they have taught us. Their weaknesses are of secondary importance.

A short biography of Józef Maria Hoene-Wroński:

1776	–	born on August 23 in Wolsztyn;
1794	–	joins the Polish army;
1795–1797	–	serves in the Russian army;
1797–1800	–	studies in Germany;
1800	–	comes to France and joins the Polish Legions in Marseilles;
1803	–	publishes his first work *Critical philosophy of Kant*;
1810	–	marries V.H. Sarrazin de Montferrier;
1853	–	dies on August 9 in Neuilly near Paris.

Translated by Jan Spaliński. This paper was originally published in the Polish journal *Wiadomości Matematyczne* (*Ann. Soc. Math. Pol.*) vol. 43 (2007). We thank the Editors of this journal for permission to reprint the paper.

One could say that the starting point of the present article is chapter XII in [6]. I read this article a long time ago, and even though I read a number of other publications about Hoene-Wroński, the content of this chapter remained present in my mind due to its balanced judgments. Here we will be mostly interested in the mathematics of Wroński, and especially in his contributions to algebra and analysis. Therefore, we shall only give the main facts from his life – the reader may find more details in [9]. Regarding philosophy, we shall restrict our attention to the most important contributions – more information can be found in [36], [37], [47], and [10]. Finally, Wroński's most important technical inventions are only mentioned here, without giving any details.

Józef Maria Hoene-Wroński
(daguerreotype from the Kórnik Library)

2. Early years in Poland. Józef Hoene was born in Wolsztyn on August 23[1] 1776. His father, Antoni, was a Czech imigrant and a well-known architect. A year later the family moved to Poznań, where the father of the future philosopher became a famous builder (in 1779 Stanisław August – the last King of Poland – gave him the title of the *royal architect*). In the years 1786–1790 Józef attended school in Poznań. Influenced by the political events of the time, he decided to join the army. His father's opposition was great, but the boy's determination was even greater. (Determination is certainly the key characteristic of Wroński's nature.) In 1792 he run away from home and changed his name, to make his father's search more difficult. From that time on he was called Józef Wroński and under this name he was drafted by the artillery corps. In the uprising of 1794 he was noted for

---

[1]Various sources give the 20 and the 24 of August.

his bravery, and was quickly promoted. During the defense of Warsaw against the Prussian army he commanded a battery – and was awarded a medal by commander in chief Tadeusz Kościuszko for his actions. He also took part in the battle near Maciejowice, during which he was taken to captivity. At that time he made the decision to join the Russian army. What was the reason for such a decision we do not know; while consulting various materials on the life of Wroński, I haven't found a trace of an explanation. Maybe – this is just a guess – he counted on the possibility of gaining an education in Russia: Wroński's main desire was a deep understanding of the laws of science, and these are universal: the same in Russia as elsewhere.... After being promoted to the rank of captain, he became an advisor of the General Staff of Suworow. In the years 1795–1797 he serves in the Russian army and is promoted to the rank of lieutenant-colonel.

3. Departure from Poland. The information about the sudden death of his father changed Wroński's plans. He inherited a large sum, which allowed him to devote himself to his studies, as he wanted for a long time. He quit the army and travelled West. Greatly inspired by Kant's philosophy, he arrived at Königsberg. However, when he found out that Kant is no longer giving lectures, Wroński left for Halle and Göttingen. In 1800 he visited England, and afterwards came to France. Fascinated by Dąbrowski's Legions, he asked the general for permission to join them. Dąbrowski agreed (however he did not honor the rank Wroński gained in Tzar's army) and sent him to Marseilles. There, Wroński could combine his service with his love for science. He became a member of Marseilles' Academy of Science and Marseilles' Medical Society.

In Marseilles Wroński underwent an enlightenment. This turning point in his life was a vision, which he had on August 15, 1803 at a ball on Napoleon's birthday. As he had described it, he had a feeling of anxiety and of certainty, that he would discover the "essence of the Absolute". Later he held that he understood the mystery of the beginning of the universe and the laws which govern it. From that time on he decided to reform human thought and create a universal philosophical system. In remembrance of that day he took the name of Maria and went down in history of science as Józef Maria Hoene-Wroński. Wroński's reform of human knowledge was to be based on a deep reform of mathematics by discovering its fundamental laws and methods. At the same time he posed the problem of solving the following three key issues in (applied) mathematics:

1. Discovering the relation between matter and energy (note Wroński's incredibly deep insight here);

2. the formation of celestial objects;

3. the formation of the universe from the celestial objects.

The most visible characteristic of Wroński's work is his determination to base all knowledge on philosophy, by finding the general principle, from which all other knowledge would follow.

Resources needed to publish papers have quickly run out, and Wroński started to support himself by giving private lessons of mathematics. Among his students was Victoria Henriette Sarrazin de Montferrier. The teacher liked this student so much, that in 1810 she became his wife. In September of the same year Wroński heads for the conquest of Paris.

4. Paris: solving equations, algorithms, continued fractions and struggles with the Academy. In 1811 Wroński publishes *Philosophy of Mathematics* [14] (see also [24]). Even earlier he has singled out two aspects of mathematical endeavor:

1. theories, whose aim is the study of the essence of mathematical notions;

2. algorithmic techniques, which comprise all methods leading to the computation of mathematical unknowns.

The second point above shows that Wroński was a pioneer of "algorithmic" thinking in mathematics. He gave many clever algorithms for solving important mathematical problems.

In 1812 Wroński publishes an article about solving equations of all degrees [15] (see also [20]). It seems that, without this paper, Wroński's scientific position would be clearer. In this paper Wroński holds that he has found *algebraic* methods to find solutions of equations of arbitrary degree. However, since 1799 it has been believed that Ruffini has proved the impossibility of solving equations of degree greater than 4 by radicals (Ruffini's proof – considered as essentially correct nowadays – at that time has lead to controversy[2] and the mathematical community has accepted this result only after Abel has published it in 1824). So did Wroński question the Ruffini–Abel theorem? Or did he not know it? As much as in the later years Wroński really did not systematically study the mathematical literature, in the first decade of the nineteenth century he has kept track of the major contributions. If one studies carefully the (difficult to understand) deliberations and calculations, it seems that, Wroński's method leads to approximate solutions, in which the error can be made arbitrarily small[3]. In his arguments besides algebraic methods, we find analytic and transcendental ones. This is the nature, for example, of his solution of the factorization problem coming from the work cited above, which we describe below. This type of approach is not quite original, it has been used by Newton for example[4]. Since Wroński considered this work so important (it was reprinted again towards the end of 1840) – in order to gain a true picture of the situation – it would be better to publish a new version with appropriate comments of someone competent, explaining what Wroński does and what he *does not* do.

---

[2]Ruffini published his results in a book and in 1801 sent a copy to Lagrange, however he did not receive any response. Legendre and other members of the Paris Academy did not consider this work as worthy of attention. Only in 1821 – a year before his death – Ruffini received a letter from Cauchy who wrote that he considers Ruffini's result as very important.

[3]It is interesting that the authors of [6] have reached a similar opinion, but without further details.

[4]Methods of Newton–Raphson and Laguerre are known.

I think that such a competent person could have been Alain Lascoux, who, reading this and other works of Wroński, could see, that he was addressing the following three algebraic problems, connected with polynomials of one variable and Euclid's algorithm for such polynomials:

1. Consider two normed polynomials $F(x)$ and $G(x)$. Suppose that $\deg(F) \geq \deg(G)$. Performing multiple division of $F(x)$ and $G(x)$:

$$F = * \, G + c_1 R_1, \quad G = * \, R_1 + c_2 R_2, \quad R_1 = * \, R_2 + c_3 R_3, \quad \ldots .$$

Successive coefficients " $*$ " are uniquely determined polynomials of the variable x such that

$$\deg G(x) > \deg R_1(x) > \deg R_2(x) > \deg R_3(x) > \cdots .$$

Instead of the "ordinary" Euclid's algorithm, were $c_1 = c_2 = c_3 = \cdots = 1$ and where $R_i(x)$ are *rational* functions of the variable x and roots of $F(x)$ and $G(x)$, one can choose c_i in such a way that the successive remainders $R_i(x)$ are polynomials of the variable x and of those roots. These remainders are called *normed polynomial remainders* or *subresultants*. Wroński constructed a clever algorithm for finding $R_i(x)$ (see [28], [29], [30]). Note that J.J. Sylvester has found other formulas for these remainders in [43] – although their validity has been established only very recently, see [31].

2. Using the algorithm in 1. and passing to the limit, Wroński [15] (see also [20]) also solved the following important *factorization problem*:

Suppose that we are given a normed polynomial $W(x) \in \mathbb{C}[x]$, which does not have roots of absolute value 1. Let

$$A := \{a \in \mathbb{C} : W(a) = 0, \, |a| > 1\}, \qquad B := \{b \in \mathbb{C} : W(b) = 0, \, |b| < 1\} .$$

Extract a factor $\prod_{b \in B}(x - b)$ from $W(x)$.

We give – following Lascoux [29] – Wroński's solution in terms of the *Schur functions* (here we use the definitions and notation for the Schur functions from [29] and [30]). The coefficients of the polynomial $W(x)$, from which we wish to extract a factor corresponding to roots of absolute value smaller than 1, are the elementary symmetric functions of $A \cup B$ – the sum of (multi)sets A and B. Therefore the problem boils down to expressing elementary symmetric functions of the variable B, in terms of the Schur functions of $A \cup B$, denoted by $S_J(A+B)$. Let the cardinality of the (multi)set A be equal to m. For $I \in \mathbb{N}^m$ i $k, p \in \mathbb{N}$ we define

$$I(k) := (i_1 + k, \ldots, i_m + k), \qquad 1^p I(k) := (1, \ldots, 1, i_1 + k, \ldots, i_m + k)$$

(where 1 is present p times). Let the cardinality of the (multi)set B be equal n. Wroński's theorem (in Lascoux's interpretation [29]) states that

$$\prod_{b \in B}(x - b) = \lim_{k \to \infty} \left(\sum_{0 \leq p \leq n} (-1)^p x^{n-p} \frac{S_{1^p I(k)}(A + B)}{S_{I(k)}(A + B)} \right)$$

(here I is an arbitrary sequence in $\mathbb{N}^m$). Notice that the solution uses a passage to the limit; therefore besides algebraic arguments, transcendental arguments are also used. One can find the proof of this result in [29]. Therefore, we see that Wroński, looking for roots of algebraic equation, *did not* limit himself to using radicals.

3. Assuming that $\deg(F) = \deg(G)+1$, Wroński also found interesting formulas for the remainders $R_i(x)$ in terms of *continued fractions* (see [30], where his formulas are also expressed in terms of the Schur functions).

We note that Wroński also used symmetric functions of the variables $x_1, x_2, \ldots,$ and especially *the aleph functions*.

$$\aleph[X_3]^{(1)} = x_1 + x_2 + x_3\,,$$

$$\aleph[X_3]^{(2)} = x_1^2 + x_2^2 + x_3^2 + x_1 x_2 + x_1 x_3 + x_2 x_3\,,$$

$$\aleph[X_3]^{(3)} = x_1^3 + x_2^3 + x_3^3 + x_1^2 x_2 + x_1^2 x_3 + x_2^2 x_3 +$$
$$+\ x_1 x_2^2 + x_1 x_3^2 + x_2 x_3^2 + x_1 x_2 x_3\,,$$

The aleph functions in three variables of degree 1, 2 and 3
from Wroński's manuscript

More generally, for $n \in \mathbb{N}$ we let $X_n = \{x_1, \ldots, x_n\}$ and define functions $\aleph[X_n]^i$ by the formula

$$\sum_{i \geq 0} \aleph[X_n]^i = \prod_{j=1}^{n} (1 - x_j)^{-1}\,,$$

i.e., $\aleph[X_n]^i$ is the sum of all monomials of degree i. Wroński considered these functions as "more important" than the "popular" *elementary* symmetric functions. This intuition of Wroński has gained – let's call it – justification in the theory of *symmetrization operators* [30] – in the theory of Gröbner bases – so important in computer algebra (see, e.g., [39]), as well as in the modern *intersection theory* in algebraic geometry [11], using rather *Segre classes*, which correspond to aleph functions, than *Chern classes*, corresponding to elementary symmetric polynomials. Here we quote one of the main creators of intersection theory – W. Fulton [11], p. 47:

> *Segre classes for normal cones have other remarkable properties not shared by Chern classes.*

All this shows that Wroński had an unusually deep intuition regarding mathematics.

In Wroński's time there was a fascination with *continued fractions*[5]. As much as the earlier generations of mathematicians (Bombelli, Cataldi, Wallis, Huygens, Euler, Lambert, Lagrange...) were interested *mainly* in expressing irrational numbers as continued fractions, obtaining such spectacular results as:

$$\sqrt{2} = 1 + \cfrac{1}{2 + \cfrac{1}{2 + \cfrac{1}{2 + \cfrac{1}{\ddots}}}} \,, \qquad e = 2 + \cfrac{1}{1 + \cfrac{1}{2 + \cfrac{2}{3 + \cfrac{3}{\ddots}}}} \,,$$

$$\pi = 3 + \cfrac{1}{6 + \cfrac{3^2}{6 + \cfrac{5^2}{6 + \cfrac{7^2}{\ddots}}}}$$

– in Wroński's time attempts were made *mainly* to express functions of one variable as continued fractions. Already in *Philosophy of Mathematics* [14] from 1811 Wroński considered *the problem of interpolation* of a function of one variable $f(x)$ by continued fractions. Let $g(x)$ be an auxiliary function vanishing at 0, and ξ – an auxiliary parameter. Wroński gives the expansion of $f(x)$ as a continued fraction

$$f(x) = c_0 + \cfrac{g(x)}{c_1 + \cfrac{g(x - \xi)}{c_2 + \cfrac{g(x - 2\xi)}{c_3 + \cfrac{g(x - 3\xi)}{\ddots}}}}$$

expressing unknown parameters c_0, c_1, c_2, ... in terms of $f(0)$, $f(\xi)$, $f(2\xi)$, ... This is connected with the *Thiele continued fractions* [44]. A few years later Wroński gave even more general continued fractions, considering instead of one auxiliary function $g(x)$ a system of functions $g_0(x), g_1(x), \ldots$, vanishing at various points:

$$0 = g_0(\alpha_0) = g_1(\alpha_1) = g_2(\alpha_2) = \cdots \,.$$

---

[5]The history of continued fractions is described in [4]. The nineteenth century can be described – without exaggeration – as the golden age of continued fractions. This was the time when this topic was known to *every* mathematician. The following are among those who were seriously involved: Jacobi, Perron, Hermite, Gauss, Cauchy and Stieltjes. Mathematicians studied continued fractions involving functions as well as those involving numbers (the same remark applies to the previous century, especially regarding the activity of Euler and Lambert). However it was Wroński who was the pioneer of functional continued fractions in interpolation theory – this fact, surprisingly, was noticed for the first time only recently by Lascoux [30].

Wroński gives determinantal formulas $f(\alpha_i)$, $i = 0, 1, \ldots$, for the coefficients c_j, $j = 0, 1, \ldots$, in the expansion

$$f(x) = c_0 + \cfrac{g_0(x)}{c_1 + \cfrac{g_1(x)}{c_2 + \cfrac{g_2(x)}{c_3 + \cfrac{g_3(x)}{\ddots}}}}.$$

These expansions are connected with the *Stieltjes continued fractions* [42] and play a key role in interpolation theory. In his book [30], Lascoux called them the *Wroński continued fractions*, therefore bringing Wroński's name for the second time (after Wrońskians) into the mathematical literature. More details, as well as specific references to Wroński's papers, can be found in [30].

In 1812 Wroński published *Criticism of Lagrange's theory of analytic functions* [16]. Wroński's views on this subject were shared by a number of other mathematicians, among others Poisson. The criticism regarded particularly the problem of interpretation of "infinitely small values" and the incomplete derivation of the Taylor formula. This is the paper where Wroński introduces for the first time "combinatorial sums" containing derivatives, today called *Wrońskians*.

In these years Wroński searched for a solid foundation for his plans; he thought that he will find it in the most distinguished scientific institution: The French Academy. In 1810 he sent to the Academy – to establish contact – the article *On the fundamental principles of algorithmic methods* containing the "Highest Law", which allows expanding functions of one variable into a series[6]. The committee judging the article had established that Wroński's formula encompasses all expansions known until that time, the Taylor formula for example, but withheld confirming the validity of formula in its most general form. Wroński insisted on a definitive answer, and – in anticipation of a dispute – declined to accept the status of a Corresponding Member of the Academy suggested by Lagrange. The Academy did not give an official response neither to Wroński's reply, nor to his further letters. On top of that, such a serious work as the earlier mentioned *Philosophy of Mathematics* was not noticed by the Academy, as well as the article *On solutions of equations*. Of course, the attitude of the Academy with respect to *Criticism of Lagrange's theory of analytic functions* could not have been different and not hostile towards Wroński. In the committee judging the article was ... Lagrange himself and his colleagues. Because of the negative opinion, Wroński withdrew his

---

[6]The name "The Highest Law" used to describe the possibility of expanding a function into a series may seem a bit pompous. We should remember, however, that mathematicians of that time were fascinated by the possibility of "passing to infinity". This fascination concerned not only infinite series, but also infinite continued fractions. Today there is nothing special about infinity: if a space needs to be compactified, one just "adds a point at infinity".... Wroński and his contemporaries treated infinity with great owe and respect as a great transcendental secret.

paper from the Academy, directing – according to his character – bitter words towards the academics from Paris (phrases: "born enemies of truth", "les savants sur brevets" are among ... the milder ones).

At this time, Wroński's material situation has become much worse. While working on his publications, he has neglected his teaching, and the illness of his wife and child forced him to sell all of his possessions. Despite all efforts, the child could not be saved, and Wroński dressed in worn out clothes and clogs. He asked Napoleon himself for funding, however Napoleon was not interested in his activity. Wroński lived on the edge of the large Polish emigration in Paris, even though – as he bitterly states in his diaries – he dedicated his treatise on equations to his Polish homeland.

A (financially) important moment was Wroński's meeting with P. Arson, a wealthy merchant and banker from Nice, to whom Wroński was introduced by his old friend Ph. Girard (by the way, Girard was the founder of Żyrardów, a Polish town). Arson, fascinated by Wroński's ideas, promised to fund his activity for a few years. In return, Wroński was to reveal him the secret of the Absolute. This strange bond of a philosopher and a banker lasted until 1816 r. Arson, Wroński's secretary, finally insisted the revealing of the secret, and when the mentor did not do so, Arson took him to court. The matter become so well known, that after a few years it was the theme of one of Balzac's books *The search for the Absolute*. Arson resigned his post, but had to pay the debts of his ex-mentor (because Wroński won in court, by convincing the judge, that he knows the mystery of the Absolute). At that time Wroński publishes *Le Sphinx*, a journal which was to popularize his social doctrines.

The years 1814–1819 bring more Wroński's publications, mostly in the area of philosophy of mathematics: *Philosophy of infinity* (1814), *Philosophy of algorithmic techniques* (1815, 1816, 1817), *Criticism of Laplace's generating functions*. The academy has neglected all these publications.

5. The stay in England. In 1820 Wroński went to England, in order to compete for an award in a contest for a method to measure distances in navigation. This trip was very unfortunate. On the boarder, the customs officials took possession of all his instruments, which Wroński never recovered. His papers were regarded as theoretical, and as such not suitable for the award. Finally, the secretary of the Board of Longitude, T. Young has made certain important modifications in the tables of his own authorship on the basis of Wroński's notes sent to him, "forgetting" to mention who should be given credit for these improvements. Of course, Wroński protested by sending a series of letters, also to the Royal Society. He had never received a response.

The very original *Introduction to the lectures of mathematics* [18] dates from this period (see also [22]), written in English and published in London in 1821. Wroński states there, that all positive knowledge is based on mathematics or in some sense draws from it. Wroński divides the development of mathematics into $4 + 1$ periods:

1. works of the scholars of East and Egypt: concrete mathematics was practiced, without the ability to raise to abstract concepts;

2. the period from Tales and Pythagoras until the Renaissance: the human mind rose to the level of high abstraction, however the discovered mathematical truths existed as unrelated facts, not connected by a general principle as, e.g., the description of the properties of conic sections;

3. the activity of Tartaglia, Cardano, Ferrari, Cavalieri, Bombelli, Fermat, Vieta, Descartes, Kepler,...: mathematics rose to the study of general laws thanks to algebra, but the achievements of mathematics are still "individual" – the "general" laws of mathematics were still unknown;

4. the discovery of differential and integral calculus by Newton and Leibniz, expansion of functions into series, continued fractions popularized by Euler, generating functions of Laplace, theory of analytic functions of Lagrange. The human mind was able to raise from the consideration of quantities themselves to the consideration of their creation in the calculus of functions, i.e., differential calculus.

The fifth period should begin with the discovery of the Highest Law and algorithmic techniques by Wroński; the development of mathematics should be based on the most general principles – "absolute ones" – encompassing all of mathematics. This is because all the methods and theories up to that time do not exhaust the essence of mathematics, as they lack a general foundation, from which everything would follow. They are relative, even though science should look for absolute principles. Therefore, the fifth period foresees a generalization of mathematics. Indeed, this will happen later, but not on the basis of philosophy, as Wroński wanted. We mention here the following mathematical theories, which appeared soon: group theory (Galois), projective geometry (Monge, Poncelet), noneuclidean geometries (Lobaczewski, Bolyai, Gauss, Riemann) and set theory (Cantor).

6. Canons of logarithms – a bestseller. In 1823 Wroński is back in Paris and is working on mathematical tables and construction of mathematical instruments: an arithmetic ring (for multiplication and division) and "arithmoscope" (for various arithmetic operations). Among Wroński's achievements in this matter, is his *Canon of logarithms* [19] (see also [23]). With the help of appropriate logarithms and cleverly devised decomposition of a number into certain parts, common for different numbers, he was able to set these parts in such a way, that these tables, even for very large numbers, fit onto one page. For logarithms with 4 decimal places the whole table can be fitted into a pocket notebook. Wroński's *Canon of logarithms* has been published many times in different languages (and shows that, besides very hard to read treatises, he could also produce works which are easier to comprehend).

In 1826 Wroński went to Belgium for a short time, where he was able to interest Belgian mathematicians in his achievements. In fact Belgian scientists were the first to bring Hoene-Wroński into worldwide scientific literature.

In 1829 Wroński, fascinated by the advances in technology, published a treatise on the steam engine.

7. Letters to the rulers of Europe. From around 1830 until the end of his life Wroński focused exclusively on the notion of messianism. At that time he published his well-known *Address to the Slavonic nations about the destiny of the World* and his most well-known works: *Messianism, Deliberations on messianism* and *Introduction to messianism.* At that time he also sent memoranda to Pope Leon XII and the Tsars, so that they would back his messianistic concept.

One should also mention, that Wroński sent letters to the rulers of Europe instructing them how they should govern. These letters contained specific mathematical formulas, how to rule. Here is an example of such a formula from *The Secret letter to his Majesty Prince Louis-Napoléon* [21] from 1851.

Let a be the degree of anarchy, d – the degree of despotism. Then

$$a = \left(\frac{m+n}{m}\cdot\frac{m+n}{n}\right)^{p-r}\cdot\left(\frac{m}{n}\right)^{p+r} = \left(\frac{m+n}{n}\right)^{2p}\cdot\left(\frac{m}{m+n}\right)^{2r},$$

$$d = \left(\frac{m+n}{m}\cdot\frac{m+n}{n}\right)^{r-p}\cdot\left(\frac{n}{m}\right)^{p+r} = \left(\frac{n}{m+n}\right)^{2p}\cdot\left(\frac{m+n}{m}\right)^{2r},$$

where m = number of members of the liberal party, p = the deviation of the philosophy of the liberal party from true religion, n = the number of members of the religious party, r = the deviation of the religious party from true philosophy. According to Wroński, for France one should take $p = r = 1$, and then

$$a = \left(\frac{m}{n}\right)^2, \qquad d = \left(\frac{n}{m}\right)^2.$$

Moreover, $\frac{m}{n} = 2$, and so $a = 4$, $d = \frac{1}{4}$. This means that, political freedom – in France of Wroński's time – is four times the normal one, and the authority of the government is one quarter of what is essential.

(The application of the above formulas to the current Polish political reality would be interesting. ...)

8. Philosophy. I. Kant's philosophy was the starting point of the philosophy of Wroński, who has transformed it into metaphysics in a way analogous to Hegel's approach. Wroński has not only created a philosophical system, but also its applications to politics, history, economy, law, psychology, music (see [38]) and education. Existence and knowledge followed from the Absolute, which he understood either as God, or as the spirit, wisdom, a thing in itself. He did not describe it, but he tried to infer from it a universal law, which he called "The Law of Creation".

In his philosophy of history he predicted reconstruction of the political system, from one full of contradictions to a completely reasonable one. In the history of philosophy he distinguished four periods, each of which imposed on itself different aims:

1. east – material aims;

2. Greak-Roman – moral aims;

12   P. Pragacz

3. medieval – religious aims;

4. modern, until the XVIII century – intellectual aims.

He treated the XIXth century as a transitional period, a time of competition of two blocs: conservative bloc whose aim is goodness and liberal bloc whose aim is the truth.

Wroński is the most distinguished Polish messianic philosopher. It is him (and not Mickiewicz nor Towiański) who introduced the notion of "messianism". Wroński held, that it is the vocation of the human race to establish a political system based on reason, in which the union of goodness and truth and religion and science will take place. The Messiah, who will bring the human race into the period of happiness, is – according to Wroński's concepts – precisely *philosophy*.

Jerzy Braun was an expert and promoter of Wroński's philosophy in Poland. His article *Aperçu de la philosophie de Wroński* published in 1967 is much valued by the French scholars of Wroński's philosophy.

9. Mathematics: The Highest Law, Wrońskians. Essentially, Wroński worked on mathematical analysis and algebra. We have already discussed Wroński's contributions to algebra. In analysis[7] he was especially interested in expanding functions in a *power series* and *differential equations*. Wroński's most interesting mathematical idea was his general method of expanding a function $f(x)$ of one variable x into a series

$$f(x) = c_1 g_1(x) + c_2 g_2(x) + c_3 g_3(x) + \cdots ,$$

when the sequence of functions $g_1(x), g_2(x), \ldots$ is given beforehand, and $c_1, c_2, \ldots$ are numerical coefficients to be determined. Notice that if

$$g_1(x), g_2(x), \ldots$$

form an orthonormal basis with respect to the standard, or any other, inner product $(\,\cdot\,,\,\cdot\,)$ on the (infinite-dimensional) vector space of polynomials of one variable, then for each i we have

$$c_i = \big(f(x), g_i(x) \big).$$

However, such a simple situation rarely happens. Wroński gave his method of finding the coefficients c_i the rank of *The Highest Law*. From today's point of view the method lacked precision and rigor (for example, Wroński did not consider the matter of convergence), however it contained – besides interesting calculations – useful ideas. These ideas were used much later by Stefan Banach, who formulated them precisely and enriched them with topological concepts, and proved that the Highest Law of Hoene-Wroński can be used in what is called today a *Banach space*, as well as in the theory of *orthogonal polynomials*. I will mention here a little known letter of Hugo Steinhaus to Zofia Pawlikowska-Brożek:

---

[7]Strictly speaking, making a distinction between algebra and analysis is not strictly correct, since Wroński often mixed algebraic and analytic methods.

> *Maybe you will find the following fact concerning two Polish mathematicians – Hoene-Wroński and Banach – interesting. In Lwów we had an edition of Wroński's work published in Paris and Banach showed me the page written by the philosopher which discussed the "Highest Law"; apparently Banach has proven to me, that Wroński is not discussing messianic philosophy – the matter concerns expanding arbitrary functions into orthogonal ones (letter of 28.06.1969).*

Banach presented a formal lecture on applying *Wroński's Highest Law* to functional analysis at a meeting in the Astronomic Institute in Warsaw, which was chaired by the well-known astronomer Tadeusz Banachiewicz. He also, as a young researcher, applied Wroński's results in one of his papers on theoretical astronomy[8]. The content of Banach's lecture appeared in print as [2]. We note that S. Kaczmarz and H. Steinhaus in their book [26] on orthogonal polynomials published in 1936 have appealed for an explanation of Wroński's contribution to the theory of those polynomials.

By developing the method of the Highest Law, Hoene-Wroński found a way to compute the coefficients of a function series. In order to achieve this, as auxiliary objects, he used certain determinants, which Thomas Muir in 1882 called *Wroński's determinants*, or *Wrońskians*. At that time, Muir worked on a treatise on the theory of determinants [33]. Looking through Wroński's papers, and especially *Criticism of Lagrange's theory of analytic functions* [16], Muir noticed that Wroński in a pioneering way introduced and systematically used "combinatorial sums" [9], denoted by the Hebrew letter *Shin* – in modern language called *determinants* – containing successive derivatives of the functions present:

$$fg' - f'g, \qquad fg'h'' + gh'f'' + hf'g'' - hg'f'' - fh'g'' - gf'h'', \qquad \ldots .$$

A fragment of page 11 of a manuscript of [16] with combinatorial sums

---

[8]T. Banachiewicz applied the ideas of Wroński's Highest Law in the calculus of the *cracovians* – see [3].

[9]In contemporary mathematics determinants are always associated with matrices – both conceptually and in notation. Historically, they were introduced earlier than matrices "as sums with signs" and as such were used in computations, finding many interesting properties, which seem to be natural only when using the language of matrices (as, for example, the Binet–Cauchy's theorem). Matrices were introduced around 1840 by Cayley, Hamilton, ..., and Sylvester created in the 1850ties a transparent calculus of determinants and minors based on the notion of a matrix (see, e.g., [43]).

In modern notation the Wrońskian of n real functions

$$f_1(x), f_2(x), \ldots, f_n(x),$$

which are $(n-1)$ times differentiable, is defined and denoted as follows:

$$W(f_1, f_2, \ldots, f_n) = \begin{vmatrix} f_1 & f_2 & \cdots & f_n \\ f_1' & f_2' & \cdots & f_n' \\ f_1'' & f_2'' & \cdots & f_n'' \\ \vdots & \vdots & \ddots & \vdots \\ f_1^{(n-1)} & f_2^{(n-1)} & \cdots & f_n^{(n-1)} \end{vmatrix}$$

(there also exists a *Wrońskian of a system of vector valued functions*). Wrońskians are one of the basic tools in the theory of differential equations ([1], [40]) and are so called in the mathematical literature from every part of the world. Probably, most often Wrońskians are used to test whether a sequence of functions is linearly independent:

> *Suppose that $f_1(x), \ldots, f_n(x)$ are $(n-1)$-fold differentiable functions. If $W(f_1, f_2, \ldots, f_n)$ is not identically zero, then the functions $f_1, \ldots, f_n$ are linearly independent*[10].

Properties and certain applications of Wrońskians were treated in [7]. The use of Wrońskians is not confined to analysis. In the classical reference work on the theory of invariants [13], the authors employ them in the algebraic theory of binary forms. Analogues of Wrońskians were constructed in other parts of mathematics. For example, *Wrońskians of linear systems* (Galbura [12], Laksov [27]), which are certain morphisms of vector bundles, are an important tool in modern algebraic geometry: they are used to the *Plücker formulas* in enumerative geometry and the theory of *Weierstrass points*. This pioneering invention, or maybe discovery of Wroński, is really very deep and lies at the heart of mathematics and proves Wroński's incredible feel for what is really important. In particular, Wroński used more general functional determinants than Wrońskians at the time when determinants of numerical matrices just begun to appear in the work of other mathematicians.

In physics, Wroński was interested in the theory of optical instruments and fluid mechanics. He improved steam engines, designed a mechanical calculator, created the concept of "moving rails", that is the contemporary *caterpillar tracks*, so once more he was ahead of his time for very many years.

10. Wroński in the eyes of people in science and art. He was a genius in many respects, and had the ability to work very hard. In the over three hundred page biography of Wroński, Dickstein [9] writes:

---

[10]The reverse implication is – in general – not true.

His iron nature required little sleep and food, he begins work early in the morning and only after a couple of hours of work he would have a meal saying: "Now I have earned my day".

and then he adds:

The seriousness of his work and the struggle against misfortune did not spoil his calm personality and cheerful character.

Wroński wrote a very large number of papers in mathematics, philosophy, physics and technical science (see [25] and [8]).

Palace in Kórnik near Poznań, containing a collection of Wroński's *original handwritten* manuscripts. They may contain interesting – unknown to the public yet – mathematical results (picture by Stanisław Nowak).

In 1875 the Kórnik Library purchased a collection of Wroński's books, articles and manuscripts [25] from Wroński's adopted daughter Bathilde Conseillant. After his death his friends (most notably Leonard Niedźwiecki – the good spirit of the Polish Emigration and a close friend of Wroński) made an effort to publish Wroński's collected works (many were left in handwritten form). It has turned out, that his works would fill 10 volumes of about 800 pages each. Thirteen years after his death, the Polish Scientific Society in Paris, whose aim was bringing together all Polish scientists, organized a contest for the evaluation of Wroński's works. Amongst the reasons why only one work was submitted to the contest, and why Wroński's works – besides a close circle of his "admirers" – were not very popular, is that they were very hard to read. This was the effect of aiming for the highest generality and of joining mathematical concepts with philosophical ones. Indeed, Wroński is not easy to read – he is a very demanding author. He was also very demanding during his life, first of all of himself, but also of others, and this did not win him friends, but rather made many enemies. It was a difficult character; in [9] we find:

> *He combined extreme simplicity in his home life with bold language and pride coming form a deep conviction about his historic mission, and the infallibility of his philosophy. He considered adversaries of his philosophy as enemies of the truth, and fought with them vehemently, often stating his arguments in a too personal way. . . .*

However, one should state clearly that during his life Wroński did not receive any *constructive* review and criticism, which – besides pointing out unclear passages in his work – would also pick out truly ingenious ideas (which were definitely present). The attitude of the French Academy does not give it credit. Wroński wrote bitterly about the academics of Paris:

> *These gentlemen are interested neither in progress nor in the truth. . . .*

Well, Wroński was many years ahead of his period. Balzac described Wroński as "the most powerful mind of Europe". The well-known Polish writer Norwid had a similar opinion of Wroński (see [38], p. 30). Wroński's political visions "anticipated" the European union – a federation of countries in a united Europe, ruled by a common parliament. Dickstein is probably correct in writing:

> *Besides versatility, the dominating feature of Wroński's mind was, so to speak, architectural ability. He himself mentioned in one of his earliest works ("Ethic philosophy"), the most beautiful privilege of the human mind is the ability to construct systems.*

Who knows whether in Germany Wroński would not have a better chance to find readers who would appreciate his works, written somewhat in "the style" of great German philosophers.

Wroński is – not only in the opinion of the author of this text – the *most unappreciated* great Polish scientist in his own country. It seems that he is much more appreciated abroad. And so, in the Museum of Science in Chicago, in a table with the names of the most prominent mathematicians in history, one can find the names of only three Polish mathematicians: Copernicus, Banach, and ... Hoene-Wroński. Also the rank of Wroński in XIXth century philosophy is high. It seems that in France Wroński – the philosopher – is valued much more than in Poland (see, e.g., [47] and [10]). The scientific legacy of such a great thinker as Wroński *should* be described in a comprehensive monograph in his homeland.

11. Non omnis moriar. Wroński's life was long and hard. Did any famous scientific authority – during his life – say a good word about what Wroński has achieved[11]?

---

[11] During Wroński's life two important works were published [41] and [32], citing his mathematical achievements. In [41], theorems and formulas of Wroński are given in a couple of dozen places, and in [32] his most important mathematical ideas are summarized. Despite of that, contemporary thinkers knew little of Wroński's achievements, and often reinvented what Wroński had discovered many years earlier.

Still in 1853 Wroński writes two papers, and prepares a third one for publication: he studies *the theory of tides*. He sends the first two to the Navy Department. He received a reply saying that Laplace's formulas are completely sufficient for the needs of the navy. This was a severe blow for the 75-year old scientist who, after 50 years of hard work, once more has not found recognition. He died on August 9, 1853 in Neuilly. Before his death he whispered to his wife:

> *Lord Almighty, I had so much more to say.*

Józef Maria Hoene-Wroński is buried at the old cemetery in Neuilly. The following words are written on his grave (in French):

$$\textit{THE SEARCH OF TRUTH IS A TESTIMONY TO THE}$$
$$\textit{POSSIBILITY OF FINDING IT.}$$

After writing this article, I have realized that it has – in fact – a lot in common with my article about A. Grothendieck in the previous volume of Impanga Lecture Notes (*Topics in cohomological studies of algebraic varieties*, Trends in Mathematics, Birkhäuser, 2005). In his diaries *Harvest and Sowings* (vol. I, p. 94), Grothendieck wrote:

> *... one night ... I realized that the DESIRE to know and the POWER to know and to discover are one and the same thing.*

Acknowledgements. I became fascinated with Hoene-Wroński thanks to Alain Lascoux – without this fascination this article could not have been written.

This text reached its final form after I have listened to the lectures and comments presented at the session of Impanga *A tribute to Józef Hoene-Wroński* described in the Introduction; I sincerely thank the speakers, as well as Jerzy Browkin and Maciej Skwarczyński. I also thank Jan Krzysztof Kowalski, Maria Pragacz and Jolanta Zaim for their help with the editorial process, as well as Wanda Karkucińska and Magdalena Marcinkowska from the Kórnik Library of the Polish Academy of Sciences for allowing me to publish the daguerreotype of Wroński and parts of his manuscripts, and also for their help in the work on those manuscripts.

Finally, I am grateful to Jan Spaliński for translating the text into English, and to Paolo Aluffi and Jacek Brodzki for their comments on an earlier version of this article.

About 20 years after Wroński's death, some notes about Wroński the mathematician and philosopher appeared in Poncelet's book *Applications d'Analyse et de Géométrie*, and also in the works of Cayley [5] and Transon [45], [46], which developed Wroński's ideas. One could say that Wrońskians "were firmly enrooted" in mathematics even before Muir coined the term. In the multi-volume history of determinants [34] and [35], works from the period 1838–1920 devoted to Wrońskians are summarized which have been written by: Liouville, Puiseaux, Christoffel, Sylvester, Frobenius, Torelli, Peano. Muir stresses there, that the interest in Wrońskians rose as time went by.

References

[1] W.I. Arnold, *Ordinary differential equations*, Nauka, Moscow 1971 (in Russian).

[2] S. Banach, *Über das "Loi suprême" von J. Hoene-Wroński*, Bulletin International de l'Académie Polonaise des sciences et de lettres, Série A (1939), 450–457.

[3] T. Banachiewicz, *Rachunek krakowianowy z zastosowaniami*, PAN, Komitet Astronomiczny, PWN, Warszawa 1959.

[4] C. Brezinski, *History of continued fractions and Padé approximants*, Springer, Berlin 1991.

[5] A. Cayley, *On Wroński's theorem*, Quart. J. Math. **12** (1873), 221–228.

[6] J. Dianni, A. Wachułka, *Tysiąc lat polskiej myśli matematycznej*, PZWS, Warszawa 1963.

[7] S. Dickstein, *Własności i niektóre zastosowania wrońskianów*, Prace Matematyczno-Fizyczne **1** (1888), 5–25.

[8] S. Dickstein, *Katalog dzieł rękopisów Hoene-Wrońskiego*, nakładem Akademii Umiejętności, Kraków 1896; also in [9], pp. 239–351.

[9] S. Dickstein, *Hoene-Wroński. Jego życie i prace*, nakładem Akademii Umiejętności, Kraków 1896.

[10] J.-C. Drouin, *Les grands thèmes de la pensée messianique en France de Wroński à Esquiros: christianisme ou laïcisme?*, in: Messianisme et slavophilie, Wyd. Uniw. Jagiell., Kraków 1987, 55–66.

[11] W. Fulton, *Intersection theory*, Springer, Berlin 1984.

[12] G. Galbura, *Il wronskiano di un sistema di sezioni di un fibrato vettoriale di rango i sopra una curva algebrica ed il relativo divisore di Brill–Severi*, Ann. Mat. Pura Appl. **98** (1974), 349–355.

[13] J.H. Grace, A. Young, *The algebra of invariants*, Cambridge University Press, Cambridge 1903; there exists a reprint: Stechert & Co., New York 1941.

[14] J.M. Hoene-Wroński, *Introduction à la Philosophie des Mathématiques et Technie de l'Algorithmique*, Courcier, Paris 1811.

[15] J.M. Hoene-Wroński, *Résolution générale des équations de tous les degrés*, Klostermann, Paris 1812.

[16] J.M. Hoene-Wroński, *Réfutation de la théorie des fonctions analytiques de Lagrange*, Blankenstein, Paris 1812.

[17] J.M. Hoene-Wroński, *Philosophie de la technie algorithmique: Loi Suprême et universelle; Réforme des Mathématiques*, Paris 1815–1817.

[18] J.M. Hoene-Wroński, *A course of mathematics, Introduction determining the general state of mathematics*, London 1821.

[19] J.M. Hoene-Wroński, *Canons de logarithms*, Didot, Paris 1824.

[20] J.M. Hoene-Wroński, *Réforme absolue et par conséquent finale du Savoir Humain. Tome I: Réforme des Mathématiques; Tome III: Résolution générale et définitive des équations algébriques de tous les degrés*, Didot, Paris 1847–1848.

[21] J.M. Hoene-Wroński, *Epitre Secrète à son Altesse le Prince Louis-Napoléon*, Dépôt des Ouvrages Messianiques, Metz 1851.

[22] J.M. Hoene-Wroński, *Wstęp do wykładu Matematyki*, transl. from French by L. Niedźwiecki, Biblioteka Polska, Quais d'Orléans 6, Paris 1880.

[23] J.M. Hoene-Wroński, *Kanony logarytmów*, transl. from French by S. Dickstein, Warszawa 1890.

[24] J.M. Hoene-Wroński, *Wstęp do Filozofii Matematyki oraz Technia Algorytmii*, transl. from French by P. Chomicz, Prace Towarzystwa Hoene-Wrońskiego, Inst. Wyd. "Biblioteka Polska", Warszawa 1937.

[25] J.M. Hoene-Wroński, *The legacy of Hoene-Wroński in the Kórnik Library* – see: www.bkpan.poznan.pl/biblioteka/index.html

[26] S. Kaczmarz, H. Steinhaus, *Theorie der Orthogonalreihen*, Warszawa–Lwów 1936.

[27] D. Laksov, *Wronskians and Plücker formulas for linear systems on curves*, Ann. Sci. École Norm. Sup. (4) **17** (1984), 45–66.

[28] A. Lascoux, *Diviser!*, in: M. Lothaire, Mots, Mélanges offerts à M.-P. Schützenberger, Hermès, Paris 1990.

[29] A. Lascoux, *Wroński's factorization of polynomials*, in: Topics in Algebra, Banach Center Publ. **26**, Part 2, PWN, Warszawa 1990, 379–386.

[30] A. Lascoux, *Symmetric functions and combinatorial operators on polynomials*, CBMS Regional Conf. Ser. in Math. **99**, Amer. Math. Soc., Providence 2003.

[31] A. Lascoux, P. Pragacz, *Double Sylvester sums for subresultants and multi-Schur functions*, J. Symbolic Comp. **35** (2003), 689–710.

[32] A.S. de Montferrier, *Dictionnaire des sciences mathématiques pures et appliquées*, Paris 1834–1840.

[33] T. Muir, *A treatise on the theory of determinants*, London 1882; there exists a reprint: Dover, New York 1960.

[34] T. Muir, *The theory of determinants in the historical order development*, 4 volumes, Macmillan & Co., London 1906, 1911, 1920, 1923; there exists a reprint: Dover, New York 1960.

[35] T. Muir, *Contributions to the history of determinants*, 1900–1920, Blackie and Son, London, Glasgow 1930.

[36] R. Murawski, *Józef Maria Hoene-Wroński – filozof i matematyk*, Materiały konferencyjne Uniwersytetu Szczecińskiego, **30** (1998), 29–46.

[37] R. Murawski, *Genius or madman? On the life and work of J.M. Hoene-Wroński*, in: European mathematics in the last centuries (ed. W. Więsław), Wrocław 2005, 77–86.

[38] C.K. Norwid, *O Szopenie*, Fundacja Narodowego Wydania Dzieł Fryderyka Chopina, Łódź 1999.

[39] H. Ohsugi, T. Wada, *Gröbner bases of Hilbert ideals of alternating groups*, J. Symb. Comp. **41** (2006), 905–908.

[40] L.S. Pontrjagin, *Ordinary differential equations*, Mir, Moscow 1974 (in Russian).

[41] F. Schweins, *Theorie der Differenzen und Differentiale*, Heidelberg 1825.

[42] T.J. Stieltjes, *Recherches sur les fractions continues*, Ann. Fac. Sc. Toulouse **8** (1894), 1–122.

[43] J.J. Sylvester, *A theory of the syzygetic relations of two rational integral functions*, Phil. Trans. Royal Soc. London **CXLIII**, Part III (1853), 407–548.

[44] T.N. Thiele, *Interpolationrechnung*, Teubner, Leipzig 1909.

[45] A. Transon, *Réflexions sur l'événement scientifique d'une formule publiée par Wroński en* 1812 *et démontrée par Cayley en* 1873, Nouvelles Annales de mathématiques **13** (1874), 161–174.

[46] A. Transon, *Lois des séries de Wroński. Sa phoronomie*, Nouvelles Annales de mathématiques **13** (1874), 305–318.

[47] F. Warrain, *L'œuvre philosophique d'Hoene-Wroński. Textes, commentaires et critique*, 3 volumes, Les Editions Vega, Paris 1933, 1936, 1938.

Piotr Pragacz
Instytut Matematyczny PAN
Śniadeckich 8
PL-00-956 Warszawa
e-mail: `P.Pragacz@impan.gov.pl`

Algebraic Cycles, Sheaves, Shtukas, and Moduli
Trends in Mathematics, 21–32

Exotic Fine Moduli Spaces of Coherent Sheaves

Jean-Marc Drézet

1. Introduction

Let X be a smooth projective irreducible algebraic variety over $\mathbb{C}$. Let $\mathcal{S}$ be a nonempty set of isomorphism classes of coherent sheaves on X. A *fine moduli space for* $\mathcal{S}$ is an integral algebraic structure M on the set $\mathcal{S}$ (i.e., $\mathcal{S}$ is identified with the set of closed points of M), such that there exists a *universal sheaf*, at least locally: there is an open cover (U_i) of M and a coherent sheaf $\mathcal{F}_i$ on each $U_i \times X$, flat on U_i, such that for every $s \in U_i$, the fiber $\mathcal{F}_{is}$ is the sheaf corresponding to s, and $\mathcal{F}_i$ is a complete deformation of $\mathcal{F}_{is}$.

The most known fine moduli spaces appear in the theory of (semi-)stable sheaves. If X is a curve and r, d are integers with $r \geq 1$, then the moduli space $M_s(r, d)$ of stable vector bundles of rank r and degree d on X is a fine moduli space if and only if r and d are relatively prime (cf. [18], [6]). Similar results have been proved on surfaces (cf. [6], [21]).

An *exotic* fine moduli space of sheaves is one that does not come from the theory of (semi-)stable sheaves. We will give several examples of exotic fine moduli spaces in this paper. Sometimes the corresponding sheaves will not be simple, contrary to stable sheaves. The detailed proofs of the results given in this paper come mainly from [3], [5] and [8].

In Section 2 we give the definitions and first properties of fine moduli spaces of sheaves. We prove that if $\mathcal{S}$ admits a fine moduli space, then the dimension of $\mathrm{End}(E)$ is independent of $E \in \mathcal{S}$, and $\mathrm{Aut}(E)$ acts trivially on $\mathrm{Ext}^1(E, E)$ (by conjugation).

In Section 3 we define *generic extensions* and build fine moduli spaces of such sheaves. They appear when one tries to construct moduli spaces of *prioritary sheaves* of given rank and Chern classes on $\mathbb{P}_2$ where there does not exist semi-stable sheaves of the same rank and Chern classes. In this case the sheaves are not simple. We give an example of fine moduli space where no global universal sheaf exists.

In Section 4 we construct fine moduli spaces of very unstable sheaves when $\dim(X) \geq 3$. In this case also the sheaves are not simple.

In Section 5 we give other examples of fine moduli spaces of sheaves, illustrating some of their properties:

- We give an example of a maximal fine moduli space of sheaves on $\mathbb{P}_2$ which is not projective.
- We show that there are simple vector bundles on $\mathbb{P}_2$ which cannot belong to an open family admitting a fine moduli space, but are limits of bundles in a fine moduli space.
- We give an example of two projective fine moduli spaces of sheaves on $\mathbb{P}_2$ which are distinct but share a nonempty open subset. One of them is the moduli space of stable sheaves of rank 6 and Chern classes -3, 8.

Notations. A *family of sheaves* on X parametrized by an algebraic variety T is a coherent sheaf $\mathcal{F}$ on $T \times X$, flat on T. If t is a closed point of T, let $\mathcal{F}_t$ denote the fiber of $\mathcal{F}$ over t.

We say that $\mathcal{F}$ is a *family of sheaves of* $\mathcal{S}$ if for every closed point t in T we have $\mathcal{F}_t \in \mathcal{S}$.

An algebraic vector bundle E on $\mathbb{P}_2$ is called *exceptional* if $\mathrm{End}(E) = \mathbb{C}$ and $\mathrm{Ext}^1(E, E) = \{0\}$. An exceptional vector bundle is stable (cf. [7]). Let Q denote the *universal quotient* on $\mathbb{P}_2$, i.e., if $\mathbb{P}_2 = \mathbb{P}(V)$ (lines in V), then we have a canonical exact sequence $0 \to \mathcal{O}(-1) \to \mathcal{O} \otimes V \to Q \to 0$.

If S, T are algebraic varieties, p_S, p_T will denote the projections $S \times T \to S$, $S \times T \to T$ respectively.

If $f : S \to T$ is a morphism of algebraic varieties and $\mathcal{E}$ a coherent sheaf on $T \times X$, let $f^\sharp(\mathcal{E}) = (f \times I_X)^*(\mathcal{E})$.

2. Fine moduli spaces of sheaves

2.1. Definitions

Let $\mathcal{S}$ be a nonempty set of isomorphism classes of coherent sheaves on X. We say that $\mathcal{S}$ is *open* if for every family $\mathcal{F}$ of sheaves parametrized by an algebraic variety T, if t is a closed point of T such that $\mathcal{F}_t \in \mathcal{S}$ then the same is true for all closed points in a suitable open neighbourhood of t in T.

Suppose that $\mathcal{S}$ is open. A fine moduli space for $\mathcal{S}$ is given by

- an integral algebraic variety M,
- an open cover $(U_i)_{i \in I}$ of M,
- for every $i \in I$, a family of sheaves of $\mathcal{F}_i$ of $\mathcal{S}$ parametrized by U_i,

such that

(i) if $s \in U_i \cap U_j$ then $\mathcal{F}_{is} \simeq \mathcal{F}_{js}$,
(ii) for every E in $\mathcal{S}$ there exists one and only one $s \in M$ such that $E = \mathcal{F}_{is}$ if $s \in U_i$,

(iii) the sheaves $\mathcal{F}_i$ satisfy the *local universal property*: for every family $\mathcal{F}$ of sheaves of $\mathcal{S}$ parametrized by an algebraic variety T, there exists a morphism $f_{\mathcal{F}} : T \to M$ such that for every $i \in I$ and $t \in f_{\mathcal{F}}^{-1}(U_i)$ there exists a neighbourhood U of t such that $f_{\mathcal{F}}(U) \subset U_i$ and $f_{\mathcal{F}}^{\sharp}(\mathcal{F}_i)_{|U \times X} \simeq \mathcal{F}_{U \times X}$.

A fine moduli space for $\mathcal{S}$ is unique in the following sense: if $(M', (U_j'), (\mathcal{F}_j'))$ is another one then M, M' are canonically isomorphic and the families $\mathcal{F}_i$, $\mathcal{F}_j'$ are locally isomorphic.

We say that $\mathcal{S}$ admits a *globally defined fine moduli space* if I can be found with just one point, i.e., if there is a *universal sheaf* on the whole $M \times X$.

2.2. Endomorphisms

Let $\mathcal{S}$ be an open set of sheaves on X admitting a fine moduli space M.

Theorem 2.2.1. *For every $E \in \mathcal{S}$, $\mathrm{Aut}(E)$ acts trivially on $\mathrm{Ext}^1(E, E)$ and $\dim(\mathrm{End}(E))$ is independent of E.*

Proof. For simplicity we assume that there is a globally defined universal sheaf $\mathcal{E}$ on $M \times X$. Using locally free resolutions we find that there exists a morphism of vector bundles $\phi : A \to B$ on M such that for every $y \in M$ there is a canonical isomorphism $\mathrm{End}(\mathcal{E}_y) \simeq \ker(\phi_y)$. Now let $p = rk(\ker(\phi))$, it is the generic dimension of $\mathrm{Aut}(E)$, for $E \in \mathcal{S}$. Now let $E \in \mathcal{S}$ and $y \in M$ such that $\mathcal{E}_y \simeq E$. Let W be the image of $\ker(\phi)_y$ in A_y. Then we have $\dim(W) \leq p$.

Let $\mathcal{F}$ be a family of sheaves in $\mathcal{S}$ parametrized by an algebraic variety, and $s \in S$ such that $\mathcal{F}_s \simeq E$. Suppose that $\mathcal{F}$ is a complete family at s, and let $\psi \in \mathrm{Aut}(\mathcal{F})$. Then we have $\psi_s \in W$.

Now it remains to prove that for any $\sigma \in \mathrm{Aut}(E)$ there exists such a family $\mathcal{F}$ and $\psi \in \mathrm{Aut}(\mathcal{F})$ such that $\psi_y = \sigma$. This will prove that $\mathrm{Aut}(E) \subset W$, and finally that $\dim(\mathrm{End}(E)) = p$.

The preceding assertion can be proved using Quot schemes. $\square$

Proposition 2.2.2. *If the sheaves in $\mathcal{S}$ are simple then there exists a universal sheaf $\mathcal{F}$ on $M \times X$. For every family $\mathcal{E}$ of sheaves of $\mathcal{S}$ parametrized by T there exists a line bundle L on T such that $f_{\mathcal{E}}^{\sharp}(\mathcal{F}) \otimes p_T^*(L) \simeq \mathcal{E}$.*

Proof. Suppose for simplicity that the sheaves in $\mathcal{S}$ are locally free. Then it is easy to build a *universal projective bundle* by gluing the projective bundles of all the local universal bundles, using the fact that the bundles of $\mathcal{S}$ are simple. Now this projective bundle comes from a universal bundle on $M \times X$ because a projective bundle which is banal on a nonempty open subset of the base is banal (cf. [12]). $\square$

3. Moduli spaces of generic extensions

3.1. Generic extensions

Let E', E be coherent sheaves on X. The extensions of E by E' are parametrized by $\mathrm{Ext}^1(E, E')$. If $\sigma \in \mathrm{Ext}^1(E, E')$ let

$$0 \longrightarrow E' \longrightarrow F_\sigma \longrightarrow E \longrightarrow 0 \qquad (*)$$

be the corresponding extension. The group $G = \mathrm{Aut}(E) \times \mathrm{Aut}(E')$ acts obviously on $\mathrm{Ext}^1(E, E')$ and if $\sigma \in \mathrm{Ext}^1(E, E')$ and $g \in G$ we have $F_{g\sigma} = F_\sigma$. Let $\sigma \in \mathrm{Ext}^1(E, E')$. The tangent map at the identity of the orbit map

$$\Phi_\sigma : G \longrightarrow \mathrm{Ext}^1(E, E')$$
$$g \longmapsto g\sigma$$

is

$$T_{\Phi_\sigma} : \mathrm{End}(E) \times \mathrm{End}(E') \longrightarrow \mathrm{Ext}^1(E, E')$$
$$(\alpha, \beta) \longmapsto \beta\sigma - \sigma\alpha.$$

We say that $(*)$ is a *generic extension* if T_{Φ_σ} is surjective. In this case $G\sigma$ is the unique open orbit in $\mathrm{Ext}^1(E, E')$. It is the generic orbit. We call F_σ the *generic extension of E by E'*.

Lemma 3.1.1. *Suppose that* $\mathrm{Hom}(E', E) = \mathrm{Ext}^1(E', E) = \mathrm{Ext}^2(E, E') = \{0\}$ *and let* $\sigma \in \mathrm{Ext}^1(E, E')$ *such that* F_σ *is a generic extension. Then we have*

$$\mathrm{Ext}^1(F_\sigma, F_\sigma) \simeq \mathrm{Ext}^1(E', E') \oplus \mathrm{Ext}^1(E, E)$$

and an exact sequence

$$0 \longrightarrow \mathrm{Hom}(E, E') \longrightarrow \mathrm{End}(F_\sigma) \longrightarrow \ker(\Phi_\sigma) \longrightarrow 0.$$

Proof. This follows easily from a spectral sequence associated to the filtration $0 \subset E' \subset F_\sigma$. $\qquad\qquad\square$

In other words, the only deformations of F_σ are generic extensions of deformations of E by deformations of E'.

3.2. Moduli spaces of generic extensions

Let $\mathcal{X}$, $\mathcal{X}'$ be open sets of sheaves admitting fine moduli spaces M, M'. Suppose that $\mathrm{Hom}(E', E) = \{0\}$ for every $E \in \mathcal{X}$, $E' \in \mathcal{X}'$. For simplicity we assume that they have globally defined universal sheaves $\mathcal{E}$, $\mathcal{E}'$ respectively. Let $Y \subset \mathcal{X} \times \mathcal{X}'$ be the set of pairs (E, E') such that

$$\mathrm{Ext}^1(E', E) = \mathrm{Ext}^2(E', E') = \mathrm{Ext}^2(E, E) = \mathrm{Ext}^2(E', E) = \mathrm{Ext}^i(E, E') = \{0\}$$

for $i \geq 2$ and that there exists a generic extension of E by E'. Let $\mathcal{Y}$ the set of such generic extensions. Then it is easy to see that the map $Y \to \mathcal{Y}$ associating to (E, E') the generic extension F_σ of E by E' is a bijection.

Theorem 3.2.1. *The set $\mathcal{Y}$ is open and admits a fine moduli space N which is isomorphic to the open subset of $M \times M'$ consisting of pairs (s,t) such that $(\mathcal{E}_s, \mathcal{E}'_t) \in Y$.*

3.3. Prioritary sheaves

A coherent sheaf E on $\mathbb{P}_2 = \mathbb{P}(V)$ is called *prioritary* if it is torsionfree and $\mathrm{Ext}^2(E, E(-1)) = \{0\}$ (or equivalently $\mathrm{Hom}(E, E(-2)) = \{0\}$). A semi-stable sheaf is prioritary but the converse is not true. A. Hirschowitz and Y. Laszlo have proved in [11] that the stack of prioritary sheaves of rank r and Chern classes c_1, c_2 is irreducible, and found when it is nonempty.

Let r, c_1, c_2 be integers, with $r \geq 1$, such that there are no semi-stable sheaves of rank r and Chern classes c_1, c_2, but there exist prioritary sheaves with these invariants. There are two cases:

(i) There exist distinct exceptional vector bundles E_1, E_2, E_3, and nonnegative integers m_1, m_2, m_3, with $m_1, m_2 > 0$, such that the generic prioritary sheaf of rank r and Chern classes c_1, c_2 is isomorphic to $(E_1 \otimes \mathbb{C}^{m_1}) \oplus (E_2 \otimes \mathbb{C}^{m_2}) \oplus (E_3 \otimes \mathbb{C}^{m_3})$.

(ii) There exist an exceptional vector bundle F, an integer $p > 0$ and a moduli space M of stable bundles M such that the generic prioritary sheaf of rank r and Chern classes c_1, c_2 is of the form $(F \otimes \mathbb{C}^p) \oplus E$, with $E \in M$.

In the first case there can be only one fine moduli space of prioritary sheaves of rank r and Chern classes c_1, c_2, consisting of one unique point.

In the second case we must assume that M is a fine moduli space of sheaves (in this case M is projective). Let μ denote the slope of sheaves in M. There is then a fine moduli space of prioritary sheaves of rank r and Chern classes c_1, c_2, consisting generic extensions $0 \to F \otimes \mathbb{C}^p \to \mathcal{E} \to E \to 0$ if $\mu \leq \mu(F)$ (resp. $0 \to E \to \mathcal{E} \to F \otimes \mathbb{C}^p \to 0$ if $\mu > \mu(F)$). It is isomorphic to an open subset of M, and to the whole of M if p is sufficiently big. In general this moduli space will not contain all the prioritary sheaves of rank r and Chern classes c_1, c_2. We will treat only one example.

3.4. Prioritary sheaves of rank 8 and Chern classes -4, 11

According to [7] there are no semi-stable sheaves of rank 8 and Chern classes -4, 11. The generic prioritary sheaf with these invariants is of the form $(Q^* \otimes \mathbb{C}^2) \oplus E$, where E is a stable sheaf of rank 4 and Chern classes -2, 4.

Let $M(4, -2, 4)$ be the moduli space of rank 4 semi-stable sheaves on $\mathbb{P}_2$ with Chern classes -2, 4. For these invariants semi-stability is equivalent to stability, hence $M(4, -2, 4)$ is a fine moduli space of sheaves. It has the following description: we have

$$M(4, -2, 4) \simeq \mathbb{P}(S^2 V) \simeq \mathbb{P}_5.$$

If D is a line in $S^2 V$ we have a canonical morphism

$$\alpha_D : \mathcal{O}(-3) \longrightarrow \mathcal{O}(-1) \otimes (S^2 V / D)$$

which is injective (as a morphism of sheaves) and the corresponding point in $M(4, -2, 4)$ is $\mathrm{coker}(\alpha_D) = \mathcal{E}_D$.

An easy computation shows that $\dim(\mathrm{Ext}^1(\mathcal{E}_D, Q^*)) = 0$ if D is not decomposable, 1 if $D = \mathbb{C}uv$, with $u \wedge v \neq \{0\}$ and 2 if $D = \mathbb{C}u^2$.

For $D \in \mathbb{P}(S^2V)$ let $0 \to Q^* \otimes \mathbb{C}^2 \to \mathcal{F}_D \to \mathcal{E}_D \to 0$ be the associated generic extension. We have $\dim(\mathrm{End}(\mathcal{F}_D)) = 5$. It follows from Theorem 3.2.1 that the set of all sheaves $\mathcal{F}_D$ admits a fine modulo space $\mathbf{M}$ isomorphic to $\mathbb{P}_5$. In particular there are universal sheaves defined at least locally.

Proposition 3.4.1. *There is no universal sheaf on* $\mathbf{M} \times \mathbb{P}_2$.

Proof. Suppose that there is a universal sheaf $\mathcal{G}$ on $\mathbf{M} \times \mathbb{P}_2$. Then

$$\mathbb{A} = p_{M*}(\mathcal{H}om(p_{\mathbb{P}_2}^*(Q^*), \mathcal{G}))$$

is a rank 2 vector bundle on $\mathbf{M}$, and there exists a line bundle L on $\mathbf{M}$ and an exact sequence of sheaves on $\mathbf{M} \times \mathbb{P}_2$

$$0 \longrightarrow p_{\mathbb{P}_2}^*(Q^*) \otimes p_{\mathbf{M}}^*(\mathbb{A}) \longrightarrow \mathcal{G} \longrightarrow \mathcal{E} \otimes p_{\mathbf{M}}^*(L) \longrightarrow 0.$$

Let $\mathbb{B} = \mathcal{E}xt^1_{p_M}(\mathcal{E}, p_{\mathbb{P}_2}^*(Q^*))$. Let $P \subset \mathbf{M}$ denote the locus of elements $\mathbb{C}u^2$, $u \in V \backslash \{0\}$, which is isomorphic to $\mathbb{P}_2$. Then it is easy to see that $\mathbb{B}_{|P}$ is isomorphic to Q_P, the bundle Q on $P \simeq \mathbb{P}_2$.

From the above extension we deduce a section of $\mathbb{B} \otimes L \otimes \mathbb{A}^*$ inducing an isomorphism $Q_P = \mathbb{B}_{|P} \simeq \mathbb{A}_{|P} \otimes L_{|P}*$, since it gives a generic extension at each point of P. But Q_P cannot be the restriction of a vector bundle on $\mathbf{M}$ since its determinant is not the restriction of a line bundle on $\mathbf{M}$. This proves that the universal sheaf $\mathcal{G}$ does not exist. $\qquad\qquad\square$

4. Fine moduli spaces of wide extensions

4.1. Introduction

Let $\mathcal{O}_X(1)$ be a very ample line bundle on X. We consider extensions of sheaves

$$0 \longrightarrow G \longrightarrow \mathcal{E} \longrightarrow F \longrightarrow \qquad\qquad (*)$$

where G, $\mathcal{E}$ are locally free and F torsionfree. We want to see what happens when $\mu(G) \gg \mu(\mathcal{E})$, say let $G = G_0(p)$ and let p tend to infinity. If we want nontrivial extensions F must not be locally free. The correct settings are as follows: F is torsionfree and $T = F^{**}/F$ is pure of codimension 2.

Using the dual of $(*)$ we obtain the exact sequence

$$0 \longrightarrow F^* \longrightarrow \mathcal{E}^* \longrightarrow G^* \xrightarrow{\ \lambda\ } \mathcal{E}xt^1(F, \mathcal{O}) \longrightarrow 0.$$

We have $\mathcal{E}xt^1(F, \mathcal{O}) \simeq \mathcal{E}xt^2(T, \mathcal{O}) = \widetilde{T}$, which pure of codimension 2, with the same support as T.

We suppose now that $\mathrm{Ext}^i(F^{**}, G) = \{0\}$ if $i \geq 1$ (which is true if $G = G_0(p)$ with $p \gg 0$). Then from $(*)$ we deduce that $\mathrm{Ext}^1(F, G) \simeq \mathrm{Ext}^2(T, G) = \mathrm{Hom}(G^*, \widetilde{T})$. It is easy to see that the element of $\mathrm{Ext}^1(F, G)$ associated to $(*)$ is λ.

Let $\Gamma = \ker(\lambda)$. We have an exact sequence

$$0 \longrightarrow F^* \longrightarrow \mathcal{E}^* \longrightarrow \Gamma \longrightarrow 0, \qquad\qquad (**)$$

$\mathrm{Ext}^1(\Gamma, F^*) \simeq \mathrm{Hom}(F^{**}, T)$, and the element of $\mathrm{Ext}^1(\Gamma, F^*)$ associated to $(**)$ is the quotient map $F^{**} \to T$.

It follows that to obtain $\mathcal{E}$ we need only the following data: two vector bundles F^{**}, G^*, a sheaf T, and two surjective morphisms $F^{**} \to T$ and $G^* \to \widetilde{T}$.

If the preceding hypotheses are satisfied we call $(*)$ (and $(**)$, $\mathcal{E}$, $\mathcal{E}^*$) *wide extensions*.

4.2. Fine moduli spaces of wide extensions

We suppose here that $\dim(X) > 2$. We start with two open families $\mathcal{X}$, $\mathcal{Y}$ of simple vector bundles on X admitting fine moduli spaces M, N, such that

- If $E \in \mathcal{X}$ (or $\mathcal{Y}$) then the trace map $\mathrm{Ext}^2(E, E) \to H^2(\mathcal{O}_X)$ is an isomorphism.
- If $A, B \in \mathcal{X}$ (or $\mathcal{Y}$) then $\mathrm{Hom}(A, B) = \{0\}$.

The family $\mathcal{X}$ will contain the F^{**} of 4.1, and $\mathcal{Y}$ the G^*.

Let $\mathcal{Z}$ be an open family of simple pure codimension 2 sheaves on X having a smooth fine moduli space Z such that the sheaves in $\mathcal{Z}$ are *perfect*, i.e., if $T \in \mathcal{Z}$ then $\mathcal{E}xt^q(T, \mathcal{O}_X) = 0$ for $q > 2$ (this is the case if T is a vector bundle on a locally complete intersection subvariety of X of codimension 2).

We suppose that

- For every $\mathbb{F} \in \mathcal{X}$, $\mathbb{G} \in \mathcal{Y}$ we have $\mathrm{Ext}^i(\mathbb{F}, \mathbb{G}^*) = \{0\}$ if $i \geq 1$, and $\mathrm{Ext}^i(\mathbb{G}^*, \mathbb{F}) = \{0\}$ if $i < \dim(X)$.
- For every $\mathbb{F} \in \mathcal{X}$, $\mathbb{G} \in \mathcal{Y}$, $T \in \mathbb{Z}$ and surjective morphisms $\mu : \mathbb{F} \to T$, $\lambda : \mathbb{G} \to \widetilde{T}$ we have $\mathrm{Ext}^i(\ker(\mu), \mathbb{G}^*) = \mathrm{Ext}^i(\ker(\lambda), \mathbb{F}^*) = \{0\}$ if $i \geq 2$, $\mathrm{Ext}^i(\mathbb{F}, T) = \mathrm{Ext}^i(\mathbb{G}, \widetilde{T}) = \{0\}$ if $i \geq 1$, and $H^0(\mathbb{F} \otimes \widetilde{T}) = H^0(\mathbb{G} \otimes T) = \{0\}$.

Let $\mathbf{F}$ (resp. $\mathbf{G}$, $\mathbf{T}$) be universal sheaves on $M \times X$ (resp. $N \times X$, $Z \times X$). Let $\mathcal{F} = \mathcal{H}om(p_M^\sharp(\mathbf{F}), \mathbf{T})$, Let $\mathcal{G} = \mathcal{H}om(p_M^\sharp(\mathbf{G}), \widetilde{\mathbf{T}})$, which are vector bundles on $M \times N \times Z$. For every $(m, n, z) \in M \times N \times Z$ we have $\mathcal{F}_{(m,n,z)} = \mathrm{Hom}(\mathbf{F}_m, \mathbf{T}_z)$, $\mathcal{G}_{(m,n,z)} = \mathrm{Hom}(\mathbf{G}_m, \widetilde{\mathbf{T}}_z)$. Let $\mathcal{F}^{\mathrm{surj}} \subset \mathcal{F}$ (resp. $\mathcal{G}^{\mathrm{surj}} \subset \mathcal{G}$) be the open subset corresponding to surjective morphisms, and $\mathcal{S}$ the set of wide extensions defined by surjective morphisms $\mathbb{F} \to T$, $\mathbb{G} \to \widetilde{T}$, with $\mathbb{F} \in \mathcal{X}$, $\mathbb{G} \in \mathcal{Y}$, $T \in \mathcal{Z}$.

Theorem 4.2.1. *The set $\mathcal{S}$ is an open family of sheaves. It has a fine moduli space* $\mathbf{M}$ *which is canonically isomorphic to* $\mathbb{P}(\mathcal{F}^{\mathrm{surj}}) \times_{M \times N \times Z} \mathbb{P}(\mathcal{G}^{\mathrm{surj}})$.

4.3. Examples

We give two examples of fine moduli spaces of wide extensions on $\mathbb{P}_3$. Let $M_{\mathbb{P}_3}(0,1)$ denote the fine moduli space of stable rank-2 vector bundles with Chern classes 0,1, i.e., *null correlation bundles*. It is canonically isomorphic to an open subset of $\mathbb{P}_5$.

4.3.1. Rank 3 bundles. Let $n > 4$ be an integer. We consider wide extensions

$$0 \longrightarrow E(n) \longrightarrow \mathcal{E} \longrightarrow \mathcal{I}_\ell \longrightarrow 0$$

where $E \in M_{\mathbb{P}_3}(0,1)$, ℓ is a line in $\mathbb{P}_3$ and $\mathcal{I}_\ell$ its ideal sheaf. The bundle $\mathcal{E}$ has rank 3 and Chern classes $2n$, $n^2 + 2$, $2n + 2$. It follows from Theorem 4.2.1 that the set of such extensions admits a fine moduli space $\mathbf{M}$ of dimension $2n + 14$. We have

$$\dim(\operatorname{End}(\mathcal{E})) = \frac{n(n+2)(n+4)}{3} + 1, \qquad \dim(\operatorname{Ext}^2(\mathcal{E},\mathcal{E})) = 2n + 10.$$

Note that $\operatorname{Ext}^2(\mathcal{E},\mathcal{E})$ does not vanish, but that $\mathbf{M}$ is smooth.

4.3.2. Rank 4 bundles. Let m, n be positive integers, with $n > \operatorname{Max}(m,4)$. We consider wide extensions

$$0 \longrightarrow E(n) \longrightarrow \mathcal{E} \longrightarrow \ker(\pi) \longrightarrow 0$$

where $E \in M_{\mathbb{P}_3}(0,1)$ and π is a surjective morphism $E' \to \mathcal{O}_\ell(m)$, ℓ being a line in $\mathbb{P}_3$ and $E' \in M_{\mathbb{P}_3}(0,1)$. The bundle $\mathcal{E}$ has rank 4 and Chern classes $2n$, $n^2 + 3$, $4n - 2m + 2$. It follows from Theorem 4.2.1 that the set of such extensions admits a fine moduli $\mathbf{M}$ space of dimension $2n + 20$. In this case also $\operatorname{Ext}^2(\mathcal{E},\mathcal{E})$ does not vanish, but $\mathbf{M}$ is smooth.

5. Further examples

5.1. Maximal fine moduli spaces

Let $\mathcal{S}$ be an open set of sheaves admitting a fine moduli space M. Then $\mathcal{S}$ (or M) is called *maximal* if there does not exist an open set of sheaves $\mathcal{T}$, such that $\mathcal{S} \subset \mathcal{T}$, $\mathcal{S} \neq \mathcal{T}$, admitting a fine moduli space. Of course if M is projective then $\mathcal{S}$ and M are maximal. We will give an example that shows that the converse is not true.

Let n be a positive integer. Let Z be a finite subscheme of $\mathbb{P}_2$ such that $h^0(\mathcal{O}_Z) = \frac{n(n+1)}{2}$, and $\mathcal{I}_Z$ be its ideal sheaf. Then it follows easily from the Beilinson spectral sequence that Z is not contained in a curve of degree $n-1$ if and only if there is an exact sequence

$$0 \longrightarrow \mathcal{O}(-n-1) \otimes \mathbb{C}^n \longrightarrow \mathcal{O}(-n) \otimes \mathbb{C}^{n+1} \longrightarrow \mathcal{I}_Z \longrightarrow 0.$$

Let $W = \operatorname{Hom}(\mathcal{O}(-n-1)\otimes\mathbb{C}^n, \mathcal{O}(-n)\otimes\mathbb{C}^{n+1})$. We consider the action of the reductive group $G = \operatorname{SL}(n) \times \operatorname{SL}(n+1)$ on $\mathbb{P}(W)$, with the obvious linearization. If $f \in W\backslash\{0\}$, then $\mathbb{C}f$ is semi-stable if and only if $\mathbb{C}f$ is stable, if and only if for any two linear subspaces $H \subset \mathbb{C}^n$, $K \subset \mathbb{C}^{n+1}$, of dimension $p > 0$, $f(\mathcal{O}(-n-1) \otimes H)$

is not contained in $\mathcal{O}(-n) \otimes K$. Let $\mathcal{N}_n$ denote the geometric quotient of the open set of stable points of $\mathbb{P}(W)$ by G. Then $\mathcal{N}_n$ is a smooth projective variety of dimension $n(n+1)$. There exist a *universal morphism*

$$\Phi : p^*_{\mathbb{P}_2}(\mathcal{O}(-n-1)) \otimes p^*_{\mathcal{N}_n}(\mathbf{H}) \longrightarrow p^*_{\mathbb{P}_2}(\mathcal{O}(-n)) \otimes p^*_{\mathcal{N}_n}(\mathbf{K}),$$

$\mathbf{H}$ (resp. $\mathbf{K}$) being a rank n (resp. $n+1$) vector bundle on $\mathcal{N}_n$.

Let $U \subset \mathcal{N}_n$ be the open subset corresponding to injective morphisms of sheaves, and $\mathcal{U}$ the set of sheaves $\mathrm{coker}(\Phi_u)$, with $u \in U$.

Theorem 5.1.1. *The set $\mathcal{U}$ admits a fine moduli space isomorphic to U, with universal sheaf $\mathrm{coker}(\Phi_{|U})$. Moreover U is maximal, and if $n \geq 5$ then $U \neq \mathcal{N}_n$.*

Hence if $n \geq 5$, U is a maximal nonprojective fine moduli space.

Let $M(1, 0, n(n+1)/2)$ denote the moduli space of rank 1 stable sheaves with Chern classes $1, n(n+1)/2$. It is a smooth projective variety which is canonically isomorphic to $\mathrm{Hilb}^{n(n+1)/2}(\mathbb{P}_2)$. The intersection $M(1, 0, n(n+1)/2) \cap U$ corresponds to length $n(n+1)/2$ subschemes of $\mathbb{P}_2$ which are not contained in a curve of degree $n-1$, and to morphisms of vector bundles ϕ in W such that the set of points $x \in \mathbb{P}_2$ such that ϕ_x is not injective is finite.

5.2. Admissible sheaves

A coherent sheaf E on X is called *admissible* if there exists an open set of sheaves containing E and admitting a fine moduli space. If E is admissible then it follows from Theorem 2.2.1 that $\mathrm{Aut}(E)$ acts trivially on $\mathrm{Ext}^1(E, E)$ by conjugation. The converse is not true. We will show that when $X = \mathbb{P}_2$ then there exists simple sheaves E such that $\mathrm{Ext}^2(E, E) = \{0\}$ and which can be deformed into stable admissible sheaves.

Let n be a positive integer. We consider morphisms of vector bundles on $\mathbb{P}_2$

$$\mathcal{O} \otimes \mathbb{C}^{2n-1} \longrightarrow Q \otimes \mathbb{C}^{2n+1}.$$

Let $W = \mathrm{Hom}(\mathcal{O} \otimes \mathbb{C}^{2n-1}, Q \otimes \mathbb{C}^{2n+1})$. The reductive group $G = \mathrm{SL}(2n-1) \times \mathrm{SL}(2n+1)$ acts on $\mathbb{P}(W)$, and there is an obvious linearization of this action. According to [1] a morphism $\phi \in W$ is semi-stable if and only if it is stable, if and only if it is injective (as a morphism of sheaves) and $\mathrm{coker}(\phi)$ is stable. We obtain in this way an isomorphism

$$\mathbb{P}(W)^s/G \;\simeq\; M(2n+3, 2n+1, (n+1)(2n+1)),$$

where $\mathbb{P}(W)^s$ denotes the set of stable points in $\mathbb{P}(W)$ and $M(2n+3, 2n+1, (n+1)(2n+1))$ the fine moduli space of stable sheaves of rank $2n+3$ and Chern classes $2n+1, (n+1)(2n+1)$.

Proposition 5.2.1. *There are injective unstable morphisms of vector bundles $\phi \in W$ such that $E = \mathrm{coker}(\phi)$ is simple. The vector bundle E is not admissible.*

Note that E can be deformed in stable bundles in $M(2n+3, 2n+1, (n+1)(2n+1))$ (which are admissible).

5.3. Deformations of fine moduli spaces of stable sheaves

We consider here rank 6 coherent sheaves on $\mathbb{P}_2$ with Chern classes -3, 8. For these invariants semi-stability is equivalent to stability, hence the moduli space $M(6, -3, 8)$ of stable sheaves of rank 6 and Chern classes -3, 8 is a fine moduli space. It is a smooth projective variety of dimension 16. We will construct another smooth projective fine moduli space $\mathbf{M}$ which has a nonempty intersection with $M(6, -3, 8)$.

Stable sheaves of rank 6 and Chern classes -3, 8 are related to morphisms

$$\mathcal{O}(-3) \otimes \mathbb{C}^2 \longrightarrow \mathcal{O}(-2) \oplus (\mathcal{O}(-1) \otimes \mathbb{C}^7).$$

Let W be the vector space of such morphisms. We consider the action of the nonreductive group

$$G = \operatorname{Aut}(\mathcal{O}(-3) \otimes \mathbb{C}^2) \times \operatorname{Aut}(\mathcal{O}(-2) \oplus (\mathcal{O}(-1) \otimes \mathbb{C}^7))$$

on $\mathbb{P}(W)$. Let

$$H = \left\{ (I, \begin{pmatrix} I & 0 \\ \alpha & I \end{pmatrix}) \;\; ; \;\; \alpha \in \operatorname{Hom}(\mathcal{O}(-2), \mathcal{O}(-1) \otimes \mathbb{C}^7) \right\}$$

be the unipotent subgroup of G, and

$$G_{\mathrm{red}} = \operatorname{Aut}(\mathcal{O}(-3) \otimes \mathbb{C}^2) \times \operatorname{Aut}(\mathcal{O}(-2)) \times \operatorname{Aut}(\mathcal{O}(-1) \otimes \mathbb{C}^7),$$

which is a reductive subgroup of G. It is possible to construct geometric quotients of G-invariant open subsets of $\mathbb{P}(W)$ using a notion of semi-stability, depending on the choice of a rational number t such that $0 < t < 1$. Let $\phi \in W$ be a nonzero morphism. Then the point $\mathbb{C}\phi$ of $\mathbb{P}(W)$ is called *semi-stable* (resp. *stable*) with respect to t if

- $\operatorname{im}(\phi)$ is not contained in $\mathcal{O}(-1) \otimes \mathbb{C}^7$,
- For every proper linear subspace $D \subset \mathbb{C}^7$, $\operatorname{im}(\phi)$ is not contained in $\mathcal{O}(-2) \oplus (\mathcal{O}(-1) \otimes D)$.
- For every 1-dimensional linear subspace $L \subset \mathbb{C}^2$, if $K \subset \mathbb{C}$, $D \subset \mathbb{C}^7$ are linear subspaces such that $\phi(\mathcal{O}(-3) \otimes L) \subset (\mathcal{O}(-2) \otimes K) \oplus (\mathcal{O}(-1) \otimes D)$, then we have

$$t \dim(K) + \frac{1-t}{7} \dim(D) \; \geq \; \frac{1}{2}$$

(resp. $>$).

Let $\mathbb{P}(W)^{ss}(t)$ (resp. $\mathbb{P}(W)^s(t)$) denote the open set of semi-stable (resp. stable) points of $\mathbb{P}(W)$ with respect to t.

According to [8], [4], if $t > \frac{3}{10}$ then there exists a good quotient $\mathbb{P}(W)^{ss}(t)//G$ and a geometric quotient $\mathbb{P}(W)^s(t)/G$. In this range $t = \frac{1}{2}$ is the only value such that $\mathbb{P}(W)^s(t) \neq \mathbb{P}(W)^{ss}(t)$. For all t such that $\frac{3}{10} < t < \frac{1}{2}$ we obtain the same geometric quotient $\mathbf{M}_1$, and for t such that $\frac{1}{2} < t < 1$ the geometric quotient $\mathbf{M}_2$. These two quotients are smooth projective varieties.

According to [2] we have an isomorphism $M(6, -3, 8) \simeq \mathbf{M}_1$. To obtain it we associate to a morphism its cokernel.

The other moduli space $\mathbf{M}_2$ is also a fine moduli space of sheaves. The corresponding open set of sheaves consists of the cokernels of the morphisms parametrized by $\mathbf{M}_2$. All these sheaves are torsion free and have at most 2 singular points. The two fine moduli spaces are distinct and have a common dense open subset.

References

[1] Drézet, J.-M. *Fibrés exceptionnels et variétés de modules de faisceaux semi-stables sur* $\mathbb{P}_2(\mathbb{C})$. Journ. Reine Angew. Math. 380 (1987), 14–58.

[2] Drézet, J.-M., *Variétés de modules extrémales de faisceaux semi-stables sur* $\mathbb{P}^2(\mathbb{C})$. Math. Ann. 290 (1991), 727–770

[3] Drézet, J.-M. *Variétés de modules alternatives.* Ann. de l'Inst. Fourier 49 (1999), 57–139.

[4] Drézet, J.-M., *Espaces abstraits de morphismes et mutations.* Journ. Reine Angew. Math. 518 (2000), 41–93

[5] Drézet, J.-M. *Déformations des extensions larges de faisceaux.* Pacific Journ. of Math. 220, 2 (2005), 201–297.

[6] Drézet, J.-M., Narasimhan, M.S. *Groupe de Picard des variétés de modules de fibrés semi-stables sur les courbes algébriques.* Invent. Math. 97 (1989), 53–94.

[7] Drézet, J.-M., Le Potier, J. *Fibrés stables et fibrés exceptionnels sur* $\mathbb{P}_2$. Ann. Ec. Norm. Sup. 18 (1985), 193–244.

[8] Drézet, J.-M., Trautmann, G. *Moduli spaces of decomposable morphisms of sheaves and quotients by nonreductive groups.* Ann. de l'Inst. Fourier 53 (2003), 107–192.

[9] Godement, R. *Théorie des faisceaux.* Actualités scientifiques et industrielles 1252, Hermann, Paris (1964).

[10] Hartshorne, R. *Algebraic Geometry.* GTM 52, Springer-Verlag (1977).

[11] Hirschowitz, A., Laszlo, Y. *Fibrés génériques sur le plan projectif.* Math. Ann. 297 (1993), 85–102.

[12] Hirschowitz, A., Narasimhan, M.S. *Fibrés de t'Hooft spéciaux et applications.* Enumerative geometry and Classical algebraic geometry, Progr. in Math. 24 (1982).

[13] Huybrechts, D., Lehn, M. *The Geometry of Moduli Spaces of Sheaves.* Aspect of Math. E31, Vieweg (1997).

[14] Maruyama, M. *Moduli of stable sheaves I.* J. Math. Kyoto Univ. 17 (1977), 91–126.

[15] Maruyama, M. *Moduli of stable sheaves II.* J. Math. Kyoto Univ. 18 (1978), 577–614.

[16] Maruyama, M., Trautmann, G. *Limits of instantons.* Intern. Journ. of Math. 3 (1992), 213–276.

[17] Okonek, C., Schneider, M., Spindler, H. *Vector bundles on complex projective spaces.* Progress in Math. 3, Birkhäuser (1980).

[18] Ramanan, S. *The moduli spaces of vector bundles over an algebraic curve.* Math. Ann. 200 (1973), 69–84.

[19] Simpson, C.T. *Moduli of representations of the fundamental group of a smooth projective variety I.* Publ. Math. IHES 79 (1994), 47–129.

[20] Siu Y., Trautmann, G. *Deformations of coherent analytic sheaves with compact supports.* Memoirs of the Amer. Math. Soc., Vol. 29, N. 238 (1981).

[21] Yoshioka, K. *A note on the universal family of moduli of stable sheaves.* Journ. Reine Angew. Math. 496 (1998), 149–161.

Jean-Marc Drézet
Institut de Mathématiques de Jussieu
UMR 7586 du CNRS
175 rue du Chevaleret
F-75013 Paris, France
e-mail: drezet@math.jussieu.fr
URL: http://www.math.jussieu.fr/~drezet

Algebraic Cycles, Sheaves, Shtukas, and Moduli
Trends in Mathematics, 33–43
© 2007 Birkhäuser Verlag Basel/Switzerland

Moduli Spaces of Coherent Sheaves on Multiples Curves

Jean-Marc Drézet

1. Introduction

Let S be a projective smooth irreducible surface over $\mathbb{C}$. The subject of this paper is the study of coherent sheaves on multiple curves embedded in S. Coherent sheaves on singular nonreduced curves and their moduli spaces have been studied (cf. [2], [3]) and some general results have been obtained in [11] by M.-A. Inaba on moduli spaces of stable sheaves on reduced varieties of any dimension. In the case of curves we may hope of course much more detailed results.

The results of this paper come mainly from [7]. We introduce new invariants for coherent sheaves on multiple curves: the *canonical filtrations*, *generalized rank* and *degree*, and prove a *Riemann-Roch theorem*. We define the *quasi locally free sheaves* which play the same role as locally free sheaves on smooth varieties. We study more precisely the coherent sheaves on double curves. In this case we can describe completely the torsion free sheaves of generalized rank 2, and give examples of moduli spaces of stable sheaves of generalized rank 3.

This work can easily be generalized to *primitive multiple curves* which have been defined and studied by C. Bănică and O. Forster in [1] and are classified in [6].

1.1. Motivations

Moduli spaces of sheaves on multiple curves behave sometimes like moduli spaces of sheaves on varieties of higher dimension. Moduli spaces can be nonreduced (this can happen only for moduli spaces of nonlocally free sheaves), we can observe the same phenomenon in the study of unstable rank 2 vector bundles on surfaces. Moduli spaces can have multiple components with nonempty intersections, this is the case also for moduli spaces of rank 2 stable sheaves on $\mathbb{P}_3$. We can hope that the study of these phenomenons on multiple curves will be simpler than in the higher-dimensional cases, and will give ideas to treat them.

2. Preliminaries

2.1. Multiple curves

Let S be a smooth projective irreducible surface over $\mathbb{C}$. Let $C \subset S$ be a smooth irreducible projective curve, and $n \geq 2$ be an integer. Let $s \in H^0(\mathcal{O}_S(C))$ be a section whose zero scheme is C and C_n be the curve defined by $s^n \in H^0(\mathcal{O}_S(nC))$. We have a filtration $C = C_1 \subset C_2 \subset \cdots \subset C_n$, hence a coherent sheaf on C_i with $i < n$ can be viewed as a coherent sheaf on C_n.

Let $\mathcal{I}_{C_i}$ denote the ideal sheaf of C_i in C_n. Then $L = \mathcal{I}_C/\mathcal{I}_{C_2}$ is a line bundle on C and $\mathcal{I}_{C_j}/\mathcal{I}_{C_{j+1}} \simeq L^j$. We have $L \simeq \mathcal{O}_C(-C)$.

Let $\mathcal{O}_n = \mathcal{O}_{C_n}$. If $0 < m < n$ we can view $\mathcal{O}_m$ as a sheaf of $\mathcal{O}_n$-modules.

2.2. Extension of vector bundles

Theorem 2.2.1. *If $1 \leq i \leq n$ then every vector bundle on C_i can be extended to a vector bundle on C_n.*

2.2.2. Parametrization. Let $\mathbb{E}$ be a vector bundle on C_n, and $\mathbb{E}_{n-1} = \mathbb{E}_{|C_{n-1}}$, $E = \mathbb{E}_{|C}$. Then we have an exact sequence

$$0 \longrightarrow \mathbb{E}_{n-1} \otimes \mathcal{O}_{n-1}(-C) \longrightarrow \mathbb{E} \longrightarrow E \longrightarrow 0 ,$$

(with $\mathcal{O}_{n-1}(-C) = \mathcal{O}_S(-C)_{|C_{n-1}}$).

Conversely, let $\mathbb{E}_{n-1}$ be a vector bundle on C_{n-1} and $E = \mathbb{E}_{n-1|C}$. Then using suitable locally free resolutions on C_n one can find canonical isomorphisms

$$\mathcal{H}om(E, \mathbb{E}_{n-1} \otimes \mathcal{O}_{n-1}(-C)) \simeq E^* \otimes E \otimes L^{n-1},$$

$$\mathcal{E}xt^1_{\mathcal{O}_n}(E, \mathbb{E}_{n-1} \otimes \mathcal{O}_{n-1}(-C)) \simeq E^* \otimes E.$$

It follows that we have an exact sequence

$$0 \longrightarrow H^1(E^* \otimes E \otimes L^{n-1}) \longrightarrow \mathrm{Ext}^1_{\mathcal{O}_n}(E, \mathbb{E}_{n-1} \otimes \mathcal{O}_{n-1}(-C)) \longrightarrow \mathrm{End}(E) \longrightarrow 0.$$

Now let $0 \to \mathbb{E}_{n-1} \otimes \mathcal{O}_{n-1}(-C) \to \mathcal{E} \to E \to 0$ be an extension, associated to $\sigma \in \mathrm{Ext}^1_{\mathcal{O}_n}(E, \mathbb{E}_{n-1} \otimes \mathcal{O}_{n-1}(-C))$. Then $\mathcal{E}$ is locally free if and only the image of σ in $\mathrm{End}(E)$ is an isomorphism. In particular if E is simple (i.e., $\mathrm{End}(E) \simeq \mathbb{C}$) then the set of vector bundles on C_n extending $\mathbb{E}_{n-1}$ can be identified with $H^1(E^* \otimes E \otimes L^{n-1})$.

2.3. Picard group

It follows from 2.2 that if $\deg(L) < 0$ we have an exact sequence of abelian groups

$$0 \longrightarrow H^1(L^{n-1}) \longrightarrow \mathrm{Pic}(C_n) \xrightarrow{r_n} \mathrm{Pic}(C_{n-1}) \longrightarrow 0,$$

where r_n is the restriction morphism. Let $\mathbf{P}_n \subset \mathrm{Pic}(C_n)$ be the subgroup consisting of line bundles whose restriction to C is $\mathcal{O}_C$. Then we have a filtration of abelian groups $O = G_0 \subset G_1 \subset \cdots \subset G_{n-1} = \mathbf{P}_n$ such that $G_i/G_{i-1} \simeq H^1(L^i)$ for $1 \leq i \leq n - 1$. Here G_i is the subgroup of $\mathbf{P}_n$ of line bundles whose restriction to C_{n-i} is trivial. It follows from this filtration that $\mathbf{P}_n$ is isomorphic to a product of groups $\mathbb{G}_a$, i.e., to a finite-dimensional vector space.

3. Canonical filtrations – generalized rank and degree

Let $P \in C$ and $z \in \mathcal{O}_{n,P}$ be a local equation of C. Let $x \in \mathcal{O}_{n,P}$ be such that x and z generate the maximal ideal of $\mathcal{O}_{n,P}$. Let M be a $\mathcal{O}_{n,P}$-module of finite type and $\mathcal{E}$ a coherent sheaf on C_n.

3.1. Canonical filtrations

3.1.1. First canonical filtration.
For $1 \leq i \leq n + 1$, let $M_i = z^{i-1} M$. The *first canonical filtration* (or simply the canonical filtration) of M is

$$M_{n+1} = \{0\} \subset M_n \subset \cdots \subset M_2 \subset M_1 = M.$$

We have

$$M_i/M_{i+1} \simeq M_i \otimes_{\mathcal{O}_{n,P}} \mathcal{O}_{C,P}, \qquad M/M_{i+1} \simeq M \otimes_{\mathcal{O}_{n,P}} \mathcal{O}_{i,P}.$$

Let $Gr(M) = \bigoplus_{i=1}^{n} M_i/M_{i+1}$. It is a $\mathcal{O}_{C,P}$-module.

Similarly one can define the first canonical filtration

$$0 = \mathcal{E}_{n+1} \subset \mathcal{E}_n \subset \cdots \subset \mathcal{E}_2 \subset \mathcal{E}_1 = \mathcal{E}$$

where the $\mathcal{E}_i$ are defined inductively: $\mathcal{E}_{i+1}$ is the kernel of the restriction $\mathcal{E}_i \to \mathcal{E}_{i|C}$. Let
$Gr(\mathcal{E}) = \bigoplus_{i=1}^{n} \mathcal{E}_i/\mathcal{E}_{i+1}$. It is concentrated on C.

3.1.2. Second canonical filtration.
The *second canonical filtration* of M

$$M^{(n+1)} = \{0\} \subset M^{(n)} \subset \cdots \subset M^{(2)} \subset M^{(1)} = M$$

is defined by $M^{(i)} = \{u; z^{n+1-i} u = 0\}$. In the same way we can define the second canonical filtration of $\mathcal{E}$

$$0 = \mathcal{E}^{(n+1)} \subset \mathcal{E}^{(n)} \subset \cdots \subset \mathcal{E}^{(2)} \subset \mathcal{E}^{(1)} = \mathcal{E}.$$

3.1.3. Basic properties.
1. We have $\mathcal{E}_i = 0$ if and only if $\mathcal{E}$ is concentrated on C_{i-1}.
2. $\mathcal{E}_i$ is concentrated on C_{n+1-i} and its first canonical filtration is
$0 = \mathcal{E}_{n+1} \subset \mathcal{E}_n \subset \cdots \subset \mathcal{E}_{i+1} \subset \mathcal{E}_i$; $\mathcal{E}^{(i)}$ is concentrated on C_{n+1-i} and its
second canonical filtration is $0 = \mathcal{E}^{(n+1)} \subset \mathcal{E}^{(n)} \subset \cdots \subset \mathcal{E}^{(i+1)} \subset \mathcal{E}^{(i)}$.
3. Canonical filtrations are preserved by morphisms of sheaves.

3.1.4. Examples.
1. If $\mathcal{E}$ is locally free and $E = \mathcal{E}_{|C}$, then $\mathcal{E}_i = \mathcal{E}^{(i)}$ and $\mathcal{E}_i/\mathcal{E}_{i+1} = E \otimes L^{i-1}$ for $1 \leq i \leq n$.
2. If $\mathcal{E}$ is the ideal sheaf of a finite subscheme T of C then $\mathcal{E}_i/\mathcal{E}_{i+1} = (\mathcal{O}_C(-T) \otimes L^{i-1}) \oplus \mathcal{O}_T$ if $1 \leq i < n$, $\mathcal{E}_n = \mathcal{O}_C(-T) \otimes L^{n-1}$, $\mathcal{E}^{(i)}/\mathcal{E}^{(i+1)} = L^{i-1}$ if $2 \leq i \leq n$ and $\mathcal{E}^{(1)}/\mathcal{E}^{(2)} = \mathcal{O}_C(-T)$.

3.2. Generalized rank and degree and Riemann–Roch theorem

The integer $R(M) = rk(Gr(M))$ is called the *generalized rank* of M.

The integer $R(\mathcal{E}) = rk(Gr(\mathcal{E}))$ is called the *generalized rank* of $\mathcal{E}$, and $\mathrm{Deg}(\mathcal{E}) = \deg(Gr(\mathcal{E}))$ is called the *generalized degree* of $\mathcal{E}$.

3.2.1. Example. If $\mathcal{E}$ is locally free and $E = \mathcal{E}_{|C}$, then $R(\mathcal{E}) = n.rk(E)$, and $\mathrm{Deg}(\mathcal{E}) = n.\deg(E) + \frac{n(n-1)}{2} rk(E) \deg(L)$.

Theorem 3.2.2 (Riemann–Roch theorem). *We have*

$$\chi(\mathcal{E}) = \mathrm{Deg}(\mathcal{E}) + R(E)(1 - g_C).$$

This follows immediately from the first canonical filtration. The generalized rank can be computed as follows

Theorem 3.2.3. *We have*

$$R(M) \;=\; \lim_{p \to \infty} \left(\frac{1}{p} \dim_{\mathbb{C}}(M \otimes_{\mathcal{O}_{n,P}} \mathcal{O}_{n,P}/(x^p)) \right).$$

This result can be used to prove that the generalized rank and degree are *additive*:

Corollary 3.2.4.
1. *Let* $0 \longrightarrow M' \longrightarrow M \longrightarrow M'' \longrightarrow 0$ *be an exact sequence of* $\mathcal{O}_{n,P}$*-modules of finite type. Then we have* $R(M) = R(M') + R(M'')$.
2. *Let* $0 \longrightarrow \mathcal{E}' \longrightarrow \mathcal{E} \longrightarrow \mathcal{E}'' \longrightarrow 0$ *be an exact sequence of coherent sheaves on* C_n*. Then we have* $R(\mathcal{E}) = R(\mathcal{E}') + R(\mathcal{E}'')$ *and* $\mathrm{Deg}(\mathcal{E}) = \mathrm{Deg}(\mathcal{E}') + \mathrm{Deg}(\mathcal{E}'')$.

Proof. This follows from Theorem 3.2.3. The assertion on degrees follows from the one on ranks and from the Riemann–Roch theorem. $\qquad\square$

The generalized rank and degree are invariant by deformation.

3.2.5. Hilbert polynomial and (semi-)stability. Let $\mathcal{D}$ be a line bundle on C_n and $D = \mathcal{D}_{|C}$. Then for every coherent sheaf $\mathcal{E}$ on C_n we have

$$R(\mathcal{E} \otimes \mathcal{D}) = R(\mathcal{E}), \quad Deg(\mathcal{E} \otimes \mathcal{D}) = Deg(\mathcal{E}) + R(\mathcal{E}) \deg(D).$$

Hence if $\mathcal{O}(1)$ is a very ample line bundle on C_n then the Hilbert polynomial of $\mathcal{E}$ with respect to $\mathcal{O}(1)$ is

$$P_{\mathcal{E}}(m) \;=\; \chi(\mathcal{E}) + R(\mathcal{E}) \deg(\mathcal{O}(1)_{|C}).m \quad .$$

It follows that a coherent sheaf $\mathcal{E}$ of positive rank is *semi-stable* (resp. *stable*) if and only if it is pure of dimension 1 (i.e., it has no subsheaf with finite support) and if for every proper subsheaf $\mathcal{F} \subset \mathcal{E}$ we have

$$\frac{\mathrm{Deg}(\mathcal{F})}{R(\mathcal{F})} \;\leq\; \frac{\mathrm{Deg}(\mathcal{E})}{R(\mathcal{E})} \qquad (\text{resp. } < \ ).$$

4. Quasi locally free sheaves

Let $P \in C$ and $z \in \mathcal{O}_{n,P}$ be a local equation of C.

Let M be a $\mathcal{O}_{n,P}$-module of finite type. Then M is called *quasi free* if there exist nonnegative integers $m_1, \dots, m_n$ and an isomorphism $M \simeq \oplus_{i=1}^{n} m_i \mathcal{O}_{i,P}$. The integers $m_1, \dots, m_n$ are uniquely determined: it is easy to recover them from the first canonical filtration of M. We say that $(m_1, \dots, m_n)$ is the *type* of M.

Let $\mathcal{E}$ be a coherent sheaf on C_n. We say that $\mathcal{E}$ is *quasi free at P* if $\mathcal{E}_P$ is quasi free, and that $\mathcal{E}$ is *quasi locally free* if it is quasi free at every point of C.

Theorem 4.0.6. *The $\mathcal{O}_{n,P}$-module M is quasi free if and only if $Gr(M)$ is a free $\mathcal{O}_{C,P}$-module, if and only if all the M_i/M_{i+1} are free $\mathcal{O}_{C,P}$-modules.*

It follows that the set of points $P \in C$ such that $\mathcal{E}$ is quasi free at P is open and nonempty, and that $\mathcal{E}$ is quasi locally free if and only if $Gr(\mathcal{E})$ is a vector bundle on C, if and only if all the $\mathcal{E}_i/\mathcal{E}_{i+1}$ are vector bundles on C.

5. Coherent sheaves on double curves

We work in this section on C_2, that we call a *double curve*. If $\mathcal{E}$ is a coherent sheaf on C_2, let $E_{\mathcal{E}} \subset \mathcal{E}$ (resp. $G_{\mathcal{E}} \subset \mathcal{E}$) be its first (resp. second) canonical filtration. Let $F_{\mathcal{E}} = \mathcal{E}/E_{\mathcal{E}}$.

For $P \in C$, let z be an equation of C in $\mathcal{O}_{2,P}$. Let $x \in \mathcal{O}_{2,P}$ such that x, z generate the maximal ideal of $\mathcal{O}_{2,P}$.

5.1. Quasi locally free sheaves

5.1.1. Locally free resolutions of vector bundles on C.
Let F be a vector bundle on C. Using Theorem 2.2.1 we find a locally free sheaf $\mathbb{F}$ on C_2 such that $\mathbb{F}_{|C} = F$, and a free resolution of F on C_2:

$$\cdots \mathbb{F} \otimes \mathcal{O}_2(-2C) \longrightarrow \mathbb{F} \otimes \mathcal{O}_2(-C) \longrightarrow \mathbb{F} \longrightarrow F \longrightarrow 0.$$

From this it follows that for every vector bundle E on C we have

$$\mathcal{E}xt^i_{\mathcal{O}_2}(F, E) \simeq \mathcal{H}om(F \otimes L^i, E)$$

for $i \geq 1$.

5.1.2. Construction of quasi locally free sheaves.
Let $\mathcal{F}$ be a quasi locally free coherent sheaf on C_2. Let $E = E_{\mathcal{F}}$, $F = F_{\mathcal{F}}$. We have an exact sequence

$$(*) \qquad 0 \longrightarrow E \longrightarrow \mathcal{F} \longrightarrow F \longrightarrow 0$$

and E, F are vector bundles on C_2. The canonical morphism $\mathcal{F} \otimes \mathcal{I}_C \to \mathcal{F}$ comes from a surjective morphism $\Phi_{\mathcal{F}} : F \otimes L \to E$.

Conversely suppose we want to construct the quasi locally free sheaves $\mathcal{F}$ whose first canonical filtration gives the exact sequence $(*)$. For this we need to compute $\mathrm{Ext}^1_{\mathcal{O}_2}(F, E)$. The Ext spectral sequence gives the exact sequence

$$0 \longrightarrow \mathrm{Ext}^1_{\mathcal{O}_C}(F, E) \longrightarrow \mathrm{Ext}^1_{\mathcal{O}_2}(F, E) \overset{\beta}{\longrightarrow} \mathrm{Hom}(F \otimes L, E) \longrightarrow 0$$

$$\| \qquad\qquad\qquad\qquad\qquad\qquad\qquad\qquad \|$$

$$H^1(\mathcal{H}om(F, E)) \qquad\qquad\qquad\qquad H^0(\mathcal{E}xt^1_{\mathcal{O}_2}(F, E))$$

Let $\sigma \in \mathrm{Ext}^1_{\mathcal{O}_2}(F, E)$ and $0 \to E \to \mathcal{E} \to F \to 0$ the corresponding extension. Then it is easy to see that this exact sequence comes from the canonical filtration of $\mathcal{E}$ if and only if $\beta(\sigma)$ is surjective. Moreover in this case we have $\Phi_{\mathcal{E}} = \beta(\sigma)$.

5.1.3. Second canonical filtration. Let $\Gamma_{\mathcal{F}} = \Gamma$ be the kernel of the surjective morphism $\Phi_{\mathcal{F}} \otimes I_{L^*} : F \to E \otimes L^*$ and G the kernel of the composition

$$\mathcal{F} \longrightarrow F \overset{\Phi_{\mathcal{F}} \otimes I_{L^*}}{\longrightarrow} E \otimes L^* \quad ,$$

which is also surjective. Then G in the maximal subsheaf of $\mathcal{F}$ which is concentrated on C. In other words, $G \subset \mathcal{F}$ is the second canonical filtration of $\mathcal{F}$, and $G = G_{\mathcal{F}}$.

5.1.4. Duality and tensor products. If M is a $\mathcal{O}_{2,P}$-module of finite type, let $M^\vee$ be the dual of M: $M^\vee = \mathrm{Hom}(M, \mathcal{O}_{2,P})$. If $\mathcal{F}$ is a coherent sheaf on C_2 let $\mathcal{E}^\vee$ denote the dual sheaf of $\mathcal{E}$, i.e., $\mathcal{E}^\vee = \mathcal{H}om(\mathcal{E}, \mathcal{O}_2)$. If N is a $\mathcal{O}_{C,P}$-module of finite type, let N^* be the dual of N: $N^* = \mathrm{Hom}(N, \mathcal{O}_{C,P})$. If E is a coherent sheaf on C let E^* be the dual of E on C. We use different notations on C and C_2 because $E^\vee \neq E^*$, we have $E^\vee = E^* \otimes L$.

Let $\mathcal{F}$ be a quasi locally free sheaf on C_2. Then $\mathcal{F}^\vee$ is also quasi locally free, and we have

$$E_{\mathcal{F}^\vee} \simeq E^*_{\mathcal{F}} \otimes L^2, \qquad F_{\mathcal{F}^\vee} \simeq G^*_{\mathcal{F}} \otimes L, \qquad G_{\mathcal{F}^\vee} \simeq F^*_{\mathcal{F}} \otimes L.$$

Let $\mathcal{E}, \mathcal{F}$ be quasi locally free sheaves on C_2. Then $\mathcal{E} \otimes \mathcal{F}$ is quasi locally free and we have $E_{\mathcal{E} \otimes \mathcal{F}} = E_{\mathcal{E}} \otimes E_{\mathcal{F}} \otimes L^*$, $F_{\mathcal{E} \otimes \mathcal{F}} = F_{\mathcal{E}} \otimes F_{\mathcal{F}}$, $G_{\mathcal{E} \otimes \mathcal{F}} = G_{\mathcal{E}} \otimes G_{\mathcal{F}} \otimes L^*$.

If $\mathcal{E}, \mathcal{F}$ are quasi locally free sheaves on C_2 then the canonical morphism $\mathcal{E}^\vee \otimes \mathcal{F} \to \mathcal{H}om(\mathcal{E}, \mathcal{F})$ is not in general an isomorphism. For instance $\mathcal{O}_C^\vee \otimes \mathcal{O}_C = L$ but $\mathcal{H}om(\mathcal{O}_C, \mathcal{O}_C) = \mathcal{O}_C$. We have an exact sequence

$$0 \longrightarrow \Gamma^*_{\mathcal{E}} \otimes \Gamma_{\mathcal{F}} \otimes L \longrightarrow \mathcal{E}^\vee \otimes \mathcal{F} \longrightarrow \mathcal{H}om(\mathcal{E}, \mathcal{F}) \longrightarrow \Gamma^*_{\mathcal{E}} \otimes \Gamma_{\mathcal{F}} \longrightarrow 0 \ .$$

The sheaves canonically associated to $\mathcal{H} = \mathcal{H}om(\mathcal{E}, \mathcal{F})$ are:

$$E_{\mathcal{H}} = \mathcal{H}om(E_{\mathcal{E}}, E_{\mathcal{F}} \otimes L), \qquad G_{\mathcal{H}} = \mathcal{H}om(F_{\mathcal{E}}, G_{\mathcal{F}}),$$

and we have exact sequences

$$0 \longrightarrow \mathcal{H}om(\Gamma_{\mathcal{E}}, E_{\mathcal{F}}) \oplus \mathcal{H}om(E_{\mathcal{E}}, \Gamma_{\mathcal{F}} \otimes L) \longrightarrow F_{\mathcal{H}}$$
$$\longrightarrow \mathcal{H}om(\Gamma_{\mathcal{E}}, \Gamma_{\mathcal{F}}) \oplus \mathcal{H}om(E_{\mathcal{E}}, E_{\mathcal{F}}) \longrightarrow 0,$$

$$0 \longrightarrow \mathcal{H}om(\Gamma_{\mathcal{E}}, E_{\mathcal{F}}) \oplus \mathcal{H}om(E_{\mathcal{E}}, \Gamma_{\mathcal{F}} \otimes L)$$
$$\longrightarrow \Gamma_{\mathcal{H}} \longrightarrow \mathcal{H}om(\Gamma_{\mathcal{E}}, \Gamma_{\mathcal{F}}) \longrightarrow 0.$$

5.2. Torsion free sheaves

A coherent sheaf on C_2 is called *torsion free* if it is pure of dimension 1, i.e., if it has no subsheaf with a zero-dimensional support.

5.2.1. First properties. Let $\mathcal{E}$ be a coherent sheaf on C_2, $E \subset \mathcal{E}$ its canonical filtration. Suppose that E is locally free, this is the case if $\mathcal{E}$ is torsion free. The quotient $\mathcal{E}/E$ may be nonlocally free. Let $\mathcal{E}/E \simeq F \oplus T$, where F is locally free on C and T supported on a finite subset of C. The kernel of the morphism $\mathcal{E} \to T$ deduced from this isomorphism is a quasi locally free subsheaf $\mathcal{F}$ of $\mathcal{E}$ containing E, and $E \subset \mathcal{F}$ is its canonical filtration. Note that $\mathcal{F}$ may not be unique, it depends on the above isomorphism. The morphism $\Phi_{\mathcal{F}} : F \otimes L \to E$ does not depend on $\mathcal{F}$ since it comes from the canonical morphism $\mathcal{E} \otimes \mathcal{I}_C \to \mathcal{E}$. So we will note $\Phi_{\mathcal{E}} = \Phi_{\mathcal{F}}$.

If T is a torsion sheaf on C, let $\widetilde{T} = \mathcal{E}xt^1_{\mathcal{O}_C}(T, \mathcal{O}_C)$, which is (noncanonically) isomorphic to T.

Lemma 5.2.2. *Let T a torsion sheaf on C and $\mathbb{F}$ a vector bundle on C_2. Let $F = \mathbb{F}_{|C}$. Then*

1. *The canonical morphism $\mathrm{Ext}^1_{\mathcal{O}_2}(T, \mathbb{F}) \to \mathrm{Ext}^1_{\mathcal{O}_2}(T, F)$ vanishes.*
2. *We have a canonical isomorphism $\mathrm{Ext}^1_{\mathcal{O}_2}(T, \mathbb{F}) \simeq \mathrm{Ext}^1_{\mathcal{O}_2}(T, F \otimes L)$, and $\mathrm{Ext}^i_{\mathcal{O}_2}(T, \mathbb{F}) = \{0\}$ if $i \geq 2$.*
3. *If $j \geq 1$ we have $\mathrm{Ext}^j_{\mathcal{O}_2}(T, F) \simeq \mathrm{Ext}^1_{\mathcal{O}_2}(T, F \otimes L^{1-j}) \simeq \mathrm{Hom}(F^* \otimes L^{j-1}, \widetilde{T})$.*

Let $\sigma_{\mathcal{E}}$ be the element of $\mathrm{Ext}^1_{\mathcal{O}_2}(T, E)$ coming from the exact sequence $0 \to E \to \mathcal{E} \to F \oplus T \to 0$. From the preceding lemma we can view $\sigma_{\mathcal{E}}$ as a morphism $E^* \to \widetilde{T}$. *This morphism is surjective if and only if $\mathcal{E}$ is torsion free.*

5.2.3. Construction of torsion free sheaves. We start with the following data: two vector bundles E, F on C, a torsion sheaf T on C and surjective morphisms $\Phi : F \otimes L \to E$ and $\sigma : E^* \to \widetilde{T}$.

Let $\mathcal{F}$ a quasi locally free sheaf on C_2 such that $\Phi_{\mathcal{F}} = \Phi$ (see 5.1.2). From $\mathcal{F}$ and σ we get an element of $\mathrm{Ext}^1_{\mathcal{O}_2}(F \oplus T, E)$ corresponding to an extension $0 \to E \to \mathcal{E} \to F \oplus T \to 0$. It is then easy to see that $E \subset \mathcal{E}$ is the canonical filtration of $\mathcal{E}$ and that $\sigma_{\mathcal{E}} = \sigma$.

5.2.4. Second canonical filtration. Let G be the kernel of the morphism

$$\mathcal{E} \dashrightarrow F \xrightarrow{\ \Phi_{\mathcal{E}} \otimes I_{L^*}\ } E \otimes L^*.$$

Then G in the maximal subsheaf of $\mathcal{E}$ which is concentrated on C. In other words, $G \subset \mathcal{E}$ is the second canonical filtration of $\mathcal{F}$.

Proposition 5.2.5. *There exist a quasi locally free sheaf V and a surjective morphism $V \to T$ such that $\mathcal{E} \simeq \ker(\alpha)$.*

Proposition 5.2.6.
1. *A $\mathcal{O}_{2,P}$-module is reflexive if and only it is torsion free.*
2. *A coherent sheaf on C_2 is reflexive if and only if it is torsion free.*

If $m \geq 1$, let $I_{m,P} =\subset \mathcal{O}_{2,P}$ be the ideal generated by x^m and z.

5.2.7. Local structure of torsion free sheaves. Let M be a torsion free $\mathcal{O}_{2,P}$-module. Then there exist integers m, q, $n_1, \ldots, n_p$ such that

$$M \simeq \left(\bigoplus_{i=1}^{p} I_{n_i,P} \right) \oplus m\mathcal{O}_{2,P} \oplus q\mathcal{O}_{C,P}.$$

5.3. Deformations of sheaves

If E is a coherent sheaf on C then the canonical morphism

$$\mathrm{Ext}^1_{\mathcal{O}_n}(E, E) \longrightarrow \mathrm{Ext}^1_{\mathcal{O}_S}(E, E)$$

is an isomorphism.

Let M be a $\mathcal{O}_{2,P}$-module, $M_2 \subset M$ its canonical filtration. Let $r_0(M) = rk(M_2)$. Then we have $R(M) \geq 2r_0(M)$. If M is quasi free then we have $R(M) = 2r_0(M)$ if and only if M is free.

Proposition 5.3.1. *Let M be a quasi free $\mathcal{O}_{2,P}$-module, and r_0 an integer such that $0 < 2r_0 \leq R(M)$. Then M can be deformed in quasi free modules N such that $r_0(N) = r_0$ if and only if $r_0 \geq r_0(M)$.*

It follows that if a quasi locally free sheaf $\mathcal{E}$ on C_2 can be deformed in quasi locally free sheaves $\mathcal{F}$ such that $r_0(\mathcal{F}) = r_0$ then we must have $r_0 \geq r_0(\mathcal{E})$. The converse is not true: if $R(\mathcal{E})$ is even, and if $\mathcal{E}$ can be deformed in locally free sheaves, then we have $\mathrm{Deg}(\mathcal{E}) \equiv \frac{R(\mathcal{E})}{2} \deg(L) \pmod 2$. Hence if this equality is not true it is impossible to deform $\mathcal{E}$ in locally free sheaves.

5.4. Rank 2 sheaves

There are 3 kinds of torsion free sheaves of generalized rank 2 on C_2:

- the rank 2 vector bundles on C,
- the line bundles on C_2,
- the sheaves of the form $\mathcal{I}_Z \otimes \mathcal{L}$, where Z is a nonempty finite subscheme of C, $\mathcal{I}_Z$ is its ideal sheaf on C_2 and $\mathcal{L}$ a line bundle on C_2.

Let Z be a finite subscheme of C and $\mathcal{L}$ a line bundle on C_2. Let $\mathcal{E} = \mathcal{I}_Z \otimes \mathcal{L}$. Then Z is uniquely determined by $\mathcal{E}$, but $\mathcal{L}$ need not be unique. The integer $i_0(\mathcal{E}) = h^0(\mathcal{O}_Z)$ is called the *index* of $\mathcal{E}$ (in particular the index of a line bundle on C_2 is 0). It is invariant by deformation of $\mathcal{E}$.

Proposition 5.4.1. *Let $p \geq 0$, d be integers and E a rank 2 vector bundle of degree d on C. Let $q = \frac{1}{2}(d + \deg(L) + p)$. Then*

1. *If E can be deformed in torsion free sheaves on C_2 that are not concentrated on C and of index p, then there exist a line bundle V on C of degree q and a nonzero morphism $\alpha : V \to E$ such that $\mathrm{Hom}((E/\mathrm{im}(\alpha)) \otimes L, V) \neq \{0\}$.*
2. *If E has a sub-line bundle V of degree q such that $\mathrm{Hom}((E/\mathrm{im}(\alpha))\otimes L, V) \neq \{0\}$ then E can be deformed in torsion free sheaves on C_2, nonconcentrated on C and of index p.*

5.5. Rank 3 sheaves

Let $l = -\deg(L)$. We suppose that $l = C^2 \geq 1$. Let $\mathcal{E}$ be a quasi locally free sheaf on C_2 not concentrated on C and of generalized rank 3. Then $\mathcal{E}$ is locally isomorphic to $\mathcal{O}_2 \oplus \mathcal{O}_C$. We have a commutative diagram with exact rows and columns

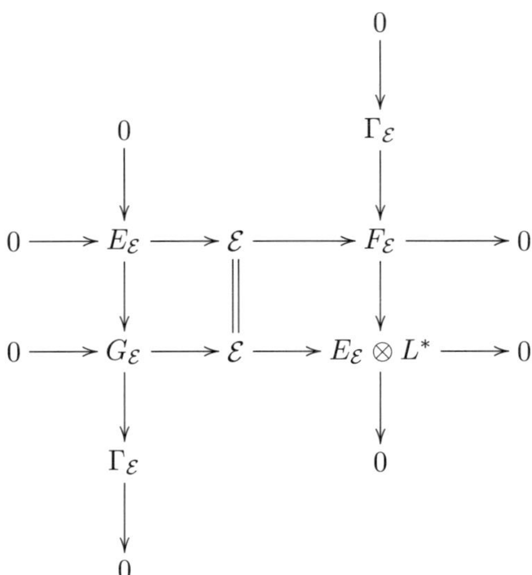

where the rows come from the first and second canonical filtrations. We have $rk(E_{\mathcal{E}}) = rk(\Gamma_{\mathcal{E}}) = 1$ and $rk(F_{\mathcal{E}}) = rk(G_{\mathcal{E}}) = 2$.

The ranks and degrees of $E_{\mathcal{E}}$, $F_{\mathcal{E}}$, $G_{\mathcal{E}}$ and $\Gamma_{\mathcal{E}}$ are invariant by deformation of $\mathcal{E}$.

Proposition 5.5.1. *The sheaf $\mathcal{E}$ is (semi-)stable if and only if*

(i) *For every sub-line bundle D' of $G_{\mathcal{E}}$ we have $\deg(D') \leq \mu(\mathcal{E})$ (resp. $<$).*
(ii) *For every quotient line bundle D'' of $F_{\mathcal{E}}$ we have $\mu(\mathcal{E}) \leq \deg(D'')$ (resp. $<$).*

It follows that if $F_{\mathcal{E}}$ and $G_{\mathcal{E}}$ are stable then so is $\mathcal{E}$. Let $\epsilon = \deg(E_{\mathcal{E}})$, $\gamma = \deg(\Gamma_{\mathcal{E}})$. We have then $\mathrm{Deg}(\mathcal{E}) = 2\epsilon + \gamma + l$. By considering the subsheaves $E_{\mathcal{E}}$, $G_{\mathcal{E}}$ of $\mathcal{E}$ we find that if $\mathcal{E}$ is semi-stable (resp. stable) then $\gamma - 2l \leq \epsilon \leq l + \gamma$ (resp. $<$).

We suppose now that $\gamma - l < \epsilon < \gamma$, which is equivalent to $\mu(E_\mathcal{E}) < \mu(G_\mathcal{E})$ and $\mu(\Gamma_\mathcal{E}) < \mu(F_\mathcal{E})$. We impose these conditions to allow $F_\mathcal{E}$ and $G_\mathcal{E}$ to be stable. In this case we get distinct components of the moduli spaces of stable sheaves corresponding to sheaves $\mathcal{E}$ such that $E_\mathcal{E}$ and $\Gamma_\mathcal{E}$ have fixed degrees. In the description of these components two moduli spaces of *Brill-Noether pairs* on C will appear: the one corresponding to $E_\mathcal{E} \subset G_\mathcal{E}$, and the one corresponding to $\Gamma_\mathcal{E} \subset F_\mathcal{E}$.

Let $M(3, 2\epsilon + \gamma + l)$ denote the moduli space of semi-stable sheaves on C_2 of generalized rank 3 and generalized degree $2\epsilon + \gamma + l$. Let $\mathcal{M}_s(\epsilon, \gamma)$ be the open subscheme of $M(3, 2\epsilon + \gamma + l)$ corresponding to quasi locally free sheaves $\mathcal{E}$ not concentrated on C, such that $\deg(E_\mathcal{E}) = \epsilon$, $\deg(\Gamma_\mathcal{E}) = \gamma$ and such that $F_\mathcal{E}$, $G_\mathcal{E}$ are stable.

Proposition 5.5.2. *The variety $\mathcal{M}_s(\epsilon, \gamma)$ is irreducible of dimension $5g + 2l - 4$. The associated reduced variety is smooth.*

The varieties $\mathcal{M}_s(\epsilon, \gamma)$ are not reduced. Let $\mathcal{M}_s^{red}(\epsilon, \gamma)$ be the reduced variety corresponding to $\mathcal{M}_s(\epsilon, \gamma)$. Then if $\mathcal{E} \in \mathcal{M}_s(\epsilon, \gamma)$ then the cokernel of the canonical map $T\mathcal{M}_s^{red}(\epsilon, \gamma)_\mathcal{E} \to T\mathcal{M}_s(\epsilon, \gamma)_\mathcal{E}$ is isomorphic to $H^0(L^*)$.

References

[1] Bănică, C., Forster, O. *Multiple structures on plane curves.* In Contemporary Mathematics 58, Proc. of Lefschetz Centennial Conf. (1984), AMS, 47–64.

[2] Bhosle Usha N. *Generalized parabolic bundles and applications to torsion free sheaves on nodal curves.* Arkiv for Matematik 30 (1992), 187–215.

[3] Bhosle Usha N. *Picard groups of the moduli spaces of vector bundles.* Math. Ann. 314 (1999) 245–263.

[4] Brambila-Paz, L., Mercat, V., Newstead, P.E., Ongay, F. *Nonemptiness of Brill-Noether loci.* Intern. J. Math. 11 (2000), 737–760.

[5] Drézet, J.-M. *Faisceaux cohérents sur les courbes multiples.* Collectanea Mathematica 57, 2 (2006), 121–171.

[6] Drézet, J.-M. *Paramétrisation des courbes multiples primitives* Preprint (2006), math.AG/0605726.

[7] Drézet, J.-M. *Déformations des extensions larges de faisceaux.* Pacific Journ. of Math. 220, 2 (2005), 201–297.

[8] Godement, R. *Théorie des faisceaux.* Actualités scientifiques et industrielles 1252, Hermann, Paris (1964).

[9] Hartshorne, R. *Algebraic Geometry.* GTM 52, Springer-Verlag (1977).

[10] Huybrechts, D., Lehn, M. *The Geometry of Moduli Spaces of Sheaves.* Aspect of Math. E31, Vieweg (1997).

[11] Inaba, M.-A. *On the moduli of stable sheaves on some nonreduced projective schemes.* Journ. of Alg. Geom. 13 (2004), 1–27.

[12] Le Potier, J. *Faisceaux semi-stables de dimension 1 sur le plan projectif.* Revue roumaine de math. pures et appliquées 38 (1993), 635–678.

[13] Maruyama, M. *Moduli of stable sheaves I.* J. Math. Kyoto Univ. 17 (1977), 91–126.

[14] Maruyama, M. *Moduli of stable sheaves II*. J. Math. Kyoto Univ. 18 (1978), 577–614.

[15] Nüssler, T., Trautmann, G. *Multiple Koszul structures on lines and instanton bundles*. Intern. Journ. of Math. 5 (1994), 373–388.

[16] Simpson, C.T. *Moduli of representations of the fundamental group of a smooth projective variety I*. Publ. Math. IHES 79 (1994), 47–129.

[17] Siu Y., Trautmann, G. *Deformations of coherent analytic sheaves with compact supports*. Memoirs of the Amer. Math. Soc., Vol. 29, N. 238 (1981).

Jean-Marc Drézet
Institut de Mathématiques de Jussieu
UMR 7586 du CNRS
175 rue du Chevaleret
F-75013 Paris, France
e-mail: `drezet@math.jussieu.fr`
URL: `http://www.math.jussieu.fr/~drezet`

Algebraic Cycles, Sheaves, Shtukas, and Moduli
Trends in Mathematics, 45–68
© 2007 Birkhäuser Verlag Basel/Switzerland

Lectures on Principal Bundles over Projective Varieties

Tomás L. Gómez

Abstract. Lectures given in the Mini-School on Moduli Spaces at the Banach center (Warsaw) 26–30 April 2005.

In these notes we will always work with schemes over the field of complex numbers $\mathbb{C}$. Let X be a scheme. A vector bundle of rank r on X is a scheme with a surjective morphism $p : \mathbb{V} \to X$ and an equivalence class of linear atlases. A linear atlas is an open cover $\{U_i\}$ of X (in the Zariski topology) and isomorphisms $\psi_i : p^{-1}(U_i) \to U_i \times \mathbb{C}^r$, such that $p = p_X \circ \psi_i$, and $\psi_j \circ \psi_i^{-1}$ is linear on the fibers. Two atlases are equivalent if their union is an atlas. These two properties are usually expressed by saying that a vector bundle is locally trivial (in the Zariski topology), and the fibers have a linear structure.

An isomorphism of vector bundles on X is an isomorphism $\varphi : \mathbb{V} \to \mathbb{V}'$ of schemes which is compatible with the linear structure. That is, $p = p' \circ \varphi$ and the covering $\{U_i\} \bigcup \{U_i'\}$ together with the isomorphisms ψ_i, $\psi_i' \circ \varphi$ is a linear structure on $\mathbb{V}$ as before.

The set of isomorphism classes of vector bundles of rank r on X is canonically bijective to the Cech cohomology set $\check{H}^1(X, \underline{\mathrm{GL}_r})$. Indeed, since the transition functions $\psi_j \circ \psi_i^{-1}$ are linear on the fibers, they are given by morphisms $\alpha_{ij} : U_i \cap U_j \to \mathrm{GL}_r$ which satisfy the cocycle condition.

Given a vector bundle $\mathbb{V} \to X$ we define the locally free sheaf E of its sections, which to each open subset $U \subset X$, assigns $E(U) = \Gamma(U, p^{-1}(U))$. This provides an equivalence of categories between the categories of vector bundles and that of locally free sheaves ([Ha, Ex. II.5.18]). Therefore, if no confusion seems likely to arise, we will use the words "vector bundle" and "locally free sheaf" interchangeably. Note that a vector bundle of rank 1 is the same thing as a line bundle. We will be interested in constructing moduli spaces of vector bundles, which can then be considered as generalizations of the Jacobian. Sometimes it will be necessary to consider also torsion free sheaves. For instance, in order to compactify the moduli space of vector bundles.

Let X be a smooth projective variety of dimension n with an ample line bundle $\mathcal{O}_X(1)$ corresponding to a divisor H. Let E be a torsion free sheaf on X. Its Chern classes are denoted $c_i(E) \in H^{2i}(X; \mathbb{C})$. We define the *degree* of E

$$\deg E = c_1(E)H^{n-1}$$

and its Hilbert polynomial

$$P_E(m) = \chi(E(m)),$$

where $E(m) = E \otimes \mathcal{O}_X(m)$ and $\mathcal{O}_X(m) = \mathcal{O}_X(1)^{\otimes m}$.

If E is locally free, we define the determinant line bundle as $\det E = \bigwedge^r E$. If E is torsion free, since X is smooth, we can still define its determinant as follows. The maximal open subset $U \subset X$ where E is locally free is *big* (with this we will mean that its complement has codimension at least two), because it is torsion free. Therefore, there is a line bundle $\det E|_U$ on U, and since U is big and X is smooth, this extends to a unique line bundle on X, which we call the determinant $\det E$ of E. It can be proved that $\deg E = \deg \det E$.

We will use the following notation. Whenever "(semi)stable" and "($\leq$)" appears in a sentence, two sentences should be read. One with "semistable" and "$\leq$", and another with "stable" and "$<$". Given two polynomials p and q, we write $p < q$ if $p(m) < q(m)$ when $m \gg 0$.

A torsion free sheaf E is *(semi)stable* if for all proper subsheaves $F \subset E$,

$$\frac{P_F}{\operatorname{rk} F} (\leq) \frac{P_E}{\operatorname{rk} E} \ .$$

A sheaf is called *unstable* if it is not semistable. Sometimes this is referred to as Gieseker (or Maruyama) stability.

A torsion free sheaf E is *slope-(semi)stable* if for all proper subsheaves $F \subset E$ with $\operatorname{rk} F < \operatorname{rk} E$,

$$\frac{\deg F}{\operatorname{rk} F} (\leq) \frac{\deg E}{\operatorname{rk} E} \ .$$

The number $\deg E / \operatorname{rk} E$ is called the *slope* of E. A sheaf is called *slope-unstable* if it is not slope-semistable. Sometimes this is referred to as Mumford (or Takemoto) stability.

Using Riemann-Roch theorem, we find

$$P_E(m) = \operatorname{rk} E \frac{m^n}{n!} + (\deg E - \operatorname{rk} E \frac{\deg K}{2}) \frac{m^{n-1}}{(n-1)!} + \cdots$$

where K is the canonical divisor. From this it follows that

$$\text{slope-stable} \implies \text{stable} \implies \text{semistable} \implies \text{slope-semistable}$$

Note that, if $n = 1$, Gieseker and Mumford (semi)stability coincide, because the Hilbert polynomial has degree 1.

Let E be a torsion free sheaf on X. There is a unique filtration, called the Harder-Narasimhan filtration,

$$0 = E_0 \subsetneq E_1 \subsetneq E_2 \subsetneq \cdots \subsetneq E_l = E$$

such that $E^i = E_i/E_{i-1}$ is semistable and

$$\frac{P_{E^i}}{\operatorname{rk} E^i} > \frac{P_{E^{i+1}}}{\operatorname{rk} E^{i+1}}$$

for all i. In particular, any torsion free sheaf can be described as successive extensions of semistable sheaves.

There is also a Harder-Narasimhan filtration for slope stability: this is the unique filtration such that $E^i = E_i/E_{i-1}$ is slope-semistable and

$$\frac{\deg E^i}{\operatorname{rk} E^i} > \frac{\deg E^{i+1}}{\operatorname{rk} E^{i+1}}$$

for all i. We denote

$$\mu_{\max}(E) = \mu(E^1) , \quad \mu_{\min}(E) = \mu(E^l).$$

Of course, in general these two filtrations will be different.

Now let E be a semistable sheaf. There is a filtration, called the Jordan-Hölder filtration,

$$0 = E_0 \subsetneq E_1 \subsetneq E_2 \subsetneq \cdots \subsetneq E_l = E$$

such that $E^i = E_i/E_{i-1}$ is stable and

$$\frac{P_{E^i}}{\operatorname{rk} E^i} = \frac{P_{E^{i+1}}}{\operatorname{rk} E^{i+1}}$$

for all i. This filtration is not unique, but the associated graded sheaf

$$\operatorname{gr}_{\mathrm{JH}}(E) = \bigoplus_{i=1}^{l} E^i$$

is unique up to isomorphism. It is easy to check that $\operatorname{gr}_{\mathrm{JH}}(E)$ is semistable. Two semistable torsion free sheaves are called *S-equivalent* if $\operatorname{gr}_{\mathrm{JH}}(E)$ and $\operatorname{gr}_{\mathrm{JH}}(E')$ are isomorphic.

There is also a Jordan-Hölder filtration for slope stability, just replacing Hilbert polynomials with degrees.

A *family of coherent sheaves* parameterized by a scheme T (also called T-family) is a coherent sheaf E_T on $X \times T$, flat over T. For each closed point $t \in T$, we get a sheaf $E_t := f^* E_T$ on $X \times t \cong X$, where $f : X \times t \to X \times T$ is the natural inclusion. We say that E_T is a family of torsion-free sheaves if E_t is torsion-free sheaf for all closed points $t \in T$. We have analogous definitions for any open condition, and hence we can talk of families of (semi)stable sheaves, of families of sheaves with fixed Chern classes $c_i(E)$, etc... Two families are *isomorphic* if E_T and E'_T are isomorphic as sheaves.

To define the notion of moduli space, we will first look at the Jacobian J of a projective scheme X. There is a bijection between isomorphism classes of line bundles L with $0 = c_1(L) \in H^2(X; \mathbb{C})$ and closed points of J.

Furthermore, if we are given a family of line bundles L_T, with vanishing first Chern class, parameterized by a scheme T, we obtain a morphism $f : T \to J$ such that for all $t \in T$, the point $f(t) \in J$ is the point corresponding to the isomorphism class of L_t. And, conversely, if we are given a morphism $f : T \to J$, we obtain a family of line bundles parameterized by T by pulling-back a Poincaré line bundle: $L_T = (\mathrm{id}_X \times f)^* \mathcal{P}$.

Note that both constructions are not quite inverse to each other. On the one hand, if M is a line bundle on T, the families L_T and $L_T \otimes p_T^* M$ give the same morphism from T to the Jacobian, and on the other hand, there is no unique Poincaré line bundle: given a linen bundle M on J, $\mathcal{P} \otimes p_J^* M$ is also a Poincaré line bundle, and the family induced by f will change to $L_T \otimes p_T^* f^* M$. This is why we declare two families of line bundles *equivalent* if they differ by the pullback of a line bundle on the parameter space T.

Using this equivalence, both constructions become inverse of each other. That is, there is a bijection between equivalence classes of families and morphisms to the Jacobian.

One could ask: is there a "better" version of the Jacobian?, i.e., some object $\mathcal{J}$ which provides a bijection between morphisms to it and families of line bundles up to isomorphism (not up to equivalence). The answer is yes, but this object $\mathcal{J}$ is not a scheme! It is an algebraic stack (in the sense of Artin): the Jacobian stack. A stack is a generalization of the notion of scheme, but we will not consider it here.

We would like to have a scheme with the same properties as the Jacobian, but for torsion-free sheaves instead of line bundles. To be able to do this, we have to consider only the semistable ones. Then there will be a moduli scheme $\mathfrak{M}(r, c_i)$ such that a family of semistable torsion-free sheaves parameterized by T, with rank r and Chern classes c_i, will induce a morphism from the parameter space T to $\mathfrak{M}(r, c_i)$. In particular, to each semistable torsion free sheaf we associate a closed point. If two stable torsion free sheaves on X are not isomorphic, they will correspond to different points of $\mathfrak{M}(r, c_i)$, but it can happen that two strictly semistable torsion free sheaves on X which are not isomorphic correspond to the same point of $\mathfrak{M}(r, c_i)$. In fact, E and E' correspond to the same point if and only if they are S-equivalent.

Another difference with the properties of the Jacobian is that in general there will be no "universal torsion free sheaf" on $X \times \mathfrak{M}(r, c_i)$, i.e., there will be no analogue of the Poincaré bundle. In other words, given a morphism $f : T \to \mathfrak{M}(r, c_i)$, there might be no family parameterized by T which induces f. If there is a universal torsion-free sheaf, we say that $\mathfrak{M}(r, c_i)$ is a *fine moduli space*, and if it does not exist, we say that it is a *coarse moduli space*.

To explain this more precisely, it is useful to use the language of representable functors. Given a scheme M over $\mathbb{C}$, we define a (contravariant) functor $\underline{M} := \mathrm{Mor}(-, M)$ from the category of $\mathbb{C}$-schemes (Sch $/\mathbb{C}$) to the category of sets (Sets) by sending an $\mathbb{C}$-scheme B to the set of morphisms $\mathrm{Mor}(B, M)$. On morphisms it is defined with composition, i.e., to a morphism $f : B \to B'$ we associate the map $\mathrm{Mor}(B', M) \to \mathrm{Mor}(B, M)$ which sends φ' to $\varphi \circ f$.

Definition 0.1 (Represents). *A functor $F : (\mathrm{Sch} /\mathbb{C}) \to (\mathrm{Sets})$ is represented by a scheme M if there is an isomorphism of functors $F \cong \underline{M}$.*

Of course, not all functors from $(\mathrm{Sch}/\mathbb{C})$ to (Sets) are representable, but if a functor F is, then the scheme M is unique up to canonical isomorphism. Given a morphism $f : M \to M'$, we obtain an natural transformation $\underline{M} \to \underline{M'}$, and, by Yoneda's lemma, every natural transformation between representable functors is induced by a morphism of schemes. In other words, the category of schemes is a full subcategory of the category of functors $(\mathrm{Sch}/\mathbb{C})'$, whose objects are contravariant functors from $(\mathrm{Sch}/\mathbb{C})$ to (Sets) and whose morphisms are natural transformation. Therefore, we will denote by the same letter a morphism of schemes and the associated natural transformation.

For instance, let $F_J : (\mathrm{Sch}/\mathbb{C}) \to$ (Sets) be the functor which sends a scheme T to the set of equivalence classes of T-families of line bundles on X, with $c_1 = 0$. This functor is represented by the Jacobian, i.e., there is an isomorphism of functors $F_J \cong \underline{J}$. This is the translation, to the language of representable functors, of the fact that there is a natural bijection between the set of equivalence classes of these families and the set of morphisms from T to J.

Definition 0.2 (Corepresents). *A functor $F : (\mathrm{Sch}/\mathbb{C}) \to$ (Sets) is corepresented by a scheme M if there is a natural transformation of functors $\phi : F \to \underline{M}$ such that given another scheme M' and natural transformation $\phi' : F \to \underline{M'}$, there is a unique morphism $\eta : M \to M'$ with $\phi' = \eta \circ \phi$.*

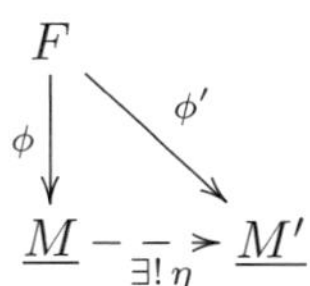

If M corepresents F, then M is unique up to canonical isomorphism. To explain why this is called "corepresentation", let $(\mathrm{Sch}/\mathbb{C})'$ be the above-defined functor category. Then it can be seen that M represents F if and only if there is a natural bijection $\mathrm{Mor}(Y, M) = \mathrm{Mor}_{(\mathrm{Sch}/\mathbb{C})'}(\underline{Y}, F)$ for all schemes Y. On the other hand, M corepresents F if and only if there is a natural bijection $\mathrm{Mor}(M, Y) = \mathrm{Mor}_{(\mathrm{Sch}/\mathbb{C})'}(F, \underline{Y})$ for all schemes Y. If M represents F, then it corepresents it, but the converse is not true.

Let X be a fixed $\mathbb{C}$-scheme. Define a functor F^{ss}_{r,c_i} from the category of schemes over $\mathbb{C}$ to the category of sets, sending a scheme T to the set $F^{ss}_{r,c_i}(T)$ of isomorphism classes of families of torsion-free sheaves on X parameterized by T, with rank r and Chern classes c_i. On morphisms it is defined by pullback, i.e., to a morphism $f : T \to T'$ we associate the map $F(T') \to F(T)$ which sends the family E'_T to $(\mathrm{id}_X \times f)^* E'_T$. Analogously, we define the functor F^s_{r,c_i} of families of stable torsion free sheaves.

It can be shown that, for any polarized smooth projective variety X, there is a scheme $\mathfrak{M}(r, c_i)$ corepresenting the above defined functor F^{ss}_{r,c_i} ([Gi, Ma, Sesh, Si]). In Section 1 we will sketch a proof of this result.

Note that the transformation of functors ϕ gives, for any T-family of semi-stable torsion free sheaves, a morphism $f : T \to \mathfrak{M}(r, c_i)$. As we mentioned before, there is a canonical bijection between closed points of $\mathfrak{M}(r, c_i)$ and S-equivalence classes of semistable torsion free sheaves.

Let $\hat{F}^{ss}_{r,c_i}$ be the functor of equivalence classes of families of semistable sheaves, where, as before, we declare two families equivalent if they differ by the pullback of a line bundle on the parameter space. There are some cases in which this functor is representable (for instance, if X is a smooth curve and the rank r and degree c_1 are coprime). In these cases, there is a universal family parameterized by the moduli space, and this universal family is unique up to the pullback of a line bundle on the moduli space.

Definition 0.3 (Moduli space). *We say that M is a moduli space for a set of objects, if it corepresents the functor of families of those objects.*

Definition 0.4 (Coarse moduli). *A scheme M is called a coarse moduli scheme for F if it corepresents F and furthermore the map*

$$\phi(\operatorname{Spec}\mathbb{C}) : F(\operatorname{Spec}\mathbb{C}) \to \operatorname{Hom}(\operatorname{Spec}\mathbb{C}, M)$$

is bijective.

Note that if a functor F is corepresented by a scheme M, then it is a coarse moduli scheme for the functor $\widetilde{F}$ of S-equivalence classes of F, i.e., the functor defined as

$$\widetilde{F}(T) = \begin{cases} F(T), & \text{if } T \neq \operatorname{Spec}\mathbb{C} \\ \text{S-equivalence classes of objects of } F(\operatorname{Spec}\mathbb{C}), & \text{if } T = \operatorname{Spec}\mathbb{C} \end{cases}$$

1. Moduli space of torsion free sheaves

In this section we will sketch the proof of the existence of the moduli space of semistable torsion free sheaves. We will start by giving a brief idea of the construction. It can be shown that there is a scheme Y classifying *semistable based sheaves*, that is, pairs (f, E), where E is a semistable sheaf and $f : V \to H^0(E(m))$ is an isomorphism between a fixed vector space V and $H^0(E(m))$. The group $\operatorname{SL}(V)$ acts on Y by "base change": an element $g \in \operatorname{SL}(V)$ sends the pair (f, E) to $(f \circ g, E)$. Two pairs (f, E) and (f', E') are in the same orbit if and only if E is isomorphic to E', therefore, the quotient of Y by the action of $\operatorname{SL}(V)$ will be a moduli space of semistable sheaves. But, does this quotient exist in the category of schemes?, i.e., is there a scheme whose points are in bijection with the $\operatorname{SL}(V)$-orbits in Y? In general the answer is no, but Geometric Invariant Theory (GIT) gives us something which is quite close to this, and is called the GIT quotient, and this will be the moduli space.

Note that we are using the group $\operatorname{SL}(V)$, and not $\operatorname{GL}(V)$. This is because if two isomorphisms f and f' only differ by multiplication with a scalar, then they correspond to the same point in Y. In other words, Y classifies pairs (f, E) up to scalar.

Let G be an algebraic group. Recall that a right action on a scheme R is a morphism $\sigma : R \times G \to R$, which we will usually denote $\sigma(z, g) = z \cdot g$, such that

$z \cdot (gh) = (z \cdot g) \cdot h$ and $z \cdot e = z$, where e is the identity element of G. A left action is analogously defined, with the associative condition $(hg) \cdot z = h \cdot (g \cdot z)$.

The orbit of a point $z \in R$ is the image $z \cdot G$. A morphism $p : R \to M$ between two schemes endowed with G-actions is called G-*equivariant* if it commutes with the actions, that is $f(z) \cdot g = f(z \cdot g)$. If the action on M is trivial (i.e., $y \cdot g = y$ for all $g \in G$ and $y \in M$), then we also say that f is G-*invariant*.

If G acts on a projective variety R, a *linearization* of the action on a line bundle $\mathcal{O}_R(1)$ consists of giving, for each $g \in G$, an isomorphism of line bundles $\tilde{g} : \mathcal{O}_R(1) \to \varphi_g^* \mathcal{O}_R(1)$, $(\varphi_g = \sigma(\cdot, g))$ which also satisfies the previous associative property. Giving a linearization is thus the same thing as giving an action on the total space $\mathbb{V}$ of the line bundle, which is linear along the fibers, and such that the projection $\mathbb{V} \to R$ is equivariant. If $\mathcal{O}_R(1)$ is very ample, then a linearization is the same thing as a representation of G on the vector space $H^0(\mathcal{O}_R(1))$ such that the natural embedding $R \to \mathbb{P}(H^0(\mathcal{O}_R(1))^\vee)$ is equivariant.

Definition 1.1 (Categorical quotient). *Let R be a scheme endowed with a G-action. A categorical quotient is a scheme M with a G-invariant morphism $p : R \to M$ such that for every other scheme M', and G-invariant morphism p', there is a unique morphism φ with $p' = \varphi \circ p$*

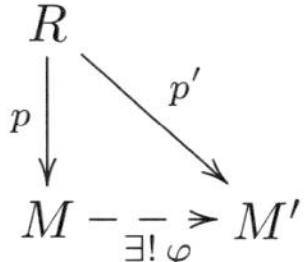

Definition 1.2 (Good quotient). *Let R be a scheme endowed with a G-action. A good quotient is a scheme M with a G-invariant morphism $p : R \to M$ such that*

1. *p is surjective and affine*
2. *$p_*(\mathcal{O}_R^G) = \mathcal{O}_M$, where $\mathcal{O}_R^G$ is the sheaf of G-invariant functions on R.*
3. *If Z is a closed G-invariant subset of R, then $p(Z)$ is closed in M. Furthermore, if Z_1 and Z_2 are two closed G-invariant subsets of R with $Z_1 \cap Z_2 = \emptyset$, then $f(Z_1) \cap f(Z_2) = \emptyset$.*

Definition 1.3 (Geometric quotient). *A geometric quotient $p : R \to M$ is a good quotient such that $p(x_1) = p(x_2)$ if and only if the orbit of x_1 is equal to the orbit of x_2.*

Clearly, a geometric quotient is a good quotient, and a good quotient is a categorical quotient.

Geometric Invariant Theory (GIT) is a technique to construct good quotients (cf. [Mu1]). Assume R is projective, and the action of G on R has a linearization on an ample line bundle $\mathcal{O}_R(1)$. A closed point $z \in R$ is called *GIT-semistable* if, for some $m > 0$, there is a G-invariant section s of $\mathcal{O}_R(m)$ such that $s(z) \neq 0$. If, moreover, the orbit of z is closed in the open set of all GIT-semistable points, it is called *GIT-polystable*, and, if furthermore, this closed orbit has the same dimension as G (i.e., if z has finite stabilizer), then z is called a *GIT-stable* point. We say that a closed point of R is *GIT-unstable* if it is not GIT-semistable.

We will use the following characterization in [Mu1] of GIT-(semi)stability. Let $\lambda : \mathbb{C}^* \to G$ be a one-parameter subgroup (by this we mean a nontrivial group homomorphism, even if λ is not injective), and let $z \in R$. Then $\lim_{t \to 0} z \cdot \lambda(t) = z_0$ exists, and z_0 is fixed by λ. Let $t \mapsto t^a$ be the character by which λ acts on the fiber of $\mathcal{O}_R(1)$. Defining $\mu(z, \lambda) = a$, Mumford proves that z is GIT-(semi)stable if and only if, for all one-parameter subgroups, it is $\mu(z, \lambda)(\leq)0$.

Proposition 1.4. *Let R^{ss} (respectively, R^s) be the subset of GIT-semistable points (respectively, GIT-stable). Both R^{ss} and R^s are open subsets. There is a good quotient $R^{ss} \to R /\!\!/ G$, the image $R^s /\!\!/ G$ of R^s is open, $R /\!\!/ G$ is projective, and the restriction $R^s \to R^s /\!\!/ G$ is a geometric quotient.*

There is one important case in which a scheme is only quasi-projective but GIT can be applied to get a projective quotient: Assume that R' is a G-acted scheme with a linearization on a line bundle $\mathcal{O}_{R'}(1)$, which is the restriction of a linearization on an ample line bundle $\mathcal{O}_R(1)$ on a projective variety R, and $R' = R^{ss}$, the open subset of GIT-semistable points of R. Then we define $R' /\!\!/ G = R /\!\!/ G$.

Now we are going to describe Grothendieck's Quot-scheme. This scheme parameterizes quotients of a fixed coherent sheaf $\mathcal{V}$ on X. That is, pairs (q, E), where $q : \mathcal{V} \twoheadrightarrow E$ is a surjective homomorphism and E is a coherent sheaf on X. An isomorphism of quotients is an isomorphism $\alpha : E \to E'$ such that the following diagram is commutative

$$
\begin{array}{ccc}
\mathcal{V} & \xrightarrow{\ q\ } & E \\
\| & & \downarrow{\scriptstyle \cong}\,\alpha \\
\mathcal{V} & \xrightarrow{\ q'\ } & E'
\end{array}
$$

A family of quotients parameterized by T is a pair $(q : p_X^* \mathcal{V} \twoheadrightarrow E_T, E_T)$ where E_T is a coherent sheaf on $X \times T$, flat over T. An isomorphism of families is an isomorphism $\alpha : E_T \to E_T'$ such that $\alpha \circ q = q'$. Recall that X is a projective scheme endowed with an ample line bundle $\mathcal{O}_X(1)$. Therefore, if E_T is flat over T then the Hilbert polynomial P_{E_t} is locally constant as a function of $t \in T$. If T is reduced, the converse is also true.

Fix a polynomial P and a coherent sheaf $\mathcal{V}$ on X. Consider the contravariant functor which sends a scheme T to the set of isomorphism classes of T-families of sheaves with Hilbert polynomial P (and it is defined as pullback on morphisms). Grothendieck proved that there is a projective scheme $\mathrm{Quot}_X(\mathcal{V}, P)$, called the Quot scheme, which represents this functor. In particular, there is a universal quotient, i.e., a tautological family of quotients parameterized by $\mathrm{Quot}_X(\mathcal{V}, P)$. We will be interested in the case $\mathcal{V} = V \otimes_{\mathbb{C}} \mathcal{O}_X(-m)$, where V is a vector space and m is sufficiently large.

Given a coherent sheaf E, there is an integer $m(E)$ such that, if $m \geq m(E)$, then $E(m)$ is generated by global sections, $h^0(E(m)) = P_E(m)$, and $h^i(E(m)) = 0$

for $i > 0$ ([H-L, Def 1.7.1]). Assume that $m \geq m(E)$ and $\dim V = P_E(m)$. An isomorphism $f : V \to H^0(E(m))$ induces a quotient

$$q : V \otimes_{\mathbb{C}} \mathcal{O}_X(-m) \xrightarrow{\cong} H^0(E(m)) \otimes_{\mathbb{C}} \mathcal{O}_X(-m) \longrightarrow E$$

as above, and this is how a scheme parameterizing based sheaves (f, E) appears as a subscheme of Grothendieck's Quot scheme.

Note that if we have a set $\mathcal{A}$ of isomorphism classes of sheaves, there might not be an integer m large enough for all sheaves. A set $\mathcal{A}$ of isomorphism classes of sheaves on X is called *bounded* if there is a family E_S of torsion free sheaves parameterized by a scheme S of finite type, such that for all $E \in \mathcal{A}$, there is at least one point $s \in S$ such that the corresponding sheaf E_s is isomorphic to E. If a set $\mathcal{A}$ is bounded, then we can find an integer m such that $m \geq m(E)$ for all $E \in \mathcal{A}$, thanks to the fact that S is of finite type.

Maruyama proved that the set $\mathcal{A}$ of semistable sheaves with fixed Hilbert polynomial is bounded, and it follows that there is an integer m_0, depending only on the polynomial P and $(X, \mathcal{O}_X(1))$, such that $m_0 \geq m(E)$ for all semistable sheaves E. This technical result is crucial in order to construct the moduli space. In fact, if $\dim X > 1$, he was able to prove it only if the base field has characteristic 0, and therefore he could only prove the existence of the moduli space in this case. Recently Langer was able to prove boundedness for characteristic $p > 0$, and therefore he was able to construct the corresponding moduli space [La].

Fix a Hilbert polynomial P, and let $m \geq m_0$. Let

$$Y \subset \mathrm{Quot}_X(V \otimes_{\mathbb{C}} \mathcal{O}_X(-m), P)$$

be the open subset of quotients such that E is torsion free and q induces an isomorphism $V \cong H^0(E(m))$. Let $\overline{Y}$ be the closure of the open set Y in $\mathrm{Quot}_X(V \otimes_{\mathbb{C}} \mathcal{O}_X(-m), P)$. Note that there is a natural action of $\mathrm{SL}(V)$ on

$$\mathrm{Quot}_X(V \otimes_{\mathbb{C}} \mathcal{O}_X(-m)),$$

which sends the quotient $q : V \otimes_{\mathbb{C}} \mathcal{O}_X(-m) \to E$ to the composition $q \circ (g \times \mathrm{id})$. It leaves Y and $\overline{Y}$ invariant, and coincides with the previously defined action for based sheaves (f, E).

To apply GIT, we also need an ample line bundle on $\overline{Y}$ and a linearization of the $\mathrm{SL}(V)$-action on it. This is done by giving an embedding of $\overline{Y}$ in $\mathbb{P}(V_1)$, where V_1 will be a vector space with a representation of $\mathrm{SL}(V)$.

There are different ways of doing this, corresponding to different representations V_1. One of them corresponds to Grothendieck's embedding of the Quot scheme. This is the method used by Simpson [Si]. Let $q : V \otimes_{\mathbb{C}} \mathcal{O}_X(-m) \twoheadrightarrow E$ be a quotient. Let $l > m$ be an integer and $W = H^0(\mathcal{O}_X(l - m))$. The quotient q induces homomorphisms

$$\begin{aligned}
q &: & V \otimes_{\mathbb{C}} \mathcal{O}_X(l - m) &\twoheadrightarrow & E(l) \\
q' &: & V \otimes W &\to & H^0(E(l)) \\
q'' &: & {\textstyle\bigwedge}^{P(l)}(V \otimes W) &\to & {\textstyle\bigwedge}^{P(l)} H^0(E(l)) \cong \mathbb{C}.
\end{aligned}$$

If l is large enough, these homomorphisms are surjective, and give Grothendieck's embedding of the Quot scheme.

$$\mathrm{Quot}_X(V \otimes_{\mathbb{C}} \mathcal{O}_X(-m), P) \longrightarrow \mathbb{P}\Big(\bigwedge{}^{P(l)}(V^\vee \otimes W^\vee) \Big).$$

The natural representation of $\mathrm{SL}(V)$ in $\bigwedge^{P(l)}(V^\vee \otimes W^\vee)$ gives a linearization of the $\mathrm{SL}(V)$ action on the very ample line bundle $\mathcal{O}_{\overline{Y}}(1)$ induced by this embedding on $\overline{Y}$.

A theorem of Simpson says that a point $(q, E) \in \overline{Y}$ is GIT-(semi)stable if and only if the sheaf E is (semi)stable and the induced linear map $f : V \to H^0(E(m))$ is an isomorphism. In other words, $Y = \overline{Y}^{ss}$. Therefore, the GIT quotient $\overline{Y} /\!\!/ \mathrm{SL}(V)$ is the moduli space $\mathfrak{M}(P)$ of semistable sheaves with Hilbert polynomial P. The Chern classes $c_i \in H^{2i}(X, \mathbb{C})$ in a family of sheaves are locally constant, therefore the moduli space $\mathfrak{M}(r, c_i)$ of semistable sheaves with fixed rank and Chern classes is a union of connected components of the scheme $\mathfrak{M}(P)$.

Another choice of representation V_1 (and therefore, of line bundle on Y and linearization of the action) is the one used by Gieseker and Maruyama. It is explained in the lectures of Schmitt.

2. Moduli space of tensors

A *tensor* of type a is a pair (E, φ) where E is a torsion free sheaf and

$$\varphi : E \overbrace{\otimes \cdots \otimes}^{a} E \longrightarrow \mathcal{O}_X$$

is a homomorphism. An isomorphism between the tensors (E, φ) and (E', φ') is a pair (f, α) where f is an isomorphism between E and E', $\alpha \in \mathbb{C}^*$, and the following diagram commutes

$$
\begin{array}{ccc}
E^{\otimes a} & \xrightarrow{\ \varphi\ } & \mathcal{O}_X \\
{\scriptstyle f^{\otimes a}}\big\downarrow & & \big\downarrow{\scriptstyle \alpha} \\
E'^{\otimes a} & \xrightarrow{\ \varphi'\ } & \mathcal{O}_X
\end{array}
$$

The definition of families of tensors and their isomorphisms are left to the reader ([G-S1, GLSS]).

To define the notion of stability for tensors, it is not enough to look at subsheaves. We have to consider filtrations $E_\bullet \subset E$. By this we always understand a $\mathbb{Z}$-indexed filtration

$$\cdots \subset E_{i-1} \subset E_i \subset E_{i+1} \subset \cdots$$

starting with 0 and ending with E (i.e., $E_k = 0$ and $E_l = E$ for some k and l). We say that the filtration is *saturated* if $E^i = E_i/E_{i-1}$ is torsion free for all i. If we delete, from 0 onward, all the non-strict inclusions, we obtain a filtration

$$0 \subsetneq E_{\lambda_1} \subsetneq E_{\lambda_2} \subsetneq \cdots \subsetneq E_{\lambda_t} \subsetneq E_{\lambda_{t+1}} = E \qquad \lambda_1 < \cdots < \lambda_{t+1}.$$

Reciprocally, from a filtration $E_{\lambda_\bullet}$ we recover the $\mathbb{Z}$-indexed filtration $E_\bullet$ by defining $E_m = E_{\lambda_{i(m)}}$, where $i(m)$ is the maximum index with $\lambda_{i(m)} \leq m$.

Let $\mathcal{I}_a = \{1, \ldots, t+1\}^{\times a}$ be the set of all multi-indexes $I = (i_1, \ldots, i_a)$ of cardinality a. Define

$$\mu_{\text{tens}}(\varphi, E_{\lambda_\bullet}) = \min_{I \in \mathcal{I}_a} \left\{ \lambda_{i_1} + \cdots + \lambda_{i_a} : \phi|_{E_{\lambda_{i_1}} \otimes \cdots \otimes E_{\lambda_{i_a}}} \neq 0 \right\}, \tag{2.1}$$

or, in terms of the $\mathbb{Z}$-indexed filtration,

$$\mu_{\text{tens}}(\varphi, E_\bullet) = \min_{I \in \mathcal{I}_a} \left\{ i_1 + \cdots + i_a : \phi|_{E_{i_1} \otimes \cdots \otimes E_{i_a}} \neq 0 \right\} \tag{2.2}$$

Definition 2.1 (Balanced filtration). *A saturated filtration $E_\bullet \subset E$ of a torsion free sheaf E is called a balanced filtration if $\sum i \operatorname{rk} E^i = 0$ for $E^i = E_i/E_{i-1}$. In terms of $E_{\lambda_\bullet}$, this is $\sum_{i=1}^{t+1} \lambda_i \operatorname{rk}(E^{\lambda_i}) = 0$ for $E^{\lambda_i} = E_{\lambda_i}/E_{\lambda_{i-1}}$.*

Definition 2.2 (Stability of tensors). *Let δ be a polynomial of degree at most $n-1$ (recall $n = \dim X$) with positive leading coefficient. We say that a tensor (E, φ) is δ-(semi)stable if φ is not identically zero and for all balanced filtrations $E_{\lambda_\bullet}$ of E, it is*

$$\left(\sum_{i=1}^{t} (\lambda_{i+1} - \lambda_i)\big(r P_{E_{\lambda_i}} - r_{\lambda_i} P\big) \right) + \mu_{\text{tens}}(\phi, E_{\lambda_\bullet}) \delta \; (\leq) \; 0 \tag{2.3}$$

We will always denote $r = \operatorname{rk} E$ and $r_i = \operatorname{rk} E_i$. The notion of stability for tensors looks complicated, but one finds that, in the applications, when the tensor has some geometric meaning, it can be simplified. We will see some examples.

A *framed bundle* is a tensor of the form $(E, \varphi : E \to \mathcal{O}_X)$. If E is a vector bundle, then taking the dual we have a section of $E^\vee$, so this is equivalent to the pairs $(E, \varphi : \mathcal{O}_X \to E)$ considered by Bradlow, García-Prada and others. In this case, it is enough to look at filtrations with one step, i.e., subsheaves $E' \subsetneq E$.

An *orthogonal sheaf* is a tensor of the form $(E, \varphi : E \otimes E \to \mathcal{O}_X)$, where E is torsion free and φ is symmetric and non-degenerate (in the sense that the induced homomorphism $\det E \to \det E^\vee$ is an isomorphism). A *symplectic sheaf* is analogously defined, requiring the tensor φ to be skew-symmetric instead of symmetric.

Given a subsheaf $E' \subset E$, its orthogonal $E'^\perp$ is defined as the kernel of the composition

$$E \xrightarrow{\; \widetilde{\varphi} \;} E^\vee \longrightarrow E'^\perp \, ,$$

where $\widetilde{\varphi}$ is induced by φ.

Definition 2.3. *An orthogonal (or symplectic) sheaf is (semi)stable if for all orthogonal filtrations, that is, filtrations with*

$$E_i^\perp = E_{-i-1}$$

for all i, the following holds

$$\sum (r P_{E_i} - r_i P_E)(\leq)0 \, .$$

It it shown in [G-S1] that an orthogonal (or symplectic) sheaf is (semi)stable if and only if it is δ-(semi)stable as a tensor, when δ has degree $n-1$.

A T-family of orthogonal sheaves is a T-family of tensors $(E_T, \varphi_T : E_T \otimes E_T \longrightarrow \mathcal{O}_{X \times T})$ such that φ_T is symmetric and non-degenerate. Note that, since being symmetric is a closed condition, it is not enough to check that φ_t is symmetric for every point $t \in T$. On the other hand, being non-degenerate is an open condition, so it is enough to check it for φ_t, for all points $t \in T$.

A *Lie algebra sheaf* is a pair (E, φ) where E is a torsion free sheaf and

$$\varphi : E \otimes E \longrightarrow E^{\vee\vee}$$

is a homomorphism such that for each point $x \in X$, where E is locally free, the induced homomorphism on the fiber $\varphi(x) : E(x) \otimes E(x) \to E(x)$ is a Lie algebra structure. An isomorphism to another Lie algebra sheaf (E', φ') is an isomorphism of sheaves $f : E \to E'$ with $\varphi' \circ (f \otimes f) = f \circ \varphi$.

At first sight, this does not seem to be included in the formalism of tensors, but, using the canonical isomorphism

$$(\bigwedge^{r-1} E)^{\vee} \otimes \det E \xrightarrow{\cong} E^{\vee\vee}, \tag{2.4}$$

a Lie sheaf becomes a tensor of the form

$$(F, \psi : F^{\otimes r+1} \longrightarrow \mathcal{O}_X), \tag{2.5}$$

with $E = F \otimes \det F$.

Definition 2.4. *A Lie tensor is a tensors of type $a = r+1$ which satisfies the following properties*

1. *ψ factors through $F \otimes F \otimes \bigwedge^{r-1} F$,*
2. *the homomorphism $\widetilde{\psi} : F \otimes F \to F^{\vee\vee} \otimes \det F^{\vee}$ associated by (2.4) is skew-symmetric.*
3. *the homomorphism $\widetilde{\psi}$ satisfies the Jacobi identity.*

There is a canonical bijection between the set of isomorphism classes of Lie sheaves $(E, \varphi : E \otimes E \to E^{\vee\vee})$ and Lie tensors $(F, \psi : F^{\otimes r+1} \longrightarrow \mathcal{O}_X)$ (with $E = F \otimes \det F$).

If the Lie algebra on the fiber $E(x)$ for all x where E is locally free is always isomorphic to a fixed semisimple Lie algebra $\mathfrak{g}$, then we say that it is a $\mathfrak{g}$-sheaf. Then, the Killing form gives an orthogonal structure $\kappa : E \otimes E \to \mathcal{O}_X$ to E.

Definition 2.5. *A $\mathfrak{g}$-sheaf is (semi)stable if for all orthogonal algebra filtrations, that is, filtrations with*

$$(1) \quad E_i^{\perp} = E_{-i-1} \quad \text{and} \quad (2) \quad [E_i, E_i] \subset E_{i+j}^{\vee\vee}$$

for all i, j, the following holds

$$\sum (r P_{E_i} - r_i P_E)(\leq)0 \,.$$

It is shown in [G-S2] that a $\mathfrak{g}$-sheaf is (semi)stable if and only if the associated tensor is δ-(semi)stable, when δ has degree $n - 1$.

We will sketch how the moduli space of tensors is constructed. The idea is similar to the construction of the moduli space of torsion free sheaves. First we construct a scheme which classifies δ-*semistable based tensors*, that is, triples (f, E, φ) where $f : V \to H^0(E(m))$ is an isomorphism, up to a constant, and (E, φ) is a δ-semistable tensor. There is a natural embedding of this scheme in a product $\mathbb{P}(V_1) \times \mathbb{P}(V_2)$, where V_1 and V_2 are representations of $\mathrm{SL}(V)$. An ample line bundle with a linearization of the $\mathrm{SL}(V)$ action is given by $\mathcal{O}_X(b_1, b_2)$. The choice of the integers b_1 and b_2 will depend on the polynomial δ, and the moduli space of δ-semistable tensors will be the GIT quotient.

To find V_2, note that the isomorphism $f : V \to H^0(E(m))$ and φ induces a linear map

$$\Phi : V^{\otimes a} \longrightarrow H^0(E(m)^{\otimes a}) \longrightarrow H^0(\mathcal{O}_X(am)) =: B \ .$$

Therefore, the semistable based tensor (f, E, φ) gives a point $(q, [\Phi])$

$$\mathcal{H} \times \mathbb{P}(V_2) := \mathrm{Quot}_X(V \otimes_{\mathbb{C}} \mathcal{O}_X(-m), P) \times \mathbb{P}\big((V^{\otimes a})^{\vee} \otimes B\big)$$

The points obtained in this way have the property that the homomorphism Φ composed with evaluation factors as

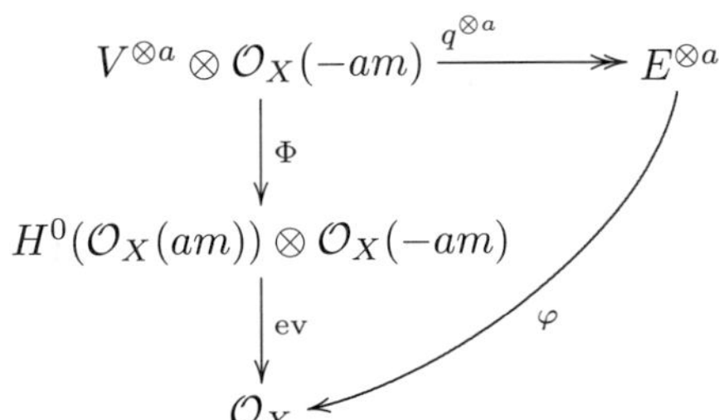

Let Z' be the closed subscheme of $\mathcal{H} \times \mathbb{P}(V_2)$ where there is a factorization as above, and let $Z \subset Z'$ be the closure of the open subset $U \subset Z'$ of points $(q : V \otimes \mathcal{O}_X(-m) \to E, [\Phi])$ such that the tensor is δ-semistable. Using Grothendieck's embedding $\mathcal{H} \to \mathbb{P}(V_1)$, explained in Section 1, we obtain a closed embedding

$$Z \longrightarrow \mathbb{P}(V_1) \times \mathbb{P}(V_2)$$

We endow Z with the polarization $\mathcal{O}_Z(b_1, b_1)$, where

$$\frac{b_2}{b_1} = \frac{P(l)\delta(m) - \delta(l)P(m)}{P(m) - a\delta(m)}$$

In other words, we use the Segre embedding

$$\mathbb{P}(V_1) \times \mathbb{P}(V_2) \longrightarrow \mathbb{P}(V_1^{\otimes b_1} \otimes V_2^{\otimes b_2})$$

and take the pullback of the ample line bundle $\mathcal{O}_{\mathbb{P}}(1)$.

It is proved in [G-S1] that a point in Z is GIT-(semi)stable if and only if the induced linear map $f : V \to H^0(E(m))$ is an isomorphism and it corresponds to a

δ-(semi)stable based tensor. Therefore, the GIT quotient $Z /\!\!/ \mathrm{SL}(V)$ is the moduli space of δ-semistable tensors.

To show how this is used to obtain moduli spaces of related objects, we will sketch the construction of the moduli space of orthogonal sheaves. First we construct the projective scheme Z as before, for tensors of type $a = 2$, i.e., of the form $(E, \varphi : E \otimes E \to \mathcal{O}_X)$. The condition of being symmetric is closed, so it defines a closed subscheme $R \subset Z^{ss}$, and the GIT quotient $R /\!\!/ \mathrm{SL}(V)$ is projective. On the other hand, the condition of being nondegenerate is open, so it defines an open subscheme $R_1 \subset R$. How can we prove that, after we remove the points corresponding to degenerate bilinear forms, the quotient is still projective? The idea is to show that, if (E, φ) is degenerate, then it is δ-unstable (we remark that, to prove this, we need the degree of δ to be $n - 1$). Therefore, $R_1 = R$, because all tensors corresponding to points in R are semistable.

In other words, the moduli space of orthogonal sheaves $R_1 /\!\!/ \mathrm{SL}(V)$ is projective because the inclusion $R_1 \hookrightarrow R$ is proper (in fact, it is the identity). Every time we impose a condition which is not closed, we have to prove a properness result of this sort, in order to show that the moduli space is projective.

The tensors defined in this section can easily be generalized to tensors of type (a, b, c), that is, pairs (E, φ) consisting of a torsion free sheaf and a homomorphism

$$\varphi : (E^{\otimes a})^{\otimes b} \longrightarrow (\det E)^{\otimes c} . \tag{2.6}$$

This more general notion will be needed in Section 5.

3. Principal bundles

Recall that, in the étale topology, an open covering of a scheme Y is a finite collection of morphisms $\{f_i : U_i \to Y\}_{i \in I}$ such that each f_i is étale, and Y is the union of the images of the f_i.

Note that an "open étale subset" of a scheme Y is not really a subset of Y, but an étale morphism $U \to Y$. If $f : X \to Y$ is a morphism, by a slight abuse of language we will denote by $f^{-1}(U)$ the pull-back

$$
\begin{array}{ccc}
f^{-1}(U) & \longrightarrow & X \\
\downarrow & & \downarrow {\scriptstyle f} \\
U & \longrightarrow & Y
\end{array}
$$

Let G be an algebraic group. A principal G-bundle on X is a scheme P with a right G-action and an invariant morphism $P \to X$ with a G-torsor structure. A G-torsor structure is given by an atlas consisting on an étale open covering $\{U_i\}$ of X and G-equivariant isomorphisms $\psi_i : p^{-1}(U_i) \to U_i \times G$, with $p = p_{U_i} \circ \psi_i$ (the G-action on $U_i \times G$ is given by multiplication on the right). Two atlases give the same G-torsor structure if their union is an atlas. An isomorphism of principal bundles is a G-equivariant isomorphism $\varphi : P \to P'$.

In short, a principal bundle is locally trivial in the étale topology, and the fibers are G-torsors. We remark that, if we were working in arbitrary characteristic, an algebraic group could be non-reduced, and we should have used the flat topology.

Given a principal G-bundle as above, we obtain an element of the étale cohomology set $\check{H}^1_{\mathrm{et}}(X, \underline{G})$, and this gives a bijection between isomorphism classes of principal G-bundles and elements of this set. Indeed, since the isomorphisms ψ_i of an atlas are required to be G-invariants, the composition $\psi_j \circ \psi_i^{-1}$ is of the form $(x, g) \mapsto (x, \alpha_{ij}(x)g)$, where $\alpha_{ij} : U_i \cap U_j \to G$ is a morphism, which satisfies the cocycle condition and defines a class in $\check{H}^1_{\mathrm{et}}(X, G)$.

Given a principal G-bundle $P \to X$ and a left action σ of G in a scheme F, we denote

$$P(\sigma, F) := P \times_G F = (P \times F)/G,$$

the associated fiber bundle. Sometimes this notation is shortened to $P(F)$ or $P(\sigma)$. In particular, for a representation ρ of G in a vector space V, $P(V)$ is a vector bundle on X, and if χ is a character of G, $P(\chi)$ is a line bundle.

If $\rho : G \to H$ is a group homomorphism, let σ be the action of G on H defined by left multiplication $h \mapsto \rho(g)h$. Then, the associated fiber bundle is a principal H-bundle, and it is denoted $\rho_* P$. We say that this principal H-bundle is obtained by *extension of structure group.*

Let $\rho : H \to G$ be a homomorphism of groups, and let P be a principal G-bundle on a scheme Y. A *reduction of structure group* of P to H is a pair (P^H, ζ), where P^H is a principal H-bundle on Y and ζ is an isomorphism between $\rho_* P^H$ and P. Two reductions (P^H, ζ) and (Q^H, θ) are isomorphic if there is an isomorphism α giving a commutative diagram

$$
\begin{array}{ccc}
P^H & \qquad \rho_* P^H \xrightarrow{\ \zeta\ } P & \qquad\qquad (3.1) \\[2pt]
{\scriptstyle\cong}\downarrow{\scriptstyle\alpha} & \quad\ \downarrow{\scriptstyle\rho_*\alpha} \qquad \| & \\[2pt]
Q^H & \qquad \rho_* Q^H \xrightarrow{\ \theta\ } P &
\end{array}
$$

The names "extension" and "reduction" come from the case in which ρ is injective, but note that these notions are still defined if the homomorphism is not injective.

If ρ is injective, giving a reduction is equivalent to giving a section σ of the associated fibration $P(G/H)$, where G/H is the quotient of G by the right action of H. Indeed, such a section gives a reduction P^H by pull-back

$$
\begin{array}{ccc}
P^H & \xrightarrow{\ i\ } & P \\[2pt]
\downarrow & & \downarrow \\[2pt]
X & \xrightarrow{\ \sigma\ } & P(G/H)
\end{array}
$$

and the isomorphism ζ is induced by i. Conversely, given a reduction (P^H, ζ), the isomorphism ζ induces an embedding $i : P^H \to P$, and the quotient by H of this morphism gives a section σ as above.

For example, if $G = \mathrm{O}(r)$ and $H = \mathrm{GL}_r$, the quotient H/G is the set of non-degenerate bilinear symmetric forms on the vector space $\mathbb{C}^r$, hence a section of $P(H/G)$ is just a non-degenerate bilinear symmetric morphism $E \otimes E \to \mathcal{O}_X$, where E is the vector bundle associated to the principal GL_r-bundle.

To construct the moduli space of principal bundles, we will assume that G is a connected reductive algebraic group. Let $G' = [G, G]$ be the commutator subgroup, and let $\mathfrak{g} = \mathfrak{z} \oplus \mathfrak{g}'$ be the Lie algebra of G, where $\mathfrak{g}'$ is the semisimple part and $\mathfrak{z}$ is the center.

Recall that, in the case of vector bundles, to obtain a projective moduli space when $\dim X > 1$, we had to consider also torsion free sheaves. Analogously, principal G-bundles are not enough if we want a projective moduli space, and this is why we also consider principal G-sheaves, which we will now define.

Definition 3.1. *A principal G-sheaf $\mathcal{P}$ over X is a triple $\mathcal{P} = (P, E, \psi)$ consisting of a torsion free sheaf E on X, a principal G-bundle P on the maximal open set U_E where E is locally free, and an isomorphism of vector bundles*

$$\psi : P(\mathfrak{g}') \xrightarrow{\ \cong\ } E|_{U_E} \, .$$

This definition can be understood from two points of view. From the first point of view, we have a torsion free sheaf E on X, together with a reduction to G, on the open set U_E, of the principal GL_r-bundle corresponding to the vector bundle $E|_{U_E}$. Indeed, the pair (P, ψ) is the same thing as a reduction to G of the principal GL_r-bundle on U_E associated to the vector bundle $E|_{U_E}$. It can be shown that, if we are given a reduction to a principal G-bundle on a big open set $U' \subsetneq U_E$, this reduction can uniquely be extended to U_E.

From the other point of view, we have a principal G-bundle on a big open set U, hence a vector bundle $P(\mathfrak{g}')$, together with a given extension of this vector bundle on U to a torsion free sheaf on the whole of X.

The Lie algebra structure of $\mathfrak{g}'$ is semisimple, hence the Killing form is non-degenerate. Correspondingly, the adjoint vector bundle $P(\mathfrak{g}')$ on U has a Lie algebra structure $P(\mathfrak{g}') \otimes P(\mathfrak{g}') \to P(\mathfrak{g}')$ and an orthogonal structure, $\kappa : P(\mathfrak{g}') \otimes P(\mathfrak{g}') \to \mathcal{O}_U$. These uniquely extend to give orthogonal and $\mathfrak{g}'$-sheaf structure to E:

$$\kappa : E \otimes E \longrightarrow \mathcal{O}_X \qquad [\,,] : E \otimes E \longrightarrow E^{\vee\vee}$$

where we have to take $E^{\vee\vee}$ in the target because an extension $E \otimes E \to E$ does not always exist. The orthogonal structure assigns an orthogonal $F^\perp = \ker(E \hookrightarrow E^{\vee} \to F^{\vee})$ to each subsheaf $F \subset E$.

Definition 3.2. *A principal G-sheaf $\mathcal{P} = (P, E, \psi)$ is said to be (semi)stable if for all orthogonal algebra filtrations $E_\bullet \subset E$, that is, filtrations with*

$$(1) \quad E_i^\perp = E_{-i-1} \quad \text{and} \quad (2) \quad [E_i, E_i] \subset E_{i+j}{}^{\vee\vee}$$

for all i, j, the following holds

$$\sum (rP_{E_i} - r_i P_E)(\leq)0$$

Replacing the Hilbert polynomials P_E and P_{E_i} by degrees, we obtain the notion of *slope (semi)-stability.*

Clearly

$$\text{slope-stable} \implies \text{stable} \implies \text{semistable} \implies \text{slope-semistable}$$

Since $G/G' \cong \mathbb{C}^{*q}$, given a principal G-sheaf, the principal bundle $P(G/G')$ obtained by extension of structure group provides q line bundles on U, and since $\operatorname{codim} X \setminus U \geq 2$, these line bundles extend uniquely to line bundles on X. Let $d_1, \ldots, d_q \in H^2(X, \mathbb{C})$ be their Chern classes. The rank r of E is clearly the dimension of $\mathfrak{g}'$. Let c_i be the Chern classes of E.

Definition 3.3 (Numerical invariants). *We call the data $\tau = (d_1, \ldots, d_q, c_i)$ the numerical invariants of the principal G-sheaf (P, E, ψ).*

Definition 3.4 (Family of semistable principal G-sheaves). *A family of (semi)stable principal G-sheaves parameterized by a scheme S is a triple (P_S, E_S, ψ_S), with E_S a family of torsion free sheaves, P_S a principal G-bundle on the open set U_{E_S} where E_S is locally free, and $\psi : P_S(\mathfrak{g}') \to E_S|_{U_{E_S}}$ an isomorphism of vector bundles, such that for all closed points $s \in S$ the corresponding principal G-sheaf is (semi)stable with numerical invariants τ.*

An isomorphism between two such families (P_S, E_S, ψ_S) and (P'_S, E'_S, ψ'_S) is a pair

$$(\beta : P_S \xrightarrow{\cong} P'_S, \gamma : E_S \xrightarrow{\cong} E'_S)$$

such that the following diagram is commutative

$$
\begin{array}{ccc}
P_S(\mathfrak{g}') & \xrightarrow{\psi} & E_S|_{U_{E_S}} \\
\beta(\mathfrak{g}') \downarrow & & \downarrow \gamma|_{U_{E_S}} \\
P'_S(\mathfrak{g}') & \xrightarrow{\psi'} & E'_S|_{U_{E_S}}
\end{array}
$$

where $\beta(\mathfrak{g}')$ is the isomorphism of vector bundles induced by β. Given an S-family $\mathcal{P}_S = (P_S, E_S, \psi_S)$ and a morphism $f : S' \to S$, the pullback is defined as $\tilde{f}^* \mathcal{P}_S = (\tilde{f}^* P_S, \overline{f}^* E_S, \tilde{f}^* \psi_S)$, where $\overline{f} = \operatorname{id}_X \times f : X \times S \to X \times S'$ and $\tilde{f} = i^*(\overline{f}) : U_{\tilde{f}^* E_S} \to U_{E_S}$, denoting $i : U_{E_S} \to X \times S$ the inclusion of the open set where E_S is locally free.

We can then define the functor of families of semistable principal G-sheaves

$$F_G^\tau : (\operatorname{Sch}/\mathbb{C}) \longrightarrow (\operatorname{Sets})$$

sending a scheme S, locally of finite type, to the set of isomorphism classes of families of semistable principal G-sheaves with numerical invariants τ. As usual, it is defined on morphisms as pullback.

Theorem 3.5. *There is a projective moduli space of semistable G-sheaves on X with fixed numerical invariants.*

This theorem is a generalization of the theorem of Ramanathan, asserting the existence of a moduli space of semistable principal bundles on a curve.

Note that in the definition of principal G-sheaf we have used the adjoint representation on the semisimple part $\mathfrak{g}'$ of the Lie algebra of G, to obtain a vector bundle $P(\mathfrak{g}')$ on a big open set of X, which we extend to the whole of X by torsion free sheaf. If we use a different representation $\rho : G \to \mathrm{GL}_r$, we have the notion of principal ρ-sheaf:

Definition 3.6. *A principal ρ-sheaf $\mathcal{P}$ over X is a triple $\mathcal{P} = (P, E, \psi)$ consisting of a torsion free sheaf E on X, a principal G-bundle P on the maximal open set U_E where E is locally free, and an isomorphism of vector bundles*

$$\psi : P(\rho) \xrightarrow{\;\cong\;} E|_{U_E} \, .$$

Now we will give some examples of principal ρ-sheaves which have already appeared:

- If $G = \mathrm{GL}_r$ and ρ is the canonical representation, then a principal ρ-sheaf is a torsion free sheaf.
- If $G = \mathrm{O}(r)$ and ρ is the canonical representation, then a principal ρ-sheaf is an orthogonal sheaf.
- If $G = \mathrm{SO}(r)$ and ρ is the canonical representation, then a principal ρ-sheaf is a special orthogonal sheaf (cf. [G-S1]), that is, a triple (E, φ, ψ) where $\varphi : E \otimes E \to \mathcal{O}_X$ symmetric and nondegenerate, and $\psi : \det E \to \mathcal{O}_X$ is an isomorphism such that $\det \varphi = \psi^{\otimes 2}$.
- If $G = \mathrm{Sp}(r)$ and ρ is the canonical representation, then a principal ρ-sheaf is a symplectic sheaf.
- If G is semisimple and ρ is injective, then giving a principal ρ-sheaf is equivalent to giving a honest singular principal bundle [Sch1, Sch2] with respect to the dual representation $\rho^\vee$ (see Section 5).

In all these cases (and also for principal G-sheaves, i.e., when $\rho : G \to \mathrm{GL}(\mathfrak{g}')$ is the adjoint representation), the stability condition is equivalent to the following:

Definition 3.7 (Stability for principal ρ-sheaves). *A principal ρ-sheaf $\mathcal{P} = (P, E, \psi)$ is said to be (semi)stable if for all reductions on any big open set $U \subset U_E$ of P to a parabolic subgroup $Q \subsetneq G$, and all dominant characters of Q, which are trivial on the center of Q, the induced filtration of saturated torsion free sheaves*

$$\cdots \subset E_{i-1} \subset E_i \subset E_{i+1} \subset \cdots$$

satisfies the following

$$\sum (r P_{E_i} - r_i P_E)(\leq)0$$

4. Construction of the moduli space of principal sheaves

In this section we will give a sketch of the construction of the moduli space in [G-S2]. The strategy is close to that of Ramanathan.

Let $r = \dim \mathfrak{g}'$, and consider the adjoint representation $\rho : G \to \mathrm{GL}_r$ of G in $\mathfrak{g}'$. The idea of Ramanathan is to start by constructing a scheme R_0 which classifies based vector bundles of rank r, and then to construct another scheme $Q \to R_0$ such that the fiber over each based vector bundle (f, E) parameterizes all reductions to G of the principal GL_r-bundle E. In other words, Q classifies tuples (f, P, E, ψ), where f is an isomorphism of a fixed vector space V with $H^0(E(m))$, P is a principal G-bundle and ψ is an isomorphism between the vector bundle $P(\rho, \mathfrak{g}')$ and E.

The problem is that ρ is not injective in general, so it is not easy to construct a reduction of structure group from GL_r to G in one step. Therefore, Ramanathan factors the representation ρ into several group homomorphisms, and then constructs reductions step by step.

Recall that $G' = [G, G]$ is the commutator subgroup. Let Z (respectively, Z') be the center of G (respectively, G'). Note that $Z' = G' \cap Z$. The adjoint representation factors as follows

$$G \xrightarrow{\rho_3} G/Z' \xrightarrow{\rho_2'} G/Z \xhookrightarrow{\rho_2} \mathrm{Aut}(\mathfrak{g}') \xhookrightarrow{\rho_1} \mathrm{GL}_r$$

and the schemes parameterizing these reductions are

$$R_3 \xrightarrow{f_3} R_2' \xrightarrow{f_2'} R_2 \xrightarrow{f_2} R_1 \longrightarrow R_0$$

In the case $\dim X = 1$ this works well because a principal G-bundle is semistable if and only if the associated vector bundle is semistable. This is no longer true if X is not a curve, and this is why, for arbitrary dimension, we do not construct the scheme R_0, but instead start directly with a scheme R_1, classifying semistable based principal $\mathrm{Aut}(\mathfrak{g}')$-sheaves.

Here $\mathrm{Aut}(\mathfrak{g}')$ denotes the subgroup of GL_r of linear automorphisms which respect the Lie algebra structure. Therefore, a based principal $\mathrm{Aut}(\mathfrak{g}')$-sheaf is the same thing as a based $\mathfrak{g}'$-sheaf.

Using the isomorphism (2.4), we can describe a $\mathfrak{g}'$-sheaf as a Lie tensor (Definition 2.4) such that the Lie algebra structure induced on the fibers of E, over points $x \in X$ where E is locally free, is isomorphic to $\mathfrak{g}'$.

Choose a polynomial δ of degree $\dim X - 1$, with positive leading coefficient. We fix the first Chern class to be zero. This is because we are interested in $\mathfrak{g}'$-sheaves, and since $\mathfrak{g}'$ is semisimple, its Killing form is nondegenerate, hence induces an orthogonal structure on the sheaf, and this forces the first Chern class to be zero.

We start with the scheme Z, defined in Section 2, classifying based tensors of type $a = r + 1$. This scheme has an open subset Z^{ss} corresponding to δ-semistable tensors. Conditions (1) to (3) in the definition of Lie tensor are closed, hence they define a closed subscheme $R \subset Z^{ss}$. Using the isomorphism (2.4), we see that the scheme R parameterizes Lie sheaves. Recall that a Lie sheaf structure induces a Killing form $\kappa : E \otimes E \to \mathcal{O}_X$.

Lemma 4.1. *There is a subscheme $R_1 \subset R$ corresponding to those Lie tensors which are $\mathfrak{g}'$-tensors.*

The family of Lie sheaves parameterized by R gives a family of Killing forms $E_R \otimes E_R \to \mathcal{O}_{X \times R}$, and hence a homomorphism $f : \det E_R \to \det E_R^\vee$. We have fixed the determinant of the tensors to be trivial, hence $\det E_R$ is the pullback of a line bundle on R, and therefore the homomorphism f is nonzero on an open set of the form $X \times W$, where W is an open set of R. The open set W is in fact the whole of R. This is because if z is a point in the complement, it corresponds to a Lie sheaf whose Killing form is non-degenerate, and hence has a nontrivial kernel. Using this, it is possible to construct a filtration which shows that this Lie sheaf is δ-unstable when $\deg \delta = \dim X - 1$, but this contradicts the fact that $R \subset Y^{ss}$.

The Killing form of a Lie algebra is semisimple if and only if it is non-degenerate. Therefore, for all points (x, t) in the open subset $\mathcal{U}_{E_R} \subset X \times R$ where E_R is locally free, the Lie algebra is semisimple.

Semisimple Lie algebras are rigid, that is, if there is a family of Lie algebras, the subset of the parameter space corresponding to Lie algebras isomorphic to a given semisimple Lie algebra is open. Therefore, since U_E is connected for all torsion free sheaves E, all points $(x, t) \in \mathcal{U}_{E_R} \subset X \times R$ where t is in a fixed connected component of R, give isomorphic Lie algebras. Let R_1 be the union of those components whose Lie algebra is isomorphic to $\mathfrak{g}'$. The inclusion

$$i : R_1 \hookrightarrow R$$

is proper, and hence, since the GIT quotient $R /\!\!/ \mathrm{SL}(V)$ is proper, also the GIT quotient $R_1 /\!\!/ \mathrm{SL}(V)$ is proper. Note that, to prove properness of i, two facts about semisimple Lie algebras were used: rigidity, and nondegeneracy of their Killing forms.

For simplicity of the exposition, to explain the successive reductions, first we will assume that for all $\mathfrak{g}'$-sheaves (E, φ), the torsion free sheaf E is locally free. In other words, $U_E = X$ (this holds, for instance, if $\dim X = 1$). At the end we will mention what has to be modified in order to consider the general case.

The group G/Z is the connected component of identity of $\mathrm{Aut}(\mathfrak{g}')$. Therefore, giving a reduction of structure group of a principal $\mathrm{Aut}(\mathfrak{g}')$-bundle P by ρ_2 is the same thing as giving a section of the finite étale morphism $P(F) \to X$, where F is the finite group $\mathrm{Aut}(\mathfrak{g}')/(G/Z)$. This implies that $R_2 \to R_1$ is a finite étale morphism, whose image is a union of connected components of R_1.

There is an isomorphism of groups $G/Z' \cong G/G' \times G/Z$, and ρ_2' is just the projection to the second factor. Therefore, a reduction to G/Z' of a principal G/Z-bundle $P^{G/Z}$ is just a pair $(P^{G/G'}, P^{G/Z})$, where $P^{G/Z}$ is the original G/Z-bundle and $P^{G/G'}$ is a G/G'-bundle. But

$$G/G' \cong \mathbb{C}^* \overbrace{\times \cdots \times}^{q} \mathbb{C}^* ,$$

hence this is just a collection of q line bundles, whose Chern classes are given by the numerical invariants which have been fixed. This implies that there is an isomorphism

$$R_2' \cong J \overbrace{\times \cdots \times J}^{q} \times R_2 \,,$$

where J is the Jacobian of X.

Finally, we have to consider reductions of a principal G/Z'-bundle to G, where Z' is a finite subgroup of the center of G. There is an exact sequence of pointed sets (the distinguished point being the trivial bundle)

$$\check{H}^1_{\mathrm{et}}(X, \underline{Z'}) \longrightarrow \check{H}^1_{\mathrm{et}}(X, \underline{G}) \longrightarrow \check{H}^1_{\mathrm{et}}(X, \underline{G/Z'}) \stackrel{\delta}{\longrightarrow} \check{H}^2_{\mathrm{et}}(X, \underline{Z'}) \,.$$

Note that Z' is abelian, therefore $H^i_{\mathrm{et}}(X, \underline{Z'})$ is an abelian group, and it is isomorphic to the singular cohomology group $H^i(X; Z')$, hence finite. A principal G/Z'-bundle admits a reduction to G if and only if the image by δ of the corresponding point is 0. This is an open and closed condition, therefore there is a subscheme $\hat{R}_2'$ of R_2', consisting of a union of connected components, corresponding to those principal G/Z-bundles admitting a reduction to G.

Let (P^G, ζ) be a reduction to G of a principal G/Z'-bundle. It can be shown that the set of isomorphism classes of all reductions to G is in bijection with the cohomology set $\check{H}^1_{\mathrm{et}}(X, \underline{Z'})$, with the unit element of this set corresponding to the chosen reduction (P^G, ζ). This cohomology set is an abelian group, because Z' is abelian. Therefore, the set of reductions of a principal G/Z'-bundle to G is an $\check{H}^1_{\mathrm{et}}(X, \underline{Z'})$-torsor, and this implies that $R_3 \to \hat{R}_2'$ is a principal $\check{H}^1_{\mathrm{et}}(X, \underline{Z'})$-bundle. Using that this cohomology set is a finite set (in fact isomorphic to the singular cohomology group $H^1(X; Z')$), and that $\hat{R}_2'$ is a union of connected components of R_2', it follows that $R_3 \to R_2'$ is finite étale.

Ramanathan [Ra, Lemma 5.1] proves that, if H is a reductive algebraic group, $f : Y \to S$ is an H-equivariant affine morphism, and $p : S \to \overline{S}$ is a good quotient, then Y has a good quotient $q : \overline{Y} \to Y$ and the induced morphism $\overline{f}$ is affine. Moreover, if f is finite, $\overline{f}$ is also finite. When f is finite and p is a geometric quotient, also q is a geometric quotient.

The group $\mathrm{SL}(V)$ acts on all the schemes R_i, and the morphisms f_2 and f_3 are equivariant and finite. Therefore, we can apply Ramanathan's lemma to those morphisms.

The morphism $f_2' : J^{\times q} \times R_2 \to R_2$ is just projection to a factor, and the group acts trivially on the fiber, therefore if $p_2 : R_2 \to \mathfrak{M}_2$ is a good quotient of R_2, $J^{\times q} \times \mathfrak{M}_2$ will give a good quotient of R_2'. Furthermore, if p_2 becomes a geometric quotient when restricting to an open set, the same will be true after taking the product with $J^{\times q}$.

Using GIT, we know that R_1 has a good quotient $\mathfrak{M}_1$, which is a geometric quotient when restricting to the open set of stable points. Therefore, the same holds for all these schemes, and the good quotient of R_3 is the moduli space of principal G-bundles.

The successive reductions in higher dimension are very similar to the reductions in the case X is a curve, except for the technical difficulty that the principal bundles in general are not defined in the whole of X, but only in a big open set. To overcome this difficulty, we need "purity" results for open sets $U \subset X$ when U is big. We will discuss them one by one.

First we consider reductions of a principal $\mathrm{Aut}(\mathfrak{g}')$-bundle P to G/Z. These are parameterized by sections of the associated fibration $P(F)$, where $F = \mathrm{Aut}(\mathfrak{g}')/(G/Z)$ is a finite group. If P is a principal bundle on a big open set U, $P(F)$ is a Galois cover of U, given by a representation of the algebraic fundamental group of $\pi(U)$ in F. Since U is a big open set, $\pi(X) = \pi(U)$ (purity of fundamental group), and hence the Galois cover $P(F)$ of U extends uniquely to a Galois cover of X. This implies that, even if $\dim X > 1$, the morphism $R_2 \to R_1$ is still finite étale, as in the curve case.

Giving a reduction of a principal G/Z-bundle on U to a principal G/Z'-bundle is equivalent to giving q line bundles on U. Since U is a big open set, the Jacobians of U and X are isomorphic (purity of Jacobian), and hence we still have $R_2' = J(X)^{\times q} \times R_2$.

Finally, we have to consider reductions of principal G/Z'-bundles to G. Using the fact that U is a big open set, there are isomorphisms $\check{H}^i_{\mathrm{et}}(X, \underline{Z}') \cong \check{H}^i_{\mathrm{et}}(U, \underline{Z}')$ for $i = 1, 2$. Therefore, the arguments used for the case $U = X$ still hold in general, and it follows that $R_3 \to R_2'$ is étale finite.

5. Construction of the moduli space of principal ρ-sheaves

In [Sch1, Sch2], A. Schmitt fixes a semisimple group G and a faithful representation ρ, defines semisimple honest singular principal bundle with respect to this data (see the definition below), and constructs the corresponding projective moduli space. Giving such an object is equivalent to giving a principal $\rho^\vee$-bundle, where $\rho^\vee$ is the dual representation in $V^\vee$. In this section we will give a sketch of Schmitt's construction.

Let G be a semisimple group, and $\rho : G \to \mathrm{GL}(V)$ a faithful representation. A honest singular principal G-bundle is a pair $(\mathcal{A}, \tau)$, where $\mathcal{A}$ is a torsion free sheaf on X and

$$\tau : \mathrm{Sym}^*(\mathcal{A} \otimes V)^G \longrightarrow \mathcal{O}_X$$

is a homomorphism of $\mathcal{O}_X$-algebras such that, if $\sigma : X \to \mathcal{H}om(V \otimes \mathcal{O}_U, \mathcal{A}|_U^\vee)/\!\!/ G$ is the induced morphism, then

$$\sigma(U) \subset \mathcal{I}som(V \otimes \mathcal{O}_U, \mathcal{A}|_U^\vee)/G \subset \mathcal{H}om(V \otimes \mathcal{O}_U, \mathcal{A}|_U^\vee)/\!\!/ G.$$

It can be shown that the points in the affine U-scheme $\mathcal{I}som(V \otimes_{\mathbb{C}} \mathcal{O}_U, \mathcal{A}|_U^\vee)$ are in the open set of GIT-polystable points of $\mathcal{H}om(V \otimes_{\mathbb{C}} \mathcal{O}_U, \mathcal{A}|_U^\vee)$, under the natural action of G, therefore the previous inclusion makes sense.

Note that the homomorphism τ is uniquely defined by its restriction to $U \subset X$, therefore, giving a honest singular principal G-bundle is equivalent to giving a principal $\rho^\vee$-sheaf (P, E, ψ), where $\rho^\vee : G \to \mathrm{GL}(V^\vee)$ is the dual representation, P is a principal GL_n-bundle, $E = \mathcal{A}$, and ψ is induced by $\sigma|_U$.

In other words, in a principal ρ-sheaf, we extend to the whole of X, as a torsion free sheaf E, the vector bundle associated to ρ, whereas, is a honest singular principal G-bundle associated to ρ, we extend the dual of the vector bundle associated to ρ.

The idea of Schmitt's construction is to transform τ into a tensor. Note that τ is an infinite collection of $\mathcal{O}_X$-module homomorphisms

$$\tau_i : \mathrm{Sym}^i(\mathcal{A} \otimes V)^G \longrightarrow \mathcal{O}_X, \tag{5.1}$$

but, since $\mathrm{Sym}^*(\mathcal{A} \otimes V)^G$ is finitely generated as a $\mathcal{O}_X$-algebra, there is an integer s such that

1. the sheaf

$$\bigoplus_{i=1}^{s} \mathrm{Sym}^i(\mathcal{A} \otimes V)^G$$

   contains a set of generators of the algebra, and
2. the subalgebra

$$\mathrm{Sym}^{(s!)}(\mathcal{A} \otimes V)^G := \bigoplus_{m=0}^{\infty} \mathrm{Sym}^{s!m}(\mathcal{A} \otimes V)^G$$

   is generated by elements in $\mathrm{Sym}^{s!}(\mathcal{A} \otimes V)^G$.

Using the homomorphisms τ_s, we construct a homomorphism of $\mathcal{O}_X$-modules

$$\bigoplus_{\sum id_i = s!} \left(\bigotimes_{i=1}^{s} \mathrm{Sym}^{d_i}\left(\mathrm{Sym}^i(\mathcal{A} \otimes V)^G \right) \right) \twoheadrightarrow \mathrm{Sym}^{s!}(\mathcal{A} \otimes V)^G \xrightarrow{\tau_s} \mathcal{O}_X \tag{5.2}$$

Note that the vector space

$$\bigoplus_{\sum id_i = s!} \left(\bigotimes_{i=1}^{s} \mathrm{Sym}^{d_i}\left(\mathrm{Sym}^i(\mathbb{C}^r \otimes V)^G \right) \right)$$

has a canonical representation of GL_n, homogeneous of degree $s!$, and hence it is a quotient of the representation

$$(\mathbb{C}^{\otimes a})^{\oplus b} \otimes (\bigwedge^r \mathbb{C}^r)^{-\otimes c}$$

for appropriate values of a, b and c. Therefore, there is a surjection

$$(\mathcal{A}^{\otimes a})^{\oplus b} \otimes (\det \mathcal{A})^{-\otimes c} \to \bigoplus_{\sum id_i = s!} \left(\bigotimes_{i=1}^{s} \mathrm{Sym}^{d_i}\left(\mathrm{Sym}^i(\mathcal{A} \otimes V)^G \right) \right) \tag{5.3}$$

and composing (5.2) with (5.3) we obtain a tensor of type (a, b, c), as in (2.6).

References

[Gi] D. Gieseker, *On the moduli of vector bundles on an algebraic surface,* Ann. Math., **106** (1977), 45–60.

[G-S1] T. Gómez and I. Sols, *Stable tensors and moduli space of orthogonal sheaves,* Preprint 2001, math.AG/0103150.

[G-S2] T. Gómez and I. Sols, *Moduli space of principal sheaves over projective varieties,* Ann. of Math. (2) **161** no. 2 (2005), 1037–1092.

[GLSS] T. Gómez, A. Langer, A. Schmitt, I. Sols, *Moduli Spaces for Principal Bundles in Large Characteristic,* Proceedings "International Workshop on Teichmüller Theory and Moduli Problems", Allahabad 2006 (India)

[Ha] R. Hartshorne, Algebraic Geometry, Grad. Texts in Math. 52, Springer Verlag, 1977.

[H-L] D. Huybrechts and M. Lehn, The geometry of moduli spaces of sheaves, Aspects of Mathematics E31, Vieweg, Braunschweig/Wiesbaden 1997.

[La] A. Langer, *Semistable sheaves in positive characteristic.* Ann. of Math. (2) **159** (2004), 251–276.

[Ma] M. Maruyama, *Moduli of stable sheaves, I and II.* J. Math. Kyoto Univ. **17** (1977), 91–126. **18** (1978), 557–614.

[Mu1] D. Mumford, Geometric invariant theory. Ergebnisse der Mathematik und ihrer Grenzgebiete, Neue Folge, Band 34. Springer-Verlag, Berlin-New York, 1965.

[Ra] A. Ramanathan, *Moduli for principal bundles over algebraic curves: I and II,* Proc. Indian Acad. Sci. (Math. Sci.), **106** (1996), 301–328, and 421–449.

[Sesh] C.S. Seshadri, *Space of unitary vector bundles on a compact Riemann surface.* Ann. of Math. (2) **85** (1967), 303–336.

[Sch1] A. Schmitt, *Singular principal bundles over higher-dimensional manifolds and their moduli spaces,* Internat. Math. Res. Notices **23** (2002), 1183–1210.

[Sch2] A. Schmitt, *A closer look at semistability for singular principal bundles,* Int. Math. Res. Not. **62** (2004), 3327–3366.

[Si] C. Simpson, *Moduli of representations of the fundamental group of a smooth projective variety I,* Publ. Math. I.H.E.S. **79** (1994), 47–129.

Tomás L. Gómez
IMAFF, Consejo Superior de Investigaciones Científicas
Serrano 113bis
E-28006 Madrid, Spain

and

Facultad de Matemáticas
Universidad Complutense de Madrid
E-28040 Madrid, Spain
e-mail: `tomas.gomez@mat.csic.es`

Algebraic Cycles, Sheaves, Shtukas, and Moduli
Trends in Mathematics, 69–103
© 2007 Birkhäuser Verlag Basel/Switzerland

Lectures on Torsion-free Sheaves and Their Moduli

Adrian Langer

Introduction

These notes come from the lectures delivered by the author at 25th Autumn School of Algebraic Geometry in Łukęcin in 2002 and the lectures delivered by the author at the IMPANGA seminar in 2004–5. The School lectures were largely based on the book [HL], whereas the IMPANGA lectures were very close to the author's papers [La1], [La2] and [La3].

So the author decide to write notes that contain more of a vision of how the subject could be lectured upon than the faithful account of the delivered lectures. This particularly refers to the last lecture that contains generic smoothness of moduli spaces of sheaves on surfaces and the proofs in the lectures were based on the different O'Grady's approach. Since this approach was already published by the author (see [La3]), there was no point in copying it. Therefore the author decided to follow the Donaldson's approach whose idea (but not necessarily technical details) is easier to understand.

The notes contain some exercises (that are not very evenly distributed) in which the author put a part of the theory that is either analogous to what is done in the lectures or it is too far away to be proven and it is quite standard.

Since the paper is relatively short, it was impossible to give full proofs of all the theorems. All the proofs are either provided or can be found in the very incomplete references at the end of the paper. The references contain either books or the references that are not contained in these books (partially because they are too new).

1. Lecture 1. Bogomolov's instability and restriction theorems

- Topological classification of vector bundles
- Semistability and its properties
- Bogomolov's instability theorem
- Restriction theorems

Let X be a smooth complex projective variety. Classification of algebraic vector bundles on X can be divided into two parts: discrete, where we distinguish vector bundles using just topological structure, and continuous, where we study holomorphic/algebraic structures on a given topological vector bundle.

1.1. Topological classification of vector bundles

This part is usually neglected in algebro-geometric papers, as fixing basic topological invariants such as rank and Chern classes distinguish algebraic vector bundles sufficiently well to study change of the algebraic structure. Nevertheless, we will recall a few results on the topological classification referring to [BP] for a nice recent account of this topic.

The set of rank r vector bundles on X is isomorphic to the set $[X, \mathrm{Gr}_r(\mathbb{C}^\infty)]$ of homotopy classes of maps from X to the infinite Grassmannian $\mathrm{Gr}_r(\mathbb{C}^\infty) = BU(r)$. Since $H^*(\mathrm{Gr}_r(\mathbb{C}^\infty))$ is a polynomial ring $\mathbb{Z}[c_1, \ldots, c_r]$, this allows to define characteristic classes of a vector bundle.

In topology, it is easier to classify vector bundles up to stable equivalence (i.e., up to adding a trivial vector bundle). Stable equivalence classes can be read off the topological Grothendieck ring $K_{\mathrm{top}}(X)$, which is the universal ring associated to all vector bundles on X with direct sum and tensor operations. Again the topological group $K_{\mathrm{top}}(X)$ is isomorphic to homotopy classes $[X, B_{\mathbb{C}} \times \mathbb{Z}]$, where $B_{\mathbb{C}} = \lim_{\rightarrow} \mathrm{Gr}_r(\mathbb{C}^{2r})$. The K-ring is easier to compute and for example one can show that

$$K^*_{\mathrm{top}}(\mathbb{CP}^n) \simeq \mathbb{Z}[\xi]/\xi^{n+1}.$$

Nevertheless, a complete topological classification of vector bundles on projective spaces is far from being complete.

It is known that a topological rank $r > n$ vector bundle on n-dimensional X is topologically equivalent to a direct sum of a rank n vector bundle and the trivial vector bundle of rank $(r - n)$. So it is sufficient to classify vector bundles of rank $r \leq n$. As an application of the Atiyah-Singer index theorem, one can show that if E is a topological vector bundle on X then

$$\int_X \mathrm{ch}(E)\, \mathrm{ch}(\xi)\, \mathrm{td}\, X \in \mathbb{Z}$$

for all classes $\xi \in K_{\mathrm{top}}(X)$. In case $H^*(X, \mathbb{Z})$ has no torsion, by the Bǎnicǎ-Putinar result (see [BP]), if Chern classes satisfy the above condition then there exists a unique topological rank $r = n$ vector bundle with such Chern classes.

In case of projective spaces, the above conditions on Chern classes of topological vector bundles can be written quite explicitly, and they are known as *Shwarzenberger conditions*. In case of lower rank $r < n$ on $\mathbb{CP}^n$, there can be several different topological structures for a given rank and collection of Chern classes. The first such example was given by M. Atiyah and E. Rees, who classified topological rank 2 vector bundles on $\mathbb{CP}^3$ and in particular showed that there are exactly two different structures if $c_1 \equiv 0 \pmod 2$. Later, vector bundles on $\mathbb{CP}^n$ were classified for $n \leq 6$, but it is not known which of these vector bundles can be realized as algebraic vector bundles. This is connected to the Hartshorne's conjecture saying that there are no indecomposable rank 2 vector bundles on $\mathbb{CP}^n$ for large n ($n \geq 7$, or maybe even $n \geq 5$ as no examples are known).

In the rest of the paper we will be mainly interested in algebraic vector bundles on surfaces. In this case to each vector bundle E one can associate its rank r, the first Chern class $c_1 E \in H^2(X, \mathbb{Z}) \cap H^{1,1}(X)$ and the second Chern class $c_2 E \in H^4(X, \mathbb{Z}) \simeq \mathbb{Z}$. These are all topological invariants and all triples (r, c_1, c_2) with $r \geq 2$ can be realized as algebraic vector bundles (R. Schwarzenberger).

1.2. Semistability and its properties

Let us fix topological invariants of a vector bundle. As usual in algebraic geometry, instead of a geometric vector bundle we study the associated locally free coherent sheaf of its sections. In general, we cannot expect that the set of all vector bundles with fixed invariants have a nice structure of algebraic variety. The necessity of restricting to some subset of the set of all vector bundles can be understood by the following standard example.

Example 1.2.1. Consider the set $\{\mathcal{O}(n) \oplus \mathcal{O}(-n)\}$ on $\mathbb{P}^1$. The topological invariants are fixed but we have infinitely many points which cannot form a nice algebraic variety. Moreover, there exists a family of vector bundles $\{E_t\}_{t \in \mathbb{C}}$ such that $E_t \simeq \mathcal{O} \oplus \mathcal{O}$ for $t \neq 0$ but $E_0 \simeq \mathcal{O}(n) \oplus \mathcal{O}(-n)$ (Exercise: construct such a family). So the point corresponding to $\mathcal{O} \oplus \mathcal{O}$ would not be closed in the moduli space.

The natural class of vector bundles which admits a nice moduli space comes, at least in the curve case, from Mumford's Geometric Invariant Theory (GIT). The corresponding vector bundles are called stable (or semistable). However, in higher dimensions if we want to get a projective moduli space then we need to add some non-locally free sheaves at the boundary of the moduli space. So we need to define semistability and stability in the more general context.

Let X be a smooth n-dimensional projective variety defined over an algebraically closed field k and let H be an ample divisor on X. For any rank $r > 0$ torsion free sheaf E we define its *slope* by

$$\mu(E) = \frac{c_1 E \cdot H^{n-1}}{r},$$

where $c_1 E$ denotes the first Chern class of the line bundle $(\bigwedge^r E)^{**}$. The *Hilbert polynomial* $P(E)$ is defined by $P(E)(k) = \chi(X, E \otimes \mathcal{O}_X(kH))$. The *reduced Hilbert polynomial* is defined as $p(E) = P(E)/r$.

Definition 1.2.2.

  1. E is called *slope H-stable* if and only if for all subsheaves $F \subset E$ with $\mathrm{rk}\, F < \mathrm{rk}\, E$ we have
  $$\mu(F) < \mu(E).$$

  2. E is called *Gieseker H-stable* if and only if for all proper subsheaves $F \subset E$
  $$p(F) < p(E)$$
  (i.e., the inequality holds for large values k).

Similarly, one can define (*slope or Gieseker*) *H-semistability* by changing the strict inequality sign $<$ to $\leq$.

Restricting our attention to semistable sheaves is not very restrictive since each (torsion-free) sheaf has a canonical filtration with semistable quotients.

Let us fix a torsion free sheaf E on X. Consider the set $\{\mu(F) : F \subset E\}$. This set has a maximal element $\mu_{\max} = \mu_{\max}(E)$ and the set

$$\{F : F \subset E, \text{ such that } E/F \text{ is torsion-free}, \mu(F) = \mu_{\max}\}$$

contains a sheaf E_1 of largest rank. This element is the largest element in this set with respect to the inclusion relation and it is called the *maximal destabilizing subsheaf* of E.

Now we consider the maximal destabilizing subsheaf E_2' in E/E_1 and set $E_2 = p^{-1}E_2'$ for the natural projection $p : E \to E/E_1$. Iterating this process we get the unique filtration

$$0 = E_0 \subset E_1 \subset \cdots \subset E_m = E$$

such that

1. all the quotients $F_i = E_i/E_{i-1}$ are semistable, and
2. $\mu_{\max}(E) = \mu(F_1) > \mu(F_2) > \cdots > \mu(F_m) = \mu_{\min}(E)$.

This filtration is called the *Harder–Narasimhan filtration* of E.

1.2.1. Properties of slope semistability in characteristic 0:

1. If E_1, E_2 are slope H-semistable torsion free sheaves then $E_1 \otimes E_2/\,\text{Torsion}$ is also slope H-semistable.
2. If $f : Y \to X$ is a finite map between smooth projective varieties then a torsion free sheaf F is slope H-semistable if and only if f^*F is slope f^*H-semistable.

1.3. Bogomolov's inequality

Now let us consider the following question: which Chern classes can be realized by semistable vector bundles?

The class $\Delta(E) = 2rc_2E - (r-1)c_1^2 E$ is called the *discriminant* of E. In the surface case we will not distinguish between $\Delta(E)$ and its degree $\int_X \Delta(E)$.

A partial answer to the above question is given by the following theorem.

Theorem 1.3.1 (Bogomolov). *Let X be a smooth complex projective surface. Then for any torsion free (slope) H-semistable sheaf E we have*

$$\Delta(E) \geq 0.$$

Proof. We can assume that E is a locally free sheaf. Indeed, for any torsion free slope H-semistable sheaf E on a smooth surface, the sheaf E^{**} is locally free, slope H-semistable and $\Delta(E^{**}) \leq \Delta(E)$.

For simplicity, we also assume that $c_1 E = 0$. The general case can be reduced to this one using either the $\mathbb{Q}$-vector bundle $E(-\frac{1}{r} \det E)$ or the vector bundle $\mathcal{E}nd E$.

Now let us note that $S^n E$ is slope H-semistable (this follows from 1.2.1.1). Hence for $C \in |kH|$ and $k > 0$ we have

$$h^0(S^n E(-C)) = 0.$$

Therefore from the short exact sequence

$$0 \to \mathcal{O}_X(-C) \to \mathcal{O}_X \to \mathcal{O}_C \to 0$$

tensored with $S^n E$ we get

$$h^0(S^n E) \le h^0(S^n E(-C)) + h^0(S^n E_C) = h^0(S^n E_C).$$

Considering $Y = \mathbb{P}(E_C) \to C$ we see that

$$h^0(S^n E_C) = h^0(Y, \mathcal{O}_{\mathbb{P}(E_C)}(n))$$

by the projection formula. Since $\dim Y = r$ there exists a constant C such that

$$h^0(S^n E) \le h^0(Y, \mathcal{O}_{\mathbb{P}(E_C)}(n)) \le C \cdot n^r$$

for all $n > 0$. Similarly, using restriction to $C \in |kH|$ for large k, one can see that there exists a constant C' such that

$$h^2(S^n E) = h^0(S^n E^* \otimes K_X) \le C' \cdot n^r$$

for all $n > 0$.

Therefore

$$\chi(X, S^n E) \le h^0(S^n E) + h^2(S^n E) \le (C + C')n^r.$$

But by Exercise 1.5.10 we have

$$\chi(X, S^n E) = -\frac{\Delta(E)}{2r} \frac{n^{r+1}}{(r+1)!} + O(n^r),$$

so $\Delta(E) \ge 0$. $\qquad \square$

The above proof of Bogomolov's theorem follows quite closely Y. Miyaoka's proof from [Mi].

Proposition 1.3.2 ([La1])**.** *Let X be a smooth complex projective surface and let H be an ample divisor on X. Then for any rank r torsion free sheaf E we have*

$$H^2 \cdot \Delta(E) + r^2(\mu_{\max} - \mu)(\mu - \mu_{\min}) \ge 0.$$

Proof. Let $0 = E_0 \subset E_1 \subset \cdots \subset E_m = E$ be the Harder–Narasimhan filtration. Set $F_i = E_i/E_{i-1}$, $r_i = \operatorname{rk} F_i$, $\mu_i = \mu(F_i)$. Then by Bogomolov's inequality and the Hodge index theorem

$$\frac{\Delta(E)}{r} = \sum \frac{\Delta(F_i)}{r_i} - \frac{1}{r} \sum_{i<j} r_i r_j \left(\frac{c_1 F_i}{r_i} - \frac{c_1 F_j}{r_j} \right)^2$$

$$\ge -\frac{1}{rH^2} \sum_{i<j} r_i r_j \left(\left(\frac{c_1 F_i}{r_i} - \frac{c_1 F_j}{r_j} \right) H \right)^2 = -\frac{1}{rH^2} \sum_{i<j} r_i r_j (\mu_i - \mu_j)^2.$$

(The Hodge index theorem says that $(DH)^2 \geq D^2 \cdot H^2$ for any divisor D.) Now the proposition follows from the following lemma:

Lemma 1.3.3. *Let r_i be positive real numbers and $\mu_1 > \mu_2 > \cdots > \mu_m$ real numbers. Set $r = \sum r_i$ and $r\mu = \sum r_i \mu_i$. Then*

$$\sum_{i<j} r_i r_j (\mu_i - \mu_j)^2 \leq r^2 (\mu_1 - \mu)(\mu - \mu_m).$$

Proof. Let us note that

$$\sum_{i<j} r_i r_j (\mu_i - \mu_j)^2 = r \left(\sum_{i=1}^{m-1} \left(\sum_{j \leq i} r_j (\mu_j - \mu) \right) (\mu_i - \mu_{i+1}) \right).$$

Using $\sum_{j \leq i} r_j \mu_j \leq (\sum_{j \leq i} r_j) \mu_1$ and simplifying yields the required inequality. $\square\square$

1.4. Restriction theorems

As an application of the above proposition we get the following effective restriction theorem (see [La1]). A weaker version of this theorem was proved by F.A. Bogomolov (see [HL, Theorem 7.3.5]).

Theorem 1.4.1. *Let E be a rank $r \geq 2$ vector bundle on a smooth complex projective surface X. Assume that E is slope H-stable. Let $D \in |kH|$ be a smooth curve. If*

$$k \geq \frac{r-1}{r} \Delta(E) + 1$$

then E_D is stable.

Proof. Assume that E_D is not stable. Then there exists a subsheaf $S \subset E_D$ such that $\mu(S) \geq \mu(E_D)$. Let us take the maximal subsheaf with this property. In this case the quotient $T = (E_D)/S$ is a vector bundle on D (it is sufficient to check that T is torsion free; if it has a torsion then the kernel S' of $E_D \to T/\text{Torsion}$ contains S and $\mu(S') \geq \mu(S)$).

Let G be the kernel of the composition $E \to E_D \to T$. The sheaf G is called an *elementary transformation* of E along T. Set $\rho = \text{rk}\, S$. Then $\text{rk}\, E = \text{rk}\, G = r$ and $\text{rk}\, T = r - \rho$ (as a vector bundle on D).

Computing $\Delta(G)$ (use Exercise 1.5.11) we get

$$\Delta(G) = \Delta(E) - \rho(r - \rho)D^2 + 2(r \deg_D T - (r - \rho)Dc_1(E)).$$

By assumption $\mu(T) \leq \mu(E_D)$, so

$$\Delta(G) \leq \Delta(E) - \rho(r - \rho)D^2.$$

Using the stability of E we get

$$\mu_{\max}(G) - \mu(G) = \mu_{\max}(G) - \mu(E) + \frac{r - \rho}{r} DH \leq \frac{r - \rho}{r} kH^2 - \frac{1}{r(r-1)}.$$

Note that we have two short exact sequences:

$$0 \to G \to E \to T \to 0$$

and
$$0 \to E(-D) \to G \to S \to 0.$$

In particular $G^* \subset (E(-D))^*$ and

$$\mu(G) - \mu_{\min}(G) = \mu(E(-D)) - \mu_{\min}(G) + \frac{\rho}{r}DH$$

$$= \mu_{\max}(G^*) - \mu((E(-D))^*) + \frac{\rho}{r}DH \leq \frac{\rho}{r}kH^2 - \frac{1}{r(r-1)}.$$

Hence, applying Proposition 1.3.2 to G we obtain

$$0 \leq H^2\Delta(G) + r^2(\mu_{\max}(G) - \mu(G))(\mu(G) - \mu_{\min}(G))$$

$$\leq H^2\Delta(E) - \rho(r-\rho)(H^2)^2k^2$$

$$+ r^2\left(\frac{r-\rho}{r}kH^2 - \frac{1}{r(r-1)}\right)\left(\frac{\rho}{r}kH^2 - \frac{1}{r(r-1)}\right).$$

Therefore

$$\frac{rH^2}{r-1}k \leq H^2 \cdot \Delta(E) + \frac{1}{(r-1)^2},$$

which contradicts our assumption on k. $\qquad\square$

Remark. Note that if E is torsion free then the restriction E_D is also torsion free for a general divisor D in a base point free linear system (see [HL, Corollary 1.1.14] for a precise statement).

As a corollary to Theorem 1.4.1 we get an effective restriction theorem for semistable sheaves.

Corollary 1.4.2. *Let E be a torsion free sheaf of rank $r \geq 2$. Assume that E is slope H-semistable. Let D be a general curve of a base point free linear system $|kH|$. If*

$$k \geq \frac{r-1}{r}\Delta(E) + 1$$

then E_D is semistable.

Proof. Let $0 = E_0 \subset E_1 \subset \cdots \subset E_m = E$ be the Jordan–Hölder filtration of E (i.e., such a filtration that all the quotients are slope H-stable torsion free sheaves; compare with the Harder–Narasimhan filtration). Set $F_i = E_i/E_{i-1}$ and $r_i = \operatorname{rk} F_i$. Let $D \in |kH|$ be any smooth curve such that all the sheaves $(F_i)_D$ have no torsion. Then the corollary follows from Theorem 1.4.1 and the following inequality

$$\frac{\Delta(E)}{r} \geq \sum \frac{\Delta(F_i)}{r_i}$$

(cf. the proof of Proposition 1.3.2). $\qquad\square$

1.5. Semistability in positive characteristic

In positive characteristic, to obtain analogues of Properties 1.2.1 we need a notion of strong semistability. Let X be defined over a characteristic p field and let $F : X \to X$ be the absolute Frobenius morphism, obtained as identity on topological spaces and raising to the pth power on sections of $\mathcal{O}_X$.

We say that a sheaf E is *strongly slope H-semistable*, if for any integer k, the pull back $(F^k)^* E$ is slope H-semistable.

S. Ramanan and A. Ramanathan showed that in positive characteristic a tensor product of strongly slope H-semistable sheaves is strongly slope H-semistable. It is easy to see that Property 1.2.1.2 also holds for strongly slope semistable sheaves.

Theorem 1.3.1 still holds with a similar proof as before. The only difference is in showing that $h^2(S^n E) = O(n^r)$, because in general $(S^n E)^*$ is no longer isomorphic to $S^n(E^*)$ for sheaves with trivial determinant. In this case one can still prove it using twice Serre's duality:

$$h^2(X, S^n E) = h^0(X, (S^n E)^* \otimes K_X) \le h^0(C, (S^n E)^* \otimes K_X \otimes \mathcal{O}_C)$$
$$= h^1(C, S^n E \otimes \mathcal{O}_C(C))$$

and Exercise 1.5.1. Alternatively, one can replace symmetric powers of E by Frobenius pull backs and suitably change the computation.

This analogue of Bogomolov's inequality and generalizations of Proposition 1.3.2 and Corollary 1.4.2 imply boundedness of H-semistable sheaves on surfaces (see Lecture 3).

Remaining properties of semistability, some of which were used before, are put into the following exercises.

Exercise 1.5.1. Let X be a projective scheme of dimension d over an algebraically closed field k. Let F be a coherent sheaf on X. Then for any line bundle L on X we have

$$h^i(X, L^{\otimes n} \otimes F) = O(n^d)$$

(Hint: use Grothendieck's method of dévissage.)

Exercise 1.5.2. Use Bogomolov's theorem to prove the Kodaira vanishing theorem on surfaces: if L is an ample line bundle on a smooth complex projective surface then $H^1(X, K_X + L) = 0$ (Hint: suppose otherwise and use Serre's duality $H^1(K_X + L) = (\operatorname{Ext}^1(L, \mathcal{O}_X))^*$ to construct a vector bundle which violates Bogomolov's inequality.)

Exercise 1.5.3. M. Raynaud constructed a smooth projective surface X over an algebraically closed field of characteristic p and an ample line bundle L on X such that $H^1(X, L^{-1}) \ne 0$. Use this and Exercise 1.5.2 to construct a counterexample to Bogomolov's theorem for slope semistable sheaves in characteristic p.

Exercise 1.5.4. Let X be a smooth projective surface X over an algebraically closed field of characteristic p. One can show that although Bogomolov's inequality fails there exists some $\alpha(r, X, H)$ depending only on r, X and H such that $\Delta(E) \geq \alpha(r, X, H)$ for any slope H-semistable torsion free sheaf E. Use this and Exercise 1.5.3 to construct a counterexample to Property 1.2.1.2 in characteristic p.

Exercise 1.5.5. Any slope semistable torsion free sheaf E on $\mathbb{P}^2$ is strongly slope semistable. Use this to show that in arbitrary characteristic Theorem 1.4.1 holds on $\mathbb{P}^2$.

Exercise 1.5.6. Show that $\Delta(T_{\mathbb{P}^2}) = 3$ and for all smooth curves $C \in |\mathcal{O}_{\mathbb{P}^2}(k)|$ with $k \leq 2$, the restriction E_C is not stable. Theorem 1.4.1 ensures that E_C is stable for any smooth curve $C \in |\mathcal{O}_{\mathbb{P}^2}(k)|$ if $k \geq 3$ (by Exercise 1.5.5 this holds in any characteristic).

Exercise 1.5.7. Let $F : X \to X$ be the absolute Frobenius morphism. Use Exercise 1.5.4 and the fact that $F^*E \subset S^p E$ for any vector bundle E to show that 1.2.1.1 also fails in characteristic p.

Exercise 1.5.8. Show that both Properties 1.2.1 fail if slope semistability is replaced by Gieseker semistability.

Exercise 1.5.9. Show that both Properties 1.2.1 fail if slope semistability is replaced by slope stability.

Partial solution: Let E be a slope H-stable with degree zero. Then E^* is also slope H-stable. If 1.2.1.1 holds for stability then $E \otimes E^*$ is slope stable. But $\mathcal{O}_X$ is a direct summand in $\mathcal{E}nd\, E = E \otimes E^*$ (at least in characteristic 0), a contradiction.

Exercise 1.5.10. Let E be a vector bundle on a smooth surface X. Use the Leray–Hirsch theorem and the Riemann–Roch formula on $\mathbb{P}(E)$ to prove that

$$\chi(X, S^n E) = -\frac{\Delta(E)}{2r}\frac{n^{r+1}}{(r+1)!} + O(n^r).$$

Exercise 1.5.11. Let E be a rank r vector bundle on a smooth surface X. Let G be an elementary transformation of E along a vector bundle T on a smooth curve D. Show that G is a rank r vector bundle on X and we have

$$c_1 G = c_1 E - \mathrm{rk}\, T \cdot D$$

and

$$c_2 G = c_2 E + \deg_D T - \mathrm{rk}\, T(D \cdot c_1 E) + \frac{\mathrm{rk}\, T(\mathrm{rk}\, T - 1)}{2} D^2.$$

(Hints: The first equality follows since T is trivial outside finitely many points of D, $c_1 T = c_1(\mathcal{O}_D^{\mathrm{rk}\, T})$ and from

$$0 \to \mathcal{O}(-D) \to \mathcal{O}_X \to \mathcal{O}_D \to 0.$$

The second equality follows from the Riemann–Roch formula.)

2. Lecture 2

- Moduli functors and moduli spaces: definition and examples
- Geometric invariant theory
- Moduli space of semistable sheaves

2.1. Moduli functors and moduli spaces

We are interested in providing the set of isomorphism classes of vector bundles on a fixed variety with a natural scheme structure. To explain what "natural" means we need a few notions from category theory. They will also be useful in describing some well-known parameter spaces like the Hilbert scheme, Grothendieck's Quot-scheme, etc.

Let $\mathcal{C}$ be a category (e.g., the category Sch/S of S-schemes of finite type) and let $\mathcal{M} : \mathcal{C} \to \mathrm{Sets}$ be a contravariant functor, called in the following a *moduli functor*. For an object X of $\mathcal{C}$ elements of $\mathcal{M}(X)$ will be called *families*. By $h_X : \mathcal{C} \to \mathrm{Sets}$ we will denote the functor of points of X defined by $h_X(Y) = \mathrm{Hom}_{\mathcal{C}}(Y, X)$.

Definition 2.1.1.

(1) $\mathcal{M}$ is *corepresented* by an object M of $\mathcal{C}$ if there is a natural transformation $\alpha : \mathcal{M} \to h_M$ such that for any natural transformation $\beta : \mathcal{M} \to h_N$ there exists a unique morphism $\varphi : M \to N$ such that $\beta = h_\varphi \alpha$.

(2) $\mathcal{M}$ is *represented* by an object M of $\mathcal{C}$ if it is isomorphic to the functor h_M.

Let $\mathcal{M}$ is a moduli functor. If the functor $\mathcal{M}$ is corepresented by M then we say that M is a *moduli space* for $\mathcal{M}$. If $\mathcal{M}$ is represented by M (or more precisely by a natural transformation $\mathcal{M} \to h_M$) then we say that M is a *fine moduli space* for the functor $\mathcal{M}$.

A moduli space M for $\mathcal{M}$ is fine if and only if there exists a *universal family* $U \in \mathcal{M}(M)$ such that the natural transformation $h_M \to \mathcal{M}$, given on $\mathrm{Hom}_{\mathcal{C}}(Y, M) \to \mathcal{M}(Y)$ via $g \to g^*U$, is an isomorphism.

We show a few basic moduli spaces in a growing order of generality.

2.1.1. Hilbert scheme. Let $\mathcal{H}ilb(X/k)$ be a moduli functor $\mathrm{Sch}/k \to \mathrm{Sets}$ defined by

$$\mathcal{H}ilb(X/k)(S) = \{ \text{ subschemes } Z \subset S \times X, \text{ flat and proper over } S\}.$$

Let X be a projective scheme with an ample line bundle $\mathcal{O}_X(1)$. Let Z be a subscheme of X. We define the Hilbert polynomial of Z as

$$P(Z)(m) := \chi(Z, \mathcal{O}_Z \otimes \mathcal{O}_X(1)^{\otimes m}).$$

Let now $Z \overset{i}{\hookrightarrow} S \times X \overset{p}{\to} S$ and let $f = p \circ i$. Let $Z_s := f^{-1}(s)$ be the fiber of f over $s \in S$. Then we define a subfunctor of $\mathcal{H}ilb(X/k)$ by

$$\mathcal{H}ilb_P(X/k)(S) = \{\text{subschemes } Z \subset S \times X, \text{ flat and}$$
$$\text{proper over } S \text{ with } P(Z_s) = P\}.$$

Theorem 2.1.2. *The functor $\mathcal{H}ilb_P(X/_k)$ is representable by a projective k-scheme. The fine moduli space of this functor is called the* Hilbert scheme *and denoted by* $\mathrm{Hilb}_P(X/k)$.

Exercise 2.1.3. Let P be a constant polynomial equal to n. Then $\mathrm{Hilb}_P(X/k)$ is the Hilbert scheme of n points in X. Show that if X is a curve then $\mathrm{Hilb}_P(X/k) = S^n X$. What is the universal object in this case?

2.1.2. Grothendieck's Quot scheme. Let $f : X \to S$ be a projective morphism of Noetherian schemes and let $\mathcal{O}_X(1)$ be an f-ample line bundle on X. Let E be an S-flat coherent sheaf on X. We want to define a scheme which parametrizes pairs (s, F) consisting of a point $s \in S$ and a coherent quotient sheaf F of $E_s = E_{X_s}$ with fixed Hilbert polynomial P. To this end let us first define the functor $\mathcal{Q}uot :$ $\mathrm{Sch}\,/S \to \mathrm{Sets}$, which to each S-scheme T associates the set of all T-flat quotients

$$E_T = \mathcal{O}_T \otimes E \to F$$

of coherent sheaves on $Y = T \times_S X$ such that for all $t \in T$ the sheaf $F_t = (F)_{Y_t}$ on the fiber $Y_t = \mathrm{Spec}\, k(t) \times_T Y$ of $Y \to T$ has Hilbert polynomial P.

Theorem 2.1.4. *The functor $\mathcal{Q}uot$ is represented by a projective S-scheme*

$$\mathrm{Quot}_{X/S}(E; P) \to S.$$

This scheme is called the Quot-scheme.

Obviously, the Quot-scheme generalizes the classical Grassmann variety. In the special case when $E = \mathcal{O}_X$ the Quot scheme gives the Hilbert scheme parameterizing S-flat subschemes of the scheme X with given Hilbert polynomial P.

2.1.3. Flag scheme. Let E be an S-flat coherent sheaf on X and let P_i, $i = 1, \ldots, k$, be some fixed polynomials.

Consider the functor $\mathcal{F}lag : \mathrm{Sch}\,/S \to \mathrm{Sets}$, which to each S-scheme T associates the set of all flags

$$0 \subset F_1 \subset \cdots \subset F_k = E_T = \mathcal{O}_T \otimes E$$

of coherent subsheaves on $Y = T \times_S X$ such that

1. the factors $\mathrm{gr}_i = F_i/F_{i-1}$ of this filtration are T-flat, and
2. for all $t \in T$ the sheaf $\mathrm{gr}_{i,t} = \mathrm{gr}_i |_{Y_t}$ has Hilbert polynomial P_i.

Theorem 2.1.5. *The functor $\mathcal{F}lag$ is represented by a projective S-scheme*

$$\mathrm{Flag}_{X/S}(E; P_1, \ldots, P_k) \to S,$$

called the flag scheme.

Exercise 2.1.6. Construct the flag scheme $\mathrm{Flag}_{X/S}(E; P_1, \ldots, P_k) \to S$ using existence of the Quot-schemes.

2.2. Geometric invariant theory (GIT)

Example 2.2.1. (*A categorical quotient*) Let a k-group G acts on a k-scheme X. Then for any k-scheme T the T-points of G, i.e., $h_G(T)$, also form a group which naturally acts on $h_X(T)$. So we can consider the functor $h_X/h_G : \mathrm{Sch}\,/k \to \mathrm{Sets}$, which to any k-scheme T associates the set of $h_G(T)$-orbits of $h_X(T)$. If this functor is corepresented by a scheme Y, then Y is called a *categorical quotient* of X by G and denoted by X/G.

Categorical quotients do not need to exist in general. But they exist if G is a reductive group acting on an affine scheme.

Let X be a k-scheme with the action σ of a k-group G:

$$\begin{array}{ccc} G \times X & \xrightarrow{\sigma} & X \\ p_2 \downarrow & & \\ X & & \end{array}$$

Definition 2.2.2. $\mathcal{F}$ is G-*linearized* if there exists an isomorphism

$$\Phi : \sigma^*\mathcal{F} \to p_2^*\mathcal{F}$$

satisfying the cocycle condition.

From now on we assume that X is a projective k-scheme and G is a reductive group (e.g., $G = \mathrm{GL}(n)$). Let L be an ample G-linearized line bundle on X.

We define (*semi*)*stable points* of the polarized scheme (X, L) as follows:

Definition 2.2.3.
1. $X^{ss}(L) := \{x \in X : \exists n \in \mathbb{N} \; \exists s \in H^0(X, nL)^G \; s(x) \neq 0\}$.
2. $X^s(L) := \{x \in X : x \in X^{ss}(L), G_x \text{ is finite, and } Gx \subset X^{ss}(L) \text{ is closed}\}$.

Theorem 2.2.4.
1. *There exists a categorical quotient* $X^{ss}(L) \to X^{ss}(L)/G$. *$k$-points of the quotient* $X^{ss}(L)/G$ *correspond to closed G-orbits in* $X^{ss}(L)$ (*but not all G-orbits need to be closed*).
2. *There exists a categorical quotient* $X^s(L) \to X^s(L)/G$. *The fibers of this map are closed G-orbits.*

To determine $X^{s(s)}(L)$ we will use the following theorem:

Theorem 2.2.5. (the Hilbert-Mumford criterion) *A point* $x \in X$ *is semistable* (*stable*) *if and only if for all non-trivial one-parameter subgroups* $\lambda : \mathbb{G}_m \to G$ *we have*

$$\mu^L(x, \lambda) \geq 0 \quad (\mu^L(x, \lambda) > 0, \text{ respectively}).$$

To define $\mu^L(x, \lambda)$ let us draw a diagram:

$$\begin{array}{ccccc} \mathbb{A}^1 \setminus \{0\} = \mathbb{G}_m & \xrightarrow{\lambda} & G & & g \\ \downarrow & & \downarrow \sigma & & \downarrow \\ \mathbb{A}^1 & \xrightarrow{f} & X & & gx = \sigma(g, x) \\ 0 & \to & f(0) = \lim_{t \to 0} \lambda(t)x & & \end{array}$$

Let Φ be the linearization of L. This linearization induces on the fibres of L maps

$$\Phi : L(f(gx)) \to L(f(x)).$$

Clearly, $f(0)$ is a fixed point of the action of $\mathbb{G}_m$, so $\mathbb{G}_m$ acts on the fibre $L(F(0))$. Let r be the weight of this action. Then we set

$$\mu^L(x, \lambda) := -r.$$

2.3. Moduli space of semistable sheaves

In these lectures we are interested in the following moduli functor.

Let $(X, \mathcal{O}_X(1))$ be a smooth polarized projective scheme over an algebraically closed field k. For a k-scheme S let p and q denote the projections of $S \times_k X$ to S and X, respectively. Let us define an equivalence relation $\sim$ on S-flat families of sheaves on X by $F_1 \sim F_2$ if and only if there exists a line bundle $\mathcal{L}$ on S such that $F_1 \simeq F_2 \otimes p^*\mathcal{L}$. Let us note that fibrewise the families F_1 and F_2 define the same family of sheaves on X. Therefore it is natural to introduce the following moduli functor $\mathcal{M}_P : \mathrm{Sch}/k \to \mathrm{Sets}$ (and $\mathcal{M}_P^s : \mathrm{Sch}/k \to \mathrm{Sets}$), which sends a k-scheme S to the set of isomorphism classes of S-flat families of Gieseker semistable (respectively, stable) sheaves on X with Hilbert polynomial P modulo the relation $\sim$.

Theorem 2.3.1. *There exists a moduli scheme M_P for the functor $\mathcal{M}_P$. It is a projective k-scheme of finite type and it contains an open subscheme M_P^s, which is the moduli scheme for $\mathcal{M}_P^s$.*

2.3.1. Geometric meaning of points of the moduli space of sheaves. Let E be a Gieseker semistable torsion free sheaf. Then either it is Gieseker stable or there exists a proper subsheaf $E_1 \subset E$ such that $p(E_1) = p(E)$. By passing to a smaller subsheaf if necessary we can assume that E_1 is Gieseker stable. Then the quotient E/E_1 is Gieseker semistable with $p(E/E_1) = p(E)$.

By induction on the rank we can therefore construct a filtration $0 = E_0 \subset E_1 \subset \cdots \subset E_m = E$ in which all the quotients E_i/E_{i-1} are Gieseker stable and $p(E_i/E_{i-1}) = p(E)$. Such a filtration is called a *Jordan–Hölder filtration* of E (but unlike the Harder–Narasimhan filtration it is not unique).

Let us set $\mathrm{gr}_{JH} E = \bigoplus E_i/E_{i-1}$.

Exercise 2.3.2. Show that $\mathrm{gr}_{JH} E$ does not depend on the choice of Jordan–Hölder filtration.

Definition 2.3.3. Two Gieseker semistable sheaves E and E' with $p(E) = p(E')$ are called *S-equivalent* if $\mathrm{gr}_{JH} E \simeq \mathrm{gr}_{JH} E'$.

An extension of a sheaf F by a sheaf G is an exact sequence of sheaves $0 \to G \to E \to F \to 0$. The set of all extensions is parametrized by a k-vector space $\mathrm{Ext}^1(F, G)$. Moreover, there exists the *universal extension* $\mathcal{E}$, i.e., such a family of sheaves on $\mathrm{Ext}^1(F, G) \times X$ that $\mathcal{E}_\eta = \mathcal{E}|_{\{\eta\} \times X}$ is the extension defined by $\eta \in \mathrm{Ext}^1(F, G)$.

If we have a Gieseker semistable sheaf E and $E_1 \subsetneq E$ is Gieseker semistable subsheaf with the same reduced Hilbert polynomial then the universal extension induces a morphism $\mathrm{Ext}^1(E/E_1, E_1) \to M$.

On the line $\mathbb{A}^1$ defined by $\eta = [E] \in \mathrm{Ext}^1(E/E_1, E_1)$ the universal extension $\mathcal{E}$ satisfies $\mathcal{E}_{t\eta} \simeq E$ for $t \neq 0$ and $\mathcal{E}_0 \simeq E_1 \oplus E/E_1$. Therefore the map $\mathbb{A}^1 \to M$ is constant (since M is separable) which shows that E defines the same point in M as $E_1 \oplus E/E_1$. Similarly, one can show that E defines the same point as $\mathrm{gr}_{JH} E$.

One can also show that two Gieseker semistable sheaves which are not S-equivalent define different points in M_P.

Corollary 2.3.4. *M_P parametrizes S-equivalence classes of Gieseker semistable sheaves. The subscheme M_P^s parametrizes isomorphism classes of Gieseker stable sheaves.*

2.3.2. Relative moduli spaces of pure sheaves. There exists a more general version of Theorem 2.3.1 which will also be useful in the following. Before formulating this more general theorem we need to generalize the notion of Gieseker semistability and stability to the so-called pure sheaves.

A sheaf E is called *pure* if it is torsion free on its scheme-theoretical support. Equivalently, for any subsheaf $F \subset E$ the dimension of the support of F is equal to the dimension of the support of E (this is denoted by $\dim E$).

The Hilbert polynomial of any coherent sheaf E can be written as a sum

$$P(E)(k) = \chi(X, E(k)) = \sum_{i=0}^{\dim E} \alpha_i(E) \frac{k^i}{i!}.$$

Then we define the *reduced Hilbert polynomial* $p(E) = \frac{P(E)}{\alpha_{\dim E}(E)}$. As in Definition 1.2.2 a pure sheaf E is called *Gieseker semistable* (*stable*) if for any proper subsheaf $F \subset E$ we have $p(F) \leq p(E)$ $(p(F) < p(E)$, respectively).

Let $f : X \to S$ be a projective morphism of k-schemes of finite type with geometrically connected fibers and let $\mathcal{O}_X(1)$ be an f-ample line bundle. Let P be a fixed polynomial.

As before we define the moduli functor $\mathcal{M}_P(X/S) : \mathrm{Sch}/S \to \mathrm{Sets}$ as

$$(\mathcal{M}_P(X/S))(T) = \left\{ \begin{array}{l} \text{equivalence classes of families of pure Gieseker} \\ \text{semistable sheaves on the fibres of } T \times_S X \to T \\ \text{which are } T\text{-flat and have Hilbert polynomial } P \end{array} \right\}.$$

Theorem 2.3.5. *There exists a projective S-scheme $M_P(X/S)$ of finite type over S, which is the moduli space for the functor $\mathcal{M}_P(X/S)$. Moreover, there is an open scheme $M_P^s(X/S) \subset M_P(X/S)$ which is the moduli space for the subfunctor of families of geometrically Gieseker stable sheaves.*

2.3.3. Picard scheme. Let $X \to T$ be a flat projective morphism of k-schemes of finite type. Assume that geometric fibres of this morphism are varieties. The *Picard scheme* $\operatorname{Pic}_{X/T}^{P}$ parametrizes line bundles with fixed Hilbert polynomial P on fibres of $X \to T$. In general, it is a quasi-projective scheme but it is not a fine moduli scheme.

However, if $X \to T$ has a section then it is a fine moduli scheme. In this case there exists a universal family of line bundles on $X \times_T \operatorname{Pic}_{X/T}^{P}$. This family is called the *Poincaré line bundle* and denoted by $\mathcal{P}$.

If X is a smooth projective variety over an algebraically closed field k then $\operatorname{Pic}_X^{P} = \operatorname{Pic}_{X/k}^{P}$ is a projective scheme. If P is the Hilbert polynomial of $\mathcal{O}_X$ then Pic_X^{P} is a group scheme with $1 = [\mathcal{O}_X]$ and tensor product as a group action. If $\operatorname{char} k = 0$ then by Cartier's theorem any group scheme is smooth, so in particular Pic_X^{P} is smooth.

Finally let us note that for any two polynomials P_1 and P_2 representing line bundles $\mathcal{L}_1$ and $\mathcal{L}_2$ the corresponding Picard schemes are isomorphic. An isomorphism between $\operatorname{Pic}_X^{P_1}$ and $\operatorname{Pic}_X^{P_2}$ is given, e.g., by tensoring with $\mathcal{L}_2 \otimes \mathcal{L}_1^*$.

Exercise 2.3.6. Prove that the Picard scheme $\operatorname{Pic}_{X/T}^{P}$ is an open subset of the moduli space $M_P(X/T)$.

Exercise 2.3.7. Let X be a smooth projective k-variety and let P be the Hilbert polynomial of a line bundle. Prove that the whole moduli space $M_P(X/k)$ of torsion-free sheaves represents a suitably defined Picard functor. In particular, in this case $\operatorname{Pic}_{X/k}^{P}$ is a projective scheme.

3. Lecture 3

- Boundedness of semistable sheaves
- Construction of the moduli space of semistable sheaves

3.1. Boundedness of semistable sheaves

Corollary 1.4.2 implies boundedness of the family of semistable sheaves on complex surfaces. More precisely, there exists a scheme S of finite type over $\mathbb{C}$ and an S-flat sheaf F on $S \times X$ such that the set $\{F_s\}_{s \in S}$, where $F_s = F \otimes \mathcal{O}_{\{s\} \times X}$, contains all slope H-semistable sheaves with fixed topological invariants.

This can be proved in a few steps. Let us first recall the Castelnuovo-Mumford criterion:

Theorem 3.1.1. *Let X be a smooth projective variety with $\mathcal{O}_X(1)$ very ample. Let E be a coherent sheaf on X. If $h^i(X, E(-i)) = 0$ for all $i > 0$ then*

1. *E is globally generated, and*
2. *$h^i(X, E(m - i)) = 0$ for all $i > 0$ and all $m \geq 0$.*

In the first step we need to prove boundedness on curves:

Boundedness for curves: Let C be a smooth curve. Let E be a sheaf on C with $\operatorname{rk} E = r$. Then

$$\mu_{\min}(E) > \mu(\omega_C(-m)) = \deg \omega_C - m \deg \mathcal{O}_C(1).$$

So if $m > \frac{\deg K_C - \mu_{\min} E}{\deg \mathcal{O}_C(1)}$ then $h^1(C, E(m)) = \operatorname{Hom}(E, \omega_C(-m)) = 0$. By Theorem 3.1.1 $E(m+1)$ is globally generated. Hence for all sheaves E with fixed Hilbert polynomial P and $\mu_{\min}(E)$ bounded from below we can find m_0 such that all $E(m_0)$ are quotients of $\mathcal{O}_X^{P(m_0)}$. So they form a bounded family as follows, e.g., from existence of Quot-schemes (see Theorem 2.1.4). In particular semistable vector bundles of fixed degree and rank form a bounded family.

Boundedness for surfaces: For simplicity we will consider only stable vector bundles (locally free sheaves). Then Theorem 1.4.1 implies that for all stable locally free sheaves E on a complex surface X, with fixed Hilbert polynomial (or rank, $c_1 E H$ and $\Delta(E)$), there exists a fixed curve C such that the restriction E_C is stable. Then it is easy to see, e.g., applying Theorem 3.1.1 and Serre's vanishing theorem that there exists m_1 such that $E(m_1)$ is globally generated and we can conclude as before.

More generally, we have the following theorem due to Maruyama in the characteristic zero case and the author in general:

Theorem 3.1.2. (*Boundedness Theorem*) *The family of slope semistable sheaves with fixed numerical data is bounded. This means that for a fixed polynomial P there exists a scheme S of finite type over k and an S-flat coherent sheaf $\mathcal{F}$ on $X \times S$, such that the set $\{\mathcal{F}_s\}_{s \in S}$ contains isomorphism classes of all slope semistable sheaves with Hilbert polynomial P.*

This implies that the moduli space of Gieseker semistable sheaves on X is a projective scheme of finite type.

For the construction of the moduli space we also need a good bound on the number of section of a sheaf:

Theorem 3.1.3. *Let X be a smooth n-dimensional projective variety with a very ample line bundle H. Then for any rank r torsion free sheaf E on X we have*

$$h^0(X, E) \leq \begin{cases} rH^n \binom{\frac{\mu_{\max}(E)}{H^n} + \ln(r+1) + n}{n} & \text{if } \mu_{\max}(E) \geq 0, \\ 0 & \text{if } \mu_{\max}(E) < 0. \end{cases}$$

In characteristic zero the above theorem was proven by C. Simpson and J. Le Potier (see [HL, Theorem 3.3.1]), using the Grauert-Mülich restriction theorem. In positive characteristic this proof does no longer work and one needs to use Bogomolov's inequality (see [La2]).

Using Theorem 3.1.3 one can prove the following characterization of semistable sheaves among torsion free sheaves:

Theorem 3.1.4. *Let P be a fixed polynomial and let m be a sufficiently large integer. Then for a torsion free sheaf with Hilbert polynomial P the following conditions are equivalent:*

1. *F is Gieseker semistable*
2. *$h^0(F(m)) = P(m)$ and for all subsheaves F' of F with rank $0 < r' = \operatorname{rk} F' < r = \operatorname{rk} F$ we have*

$$h^0(F'(m)) \leq \frac{r'P(m)}{r}$$

with equality if and only if $P(F') = P(F)$.

Exercise 3.1.5. Prove that the following implications hold:

$$\begin{array}{ccc}
E \text{ is slope stable} & \Rightarrow & E \text{ is slope semistable} \\
\Downarrow & & \Uparrow \\
E \text{ is Gieseker stable} & \Rightarrow & E \text{ is Gieseker semistable}
\end{array}$$

3.2. Construction of the moduli space M_P.

By Theorem 3.1.2 and Exercise 3.1.5 the family of all Gieseker semistable sheaves with fixed Hilbert polynomial P is bounded. This implies that there exists an integer m_0 such that for all $m \geq m_0$ and for all Gieseker semistable sheaves E with Hilbert polynomial P

1. $E(m)$ is globally generated,
2. $H^i(X, E(m)) = 0$ for $i > 0$.

In particular, these conditions imply that $h^0(E(m)) = \chi(X, E(m)) = P(m)$.

Set $V = k^{\oplus P(m)}$ and $\mathcal{H} = V \otimes \mathcal{O}_X(-m)$. Consider the Quot-scheme $Q = \operatorname{Quot}(\mathcal{H}; P)$.

Let R be a subset of Q parameterizing all quotients $[\mathcal{H} \to E]$ such that E is Gieseker semistable and the induced map $V = H^0(X, \mathcal{H}(m)) \to H^0(X, E(m))$ is an isomorphism.

For any point $[\mathcal{H} \to E] \in R$ and any isomorphism $V \to V$ we get a point in R corresponding to the composition

$$V \otimes \mathcal{O}_X(-m) \to V \otimes \mathcal{O}_X(-m) \to E.$$

Therefore $\operatorname{GL}(V)$ acts on R and, at least set theoretically, the quotient of R by $\operatorname{GL}(V)$ parameterizes Gieseker semistable sheaves on X. To get a scheme structure on this quotient we need to use Theorem 2.2.4.

To construct a GIT quotient first we need to find a $\operatorname{GL}(V)$-linearized polarization $\mathcal{L}$ on R. There are two different constructions of polarizations and both of them are used in the description of line bundles on the moduli space. One of them is due to C. Simpson and the other is due to D. Gieseker, and although it is less natural it was constructed much earlier. The problem with Gieseker's construction is that it does not easily generalize to pure sheaves.

In Simpson's construction we consider

$$\mathcal{L}_l := \det(p_*(\tilde{F} \otimes q^*\mathcal{O}_X(l))),$$

where $\tilde{F}$ is the universal quotient sheaf on the Quot-scheme. For large l this line bundle is very ample on Q. The Quot-scheme is constructed as a subscheme of certain Grassmannian and $\mathcal{L}_l$ comes from the Plücker embedding of this Grassmannian into a projective space.

In Gieseker's construction the constructed line bundle is not ample on Q but only on R.

To any family F of sheaves on a smooth projective variety X/k parametrized by S one can associate the family $\det F$ of determinant line bundles. This induces a natural transformation of functors $\mathcal{M}_P$ and $\mathcal{P}ic_X$. Since the functor $\mathcal{P}ic_X$ is represented by $\operatorname{Pic} X$ we get a map $\det : S \to \operatorname{Pic} X$ such that $\det F \simeq \det^* \mathcal{P} \otimes p^* \mathcal{L}$ for a certain line bundle $\mathcal{L}$ on S.

If we make this construction for the universal quotient $V \otimes q^*\mathcal{O}_X(-m) \to \tilde{F}$ on $Q \times X$ then we get a line bundle $\mathcal{A} = \mathcal{L}_{\overline{R}}$ on the closure of R in Q.

This line bundle $\mathcal{A}$ is also a quotient of $\det^*(\bigwedge^r V \otimes p^*\mathcal{P}^*(-rm))$ and the quotient map induces a morphism of schemes

$$\zeta : \overline{R} \to \mathbb{P}_{\operatorname{Pic} X}\left(\det^*\left(\bigwedge^r V \otimes p^*\mathcal{P}^*(-rm)\right)\right)$$

over $\operatorname{Pic} X$. One can easily see that $\zeta|_R$ is injective, so $\mathcal{A} = \zeta^*\mathcal{O}_{\mathbb{P}}(1)$ is det-ample.

Once we constructed a $\operatorname{GL}(V)$-linearized line bundle $\mathcal{L}_l$ on $\overline{R}$ one has to prove that the points of R correspond to semistable points of $\operatorname{GL}(V)$ action on $(\overline{R}, \mathcal{L}_l)$ and the points of R^s (corresponding to those points $[\mathcal{H} \to E] \in R$ for which E is Gieseker H-stable) correspond to stable points of $\operatorname{GL}(V)$ action on $(\overline{R}, \mathcal{L}_l)$. Sice the center of $\operatorname{GL}(V)$ acts on Q trivially, we need only to find (semi)stable points for the $\operatorname{SL}(V)$-action. This is the content of Theorem 3.2.2.

Once we know it by Theorem 2.2.4 there exists a good quotient $\pi : R \to M$ and M is the moduli space of Gieseker H-semistable sheaves. Moreover, the restriction $\pi|_{R^s} : R^s \to M^s$ is a geometric quotient (i.e., $\operatorname{GL}(V)$-orbits of stable sheaves are closed).

By the properties of a GIT quotient, the points of M correspond to the closed orbits of $\operatorname{GL}(V)$-action on R. Therefore it is sufficient to identify such orbits. We used a slightly different method of identifying points of the moduli space in 2.3.1.

For further use let us note the following theorem, which is not needed for the construction of the moduli space:

Theorem 3.2.1. $R^s \to M^s$ *is a principal* $\operatorname{PGL}(V)$-*bundle, locally trivial in the étale topology.*

This follows from the fact that the scheme-theoretic stabilizer of the $\operatorname{GL}(V)$-action at every point of R^s is equal to $\operatorname{Aut} E = k^*$.

Theorem 3.2.2. *The set* $\overline{R}^{ss}(\mathcal{L}_l)$ *of semistable points of the action of* $SL(V)$ *on* $(\overline{R}, (\mathcal{L}_l)_{\overline{R}})$ *is equal to* R.

Proof. Let us take a point

$$[\rho : V \otimes \mathcal{O}(-m) \to F] \in Q$$

and a 1-parameter subgroup

$$\lambda : \mathbb{G}_m \to \mathrm{SL}(V).$$

We want to find $[\bar\rho] := \lim_{t \to 0} \lambda(t)[\rho]$ and the weight of the induced $\mathbb{G}_m$-action on $\mathcal{L}_l([\rho])$.

To this end let us decompose V into weight-spaces for the induced $\mathbb{G}_m$-action: $V = \bigoplus_{n \in \mathbb{Z}} V_n$, where $\mathbb{G}_m$ acts on V_n with weight n. Let us set $V_{\leq n} = \bigoplus_{m \leq n} V_m$, $F_{\leq n} = \rho(V_{\leq n} \otimes \mathcal{O}(-m))$ and $F_n := F_{\leq n}/F_{\leq n-1}$. Then

$$[\bar\rho] = [\bigoplus V_n \to \bigoplus F_n].$$

Now we look for the weight of the action of $\mathbb{G}_m$ on the fiber $\mathcal{L}_l([\bar\rho])$.

As $\mathbb{G}_m$ acts on F_n with weight n, it acts on $H^i(F_n(l))$ with the same weight n. Since the fiber $\mathcal{L}_l([\bar\rho]) = \det H^*((\bigoplus F_n)(l))$, $\mathbb{G}_m$ acts on $\mathcal{L}_l([\bar\rho])$ with weight $\sum_{n \in \mathbb{Z}} n P(F_n, l)$. So

$$\mu^{\mathcal{L}_l}([\rho], \lambda) = -\sum_{n \in \mathbb{Z}} n P(F_n, l).$$

Let us take a subspace $V' \subset V$. It gives the filtration together with weights defined up to a multiple, so it is associated to a certain 1-parameter subgroup of $\mathrm{SL}(V)$. Let us set $F' = \rho(V' \otimes \mathcal{O}(-m))$ and

$$\Theta_F(V') = \dim V \cdot P(F', l) - \dim V' \cdot P(F, l).$$

Using the Abel transformation we can rewrite the above expression for $\mu^{\mathcal{L}_l}([\rho], \lambda)$ as

$$\mu^{\mathcal{L}_l}([\rho], \lambda) = \frac{1}{\dim V} \sum_{n \in \mathbb{Z}} \Theta_F(V_{\leq n}).$$

So we get the following corollary:

Corollary 3.2.3. *A point* $[\rho : \mathcal{H} \to F] \in Q$ *is* $\mathrm{SL}(V)$*-semistable if and only if for all* $V' \subset V$ *we have* $\Theta_F(V') \geq 0$.

Unfortunately, usually one is not able to prove that $\mathrm{SL}(V)$-semistable points of Q are in R. The main problem is to prove that an $\mathrm{SL}(V)$-semistable point of Q corresponds to a torsion free sheaf (or to a limit of such points). So we need to restrict to $\overline{R}$. Then Theorem 3.1.4 and the above corollary imply the following corollary which finishes the construction:

Corollary 3.2.4. *If* $l \gg 0$ *then* $[\rho : \mathcal{H} \to F] \in \overline{R}$ *is* $\mathrm{SL}(V)$*-semistable if and only if* F *is Gieseker semistable.* $\square$

4. Lecture 4

- Line bundles on moduli spaces
- Strange duality

4.1. Line bundles on moduli spaces

4.1.1. Grothendieck's K and K^0 groups of sheaves on varieties. Let X be an n-dimensional Noetherian scheme. The *Grothendieck group $K(X)$* of coherent sheaves on X is the quotient of the free abelian group generated by the coherent sheaves on X by the subgroup generated by $[F] - [F'] - [F'']$, where F, F' and F'' are coherent sheaves in the exact sequence

$$0 \to F' \to F \to F'' \to 0.$$

The Grothendieck group $K^0(X)$ of locally free sheaves is defined in the same way but using only locally free sheaves and short exact sequences of locally free sheaves.

If X is smooth then $K(X) \simeq K^0(X)$ (see Exercise 4.1.1) has structure of a commutative ring with $1 = [\mathcal{O}_X]$ and $[F_1] \cdot [F_2] := [F_1 \otimes F_2]$ for locally free sheaves F_1 and F_2.

In this case we can introduce a quadratic form $\chi : K(X) \times K(X) \to \mathbb{Z}$. This is defined by

$$(a, b) \to \chi(a \cdot b) = \int_X \mathrm{ch}(a)\,\mathrm{ch}(b)\,\mathrm{td}(X)$$

for $a, b \in K(X)$, where $\mathrm{ch}(x)$ is the Chern character of x and $\mathrm{td}(X)$ is the Todd class of the tangent bundle of X.

Exercise 4.1.1. Let X be a smooth variety over an algebraically closed field k. Then there exists an isomorphism $K^0(X) \simeq K(X)$ (see [Ha, Chapter 3, Exercise 6.9]).

Exercise 4.1.2. Show that for any coherent sheaves F_1 and F_2 we have

$$[F_1] \cdot [F_2] = \sum_i (-1)^i [\mathcal{T}or_i^{\mathcal{O}_X}(F_1, F_2)]$$

Exercise 4.1.3. Let us set $[F]^* = \sum (-1)^i [\mathcal{E}xt^i(F, \mathcal{O}_X)]$ for a coherent sheaf F (note that if F is locally free then $[F]^* = [F^*]$). Show that

$$[F_1] \cdot [F_2]^* = \sum (-1)^i [\mathcal{E}xt^i(F_2, F_1)].$$

Exercise 4.1.4. For any two coherent sheaves F_1 and F_2 let us set

$$\chi(F_1, F_2) = \sum (-1)^i \dim \mathrm{Ext}^i(F_1, F_2).$$

Show that $\chi(F_1, F_2) = \chi(X, [F_2] \cdot [F_1]^*)$.

Exercise 4.1.5. Let E and F be torsion free sheaves on a smooth surface X. Set $r_E = \mathrm{rk}\,E$, $r_F = \mathrm{rk}\,F$ and $\xi_{E,F} = \frac{c_1 F}{r_F} - \frac{c_1 E}{r_E}$. Show that

$$\chi(E, F) = -\left(r_E \frac{\Delta(F)}{2r_F} + r_F \frac{\Delta(E)}{2r_E}\right) + r_E r_F \left(\frac{1}{2}\xi_{E,F}^2 - \frac{1}{2}\xi_{E,F} K_X + \chi(\mathcal{O}_X)\right).$$

4.1.2. Line bundles defined by families of sheaves. A projective morphism $f : X \to S$ induces a homomorphism $f_! : K(X) \to K(S)$ defined by

$$f_!([F]) = \sum_{i \geq 0} (-1)^i [R^i f_* F].$$

Proposition 4.1.6. *If $f : X \to S$ is a smooth projective morphism, S is a k-scheme of finite type and E is an S-flat coherent sheaf on X then $[E] \in K^0(X)$ and $f_!([E]) \in K^0(S)$. More precisely, E has a locally free resolution $E_\bullet$ such that all sheaves $R^i f_* E_j$ are locally free.*

Let p and q denote natural projections of $S \times X$ to S and X, respectively. The above proposition implies that if E is an S-flat family of sheaves on a smooth projective variety X then $[E] \in K^0(S \times X)$ and we have a map $p_! : K^0(S \times X) \to K^0(S)$. Therefore the following definition makes sense:

Definition 4.1.7. For each such family E we define the homomorphism

$$\lambda_E : K(X) \to \operatorname{Pic} S \quad \text{by} \quad \lambda_E(u) = \det p_!(q^* u \cdot [E]).$$

In this way we constructed line bundles on a scheme parametrizing a given family of sheaves on X.

4.1.3. Line bundles on moduli spaces. We want to make a similar construction as above to construct line bundles on the moduli space.

If the moduli space M_P is fine then there exists the universal sheaf $\mathcal{E}$ on $M \times X$ and $\lambda_{\mathcal{E}}$ produces line bundles on M_P for each class $u \in K(X)$. In general M_P is not fine and we cannot expect existence of a homomorphism $\lambda : K(X) \to \operatorname{Pic} M_P$. However, such a homomorphism exists if we restrict to some subspace of $K(X)$.

Let c be a class in $K(X)$. Then we can define the moduli space $M(c)$ of Gieseker H-semistable sheaves of class c. This is a well-defined open and closed subscheme of M_P for P defined by $P(m) = \chi(X, c \cdot [\mathcal{O}_X(mH)])$. We also set $M^s(c) = M_P^s \cap M(c)$. We will produce line bundles on $M(c)$ and $M^s(c)$.

Set

$$K_c = c^\perp = \{u \in K(X) : \chi(c \cdot u) = 0\}$$

and

$$K_{c,H} = c^\perp \cap \{1, h, h^2, \ldots, h^{\dim X}\}^{\perp\perp}.$$

Theorem 4.1.8.
(1) *There exists a group homomorphism $\lambda^s : K_c \to \operatorname{Pic} M^s(c)$ such that for any S-flat family $\mathcal{E}$ of Gieseker H-stable sheaves of class c on X if $\Phi_{\mathcal{E}} : S \to M^s(c)$ denotes the classifying morphism then $\Phi_{\mathcal{E}}^*(\lambda^s(u)) = \lambda_{\mathcal{E}}(u)$ for $u \in K_c$.*
(2) *(char $k = 0$) There exists a group homomorphism $\lambda : K_{c,H} \to \operatorname{Pic} M(c)$ such that for any S-flat family $\mathcal{E}$ of Gieseker H-semistable sheaves of class c on X if $\Phi_{\mathcal{E}} : S \to M(c)$ denotes the classifying morphism then $\Phi_{\mathcal{E}}^*(\lambda(u)) = \lambda_{\mathcal{E}}(u)$ for $u \in K_{c,H}$.*
(3) *(char $k = 0$) For any $u \in K_{c,H}$ the restriction of $\lambda(u)$ to $M^s(c)$ gives $\lambda^s(u)$.*

Proof. We will prove only the first part of the theorem. The rest can be proved in a similar way.

Let $R^s(c) \subset R$ be an open subset of R used in the construction of the moduli space, corresponding to stable sheaves of class c. Let $u \in K(X)$ and set $\mathcal{L} = \lambda_{\tilde{F}}(u)$, where $\tilde{F}$ is the universal quotient sheaf on $R^s(c) \times X$. $\mathcal{L}$ has a natural $\mathrm{GL}(V)$ linearization coming from $\tilde{F}$.

We want to check that if $u \in K_c$ then $\mathcal{L}$ descends to a line bundle on $M^s(c)$. To check it, it is sufficient to check that for any point $[\rho] = [\rho : \mathcal{H} \to E] \in R^s(c)$ the stabilizer of $\mathrm{GL}(V)$ action at $[\rho]$ acts trivially on the fibre $\mathcal{L}([\rho])$.

The stabilizer of $\mathrm{GL}(V)$ at ρ is equal to the image of a natural homomorphism $\mathrm{Aut}\, E \to \mathrm{GL}(V)$ sending φ to $H^0(\rho(m))^{-1} \circ H^0(\varphi(m)) \circ H^0(\rho(m))$ (this is well defined since by our assumption $H^0(\rho(m))$ is an isomorphism).

So we need to understand the action of $\mathrm{Aut}\, E \simeq k^*$ on $\mathcal{L}([\rho])$. Since higher direct images commute with base change we have $\mathcal{L}([\rho]) \simeq \det H^\bullet(X, E \otimes u)$. Therefore $A \in \mathrm{Aut}\, E$ acts by $A^{\chi(c \cdot u)}$ and this action is trivial if $u \in K_c$. $\qquad\square$

4.2. Strange duality

In the remaining part of this lecture we assume that the base field has characteristic zero.

4.2.1. The strange duality morphism.

Let us take two classes c and c^* in $K(X)$ such that $c \in K_{c^*,H}$ and $c^* \in K_{c,H}$ (in particular $\chi(c \cdot c^*) = 0$). Then we can define $\mathcal{D}_{c,c^*} = \lambda_c(c^*)$ and $\mathcal{D}_{c^*,c} = \lambda_{c^*}(c)$. Let p_1 and p_2 denote the projections of $M(c) \times M(c^*)$ onto the first and the second factor, respectively. Let us take a line bundle $\mathcal{D} = p_1^* \mathcal{D}_{c,c^*} \otimes p_2^* \mathcal{D}_{c^*,c}$.

We will assume that X is either a curve or a surface. We will also make some mild assumptions on the classes c and c^*. More precisely, we assume that the rank of c is positive and semistable sheaves of class c^* are pure of dimension $(\dim X - 1)$ (in [LP2] one can find a different set of assumptions that allows to consider also other cases). Then the line bundle $\mathcal{D}$ has the following universal property:

Proposition 4.2.1 (Le Potier, [LP2]). *For all S-flat families $\mathcal{F}$ of Gieseker H-semistable sheaves of class c and $\mathcal{G}$ of Gieseker H-semistable sheaves of class c^* let $\Phi_\mathcal{F} : S \to M(c)$ and $\Phi_\mathcal{G} : S \to M(c^*)$ be the corresponding classifying morphisms. Then for*

$$\Phi = (\Phi_\mathcal{F}, \Phi_\mathcal{G}) : S \to M(c) \times M(c^*)$$

we have $\Phi^* \mathcal{D} = \det(p_!(\mathcal{F} \otimes \mathcal{G}))$.

One can show that our assumptions imply that $\mathcal{F} \otimes \mathcal{G}$ is S-flat so by Proposition 4.1.6 $\det(p_!(\mathcal{F} \otimes \mathcal{G}))$ makes sense.

Our assumptions imply also that $H^2(F \otimes G) = 0$ for all Gieseker H-semistable sheaves F of class c and G of class c^*. Hence we can construct a canonical section $\sigma_{c,c^*} \in H^0(M(c) \times M(c^*), \mathcal{D})$ such that its zero set is equal to $\{([F], [G]) : H^1(F \otimes G) \neq 0\}$.

This section gives an element of $H^0(M(c), \mathcal{D}_{c,c^*}) \otimes H^0(M(c^*), \mathcal{D}_{c^*,c})$ so we have a linear map

$$D_{c,c^*} : H^0(M(c^*), \mathcal{D}_{c^*,c})^* \to H^0(M(c), \mathcal{D}_{c,c^*}),$$

which is called the *strange duality map*.

Conjecture 4.2.2 (Strange Duality Conjecture). *Whenever defined and non-zero the map D_{c,c^*} is an isomorphism.*

Geometric interpretation of this conjecture is the following. We have a rational map

$$\Psi : M(c) \dashrightarrow \mathbb{P}(H^0(M(c^*), \mathcal{D}_{c^*,c}))$$

which sends $[F] \in M(c)$ to the divisor $\{[G] \in M(c^*) : H^1(F \otimes G) \neq 0\}$. One can check that $\Psi^* \mathcal{O}(1) = \mathcal{D}_{c,c^*}$. The Strange Duality Conjecture asks if Ψ is a morphism and if the image of Ψ is not contained in any hyperplane.

4.2.2. Strange duality on curves. Let C be a smooth projective curve. We have $K(C) = \mathbb{Z} \oplus \operatorname{Pic} C$, where the first factor corresponds to the rank and the second to the determinant. We also have a numerical K-group $K(C)_{\mathrm{num}} = \mathbb{Z} \oplus \mathbb{Z}$, where the first factor corresponds to the rank and the second one to the degree of the determinant.

Let us take $c = [\mathcal{O}_C^r] \in K(C)$ and $c^* = (1, g-1) \in K(C)_{\mathrm{num}}$. One can easily check that $\chi(c \cdot c^*) = 0$ and assumptions needed to define the strange duality map are satisfied.

In this case $\mathcal{D}_{c,c^*}$ is the generator of $\operatorname{Pic} M(c)$ and $M(c^*) = J^{g-1}$ is the Jacobian parametrizing line bundles of degree $g-1$.

Since we have the Poincaré line bundle $\mathcal{P}$ on J^{g-1} (see 2.3.3) we get

$$\mathcal{D}_{c^*,c} = \det(\pi_!(\mathcal{P} \otimes c)) = r\det(\pi_!\mathcal{P}) = r\Theta,$$

where Θ is the theta divisor defined by $\Theta = \{L \in J^{g-1} : h^0(C, L) \neq 0\}$.

The strange duality map $\Psi : M \dashrightarrow \mathbb{P}(H^0(J^{g-1}, r\Theta))$ sends $[E]$ to $\{L \in J^{g-1} : h^0(C, E \otimes L) \neq 0\}$ and the strange duality implies that

$$h^0(M(c), \mathcal{D}_{c,c^*}) = h^0(J^{g-1}, r\Theta).$$

This is a special case of the so-called *Verlinde formula* computing the number of sections of tensor powers of $\mathcal{D}_{c,c^*}$ on $M(c)$.

Theorem 4.2.3. *Strange duality conjecture holds for curves.*

This theorem was first proven for generic curves by P. Belkale in [Be] and then for all curves by A. Marian and M. Oprea in [MO]. Unfortunately, the proof uses the Verlinde formula, so we do not get a new proof of the Verlinde formula.

4.2.3. Strange duality on surfaces. Strange duality on surfaces is known only in very special cases. We show essentially all the known cases except for those presented in [LP2] where both c and c^* correspond either to sheaves of rank 0 or to ideal sheaves of points.

The first case is due to G. Danila (see [Da]). We consider the projective plane $\mathbb{P}_2$. Let c be the class of a rank 2 vector bundle with $c_1 = 0$ and $c_2 = n$ and let c^* be the class of $\mathcal{O}_L(-1)$ for some line L. Then $\chi(c \cdot c^*) = 0$ and one can check that all our assumptions used in the definition of the strange duality map are satisfied.

Let us note that $M(c^*)$ is the dual $\mathbb{P}_2$ (it parametrizes lines in $\mathbb{P}_2$) and $M(c)$ parametrizes semistable sheaves of rank 2 on $\mathbb{P}_2$ with $c_1 = 0$ and $c_2 = n$.

By the Grauert–Müllich theorem (false if char $k > 0$) the restriction of any such semistable sheaf to a general line is trivial. The lines on which the restriction is non-trivial (the so-called *jumping lines*) form a curve of degree n in $\mathbb{P}_2^*$ (this fact is called *Barth's theorem*).

In this case Ψ is defined everywhere on $M(c)$ and it is known as *the Barth morphism*. However, even in this case the strange duality conjecture is known only for small n:

Theorem 4.2.4. ([Da, Theorem 1.1]) *The strange duality map is an isomorphism for $2 \leq n \leq 19$.*

The other case when strange duality is known is due to K. O'Grady (see [O, 5.3]). Let $X \subset \mathbb{P}^g$ be a non-degenerate K3 surface with Picard group generated by $L = \mathcal{O}_X(1)$ with $L^2 = 2g - 2$. Let c be the class of an ideal sheaf of 2 points and let c^* be the class of a rank 2 sheaf with $c_1 = c_1(L)$ and $c_2 = g - 1$. Then $M(c) = X^{[2]}$ is the Hilbert scheme of 2 points on X. Note that the above classes do not satisfy our assumptions and one needs the general setting of [LP2] to define the strange duality.

Theorem 4.2.5. ([O, Proposition 5.1 and Section 5.3]) *The strange duality map is an isomorphism for $g \leq 8$.*

5. Lecture 5

- Deformation theory of Quot schemes and local structure of moduli spaces
- e-stability
- Examples of unobstructed moduli spaces

This lecture starts with a local study of moduli spaces.

5.1. Deformation theory of Quot schemes and local structure of moduli spaces

Theorem 5.1.1 (see [HL], Proposition 2.2.8). *Let X be a projective scheme over k, $\mathcal{H}$ a coherent sheaf and P a Hilbert polynomial. Take a k-rational point $[\sigma : \mathcal{H} \to E] \in \mathrm{Quot}(\mathcal{H}; P)$ and set $K = \ker \sigma$. Then*

$$\dim \mathrm{Hom}(K, E) - \dim \mathrm{Ext}^1(K, E) \leq \dim_{[\sigma]} \mathrm{Quot}(\mathcal{H}; P) \leq \dim \mathrm{Hom}(K, E).$$

Moreover, if $\mathrm{Ext}^1(K, E) = 0$ then $\mathrm{Quot}(\mathcal{H}; P)$ is smooth at $[\sigma]$.

Corollary 5.1.2. *Let E be a Gieseker stable sheaf with Hilbert polynomial P. If $\mathrm{Ext}^2(E, E) = 0$ then M_P^s is smooth at $[E]$ and it has dimension $\dim \mathrm{Ext}^1(E, E)$.*

Proof. Take a point $[\mathcal{H} \to E] \in R^s$ as in the construction of M_P^s. From the short exact sequence

$$0 \to K \to \mathcal{H} \to E \to 0$$

we get

$$0 \to \mathrm{Hom}(E, E) \to \mathrm{Hom}(\mathcal{H}, E) \to \mathrm{Hom}(K, E) \to \overset{1}{\mathrm{Ext}}(E, E) \to \overset{1}{\mathrm{Ext}}(\mathcal{H}, E)$$

$$\to \overset{1}{\mathrm{Ext}}(K, E) \to \overset{2}{\mathrm{Ext}}(E, E) \to \overset{2}{\mathrm{Ext}}(\mathcal{H}, E).$$

Now by construction $\mathrm{Ext}^i(\mathcal{H}, E) = H^i(E(m))^{\dim V}$ vanishes for $i > 0$ and it has dimension $(\dim V)^2$ for $i = 0$. Since E is stable, $\dim \mathrm{Hom}(E, E) = 1$. Then $\mathrm{Ext}^1(K, E) = \mathrm{Ext}^2(E, E) = 0$ and by the above theorem R^s is smooth at $[\mathcal{H} \to E]$. Moreover, at this point R^s has dimension $\dim \mathrm{Hom}(K, E) = \dim \mathrm{Ext}^1(E, E) + \dim \mathrm{PGL}(V)$.

Since $R^s \to M^s$ is a $\mathrm{PGL}(V)$-principal bundle (see Theorem 3.2.1), by descent R^s is smooth if and only if M^s is smooth. So the above dimension count and Theorem 5.1.1 imply the required assertions. $\qquad\square$

In fact, one can prove a more precise version of this theorem. Namely, for any Gieseker stable sheaf E there exists a map

$$\mathrm{Ext}^1(E, E) \supset U \overset{\Phi}{\to} \mathrm{Ext}^2(E, E)$$

well defined on an open neighbourhood of 0 such that the germ of M^s at $[E]$ is the germ of $\Phi^{-1}(0)$ at 0. The map Φ is called a *Kuranishi map* but it is not unique in algebraic category.

Let X be a smooth projective variety. Then we have a map

$$\det : M \to \mathrm{Pic}\, X$$

sending $[E]$ to $[\det E]$ (it is well defined even at points corresponding to strictly semistable sheaves).

Theorem 5.1.3. *Let E be a Gieseker stable sheaf. Then after canonical isomorphisms the tangent map $T_{[E]}M \to T_{[\det E]}\mathrm{Pic}\, X$ can be identified with the trace map*

$$\mathrm{Tr} : \mathrm{Ext}^1(E, E) \to \mathrm{Ext}^1(\det E, \det E) \simeq H^1(\mathcal{O}_X).$$

Moreover, obstructions for $[E]$ to be a smooth point of M map to the obstructions for $[\det E]$ to be a smooth point of $\mathrm{Pic}\, X$ via the trace map

$$\mathrm{Tr} : \mathrm{Ext}^2(E, E) \to \mathrm{Ext}^2(\det E, \det E) \simeq H^2(\mathcal{O}_X).$$

Let us denote the kernel of the trace map

$$\mathrm{Tr} : \mathrm{Ext}^i(E, E) \to \mathrm{Ext}^i(\det E, \det E) \simeq H^i(\mathcal{O}_X)$$

by $\mathrm{Ext}^i(E, E)_0$.

If the characteristic of the base field is 0 then by Cartier's theorem any group scheme is smooth, so $\mathrm{Pic}\, X$ is smooth. In this case, if E is a Gieseker stable sheaf then M^s is smooth at $[E]$ if $\mathrm{Ext}^2(E, E)_0 = 0$.

More precisely, if we set $M(\mathcal{L}) = \det^{-1}([\mathcal{L}])$ (this definition needs alteration in positive characteristic) then the germ of $M^s(\mathcal{L})$ at any point $[E]$ is the germ at 0 of $\Psi^{-1}(0)$ for some map

$$\mathrm{Ext}^1(E, E)_0 \supset U \xrightarrow{\Psi} \mathrm{Ext}^2(E, E)_0$$

defined on an open neighbourhood U of 0.

5.2. e-stability

Let X be a smooth projective n-dimensional variety and let H be an ample divisor on X. Let us set $|H| = \sqrt[n]{H^n}$.

Definition 5.2.1. Let $e \geq 0$ be a real number. A torsion-free sheaf E on X is called e-(semi)stable if for every subsheaf $E' \subset E$ of rank $r' < r = \mathrm{rk}\, E$ we have

$$\mu(E') \,(\leq)\, \mu(E) - \frac{e|H|}{r'}.$$

Note that any e-(semi)stable sheaf is slope (semi)stable.

Lemma 5.2.2. *Let E be a rank r torsion free sheaf on X. Assume that there exists a non-zero $s \in \mathrm{Hom}(E, E \otimes \mathcal{A})$ with trivial determinant.*

1. *If $c_1\mathcal{A} \cdot H < 0$ then E is not slope semistable.*
2. *E is not e-stable for all*

$$e \geq \frac{(r-1)[c_1\mathcal{A} \cdot H]_+}{2|H|}$$

Proof. Since $\det s = 0$, the subsheaves $\ker s \subset E$ and $\mathrm{im}\, s \subset E \otimes \mathcal{A}$ are non-trivial of ranks $k < r$ and $r - k < r$, respectively. Hence, if E is e-stable then

$$\mu(\ker s) < \mu(E) - \frac{e|H|}{k}$$

and

$$\mu(\mathrm{im}\, s) < \mu(E) + c_1\mathcal{A} \cdot H - \frac{e|H|}{r - k}.$$

Then

$$r\mu(E) = k\mu(\ker s) + (r-k)\mu(\mathrm{im}\, s) < r\mu(E) + (r-k)c_1\mathcal{A} \cdot H - 2e|H|$$

and hence

$$e < \frac{(r-k)c_1\mathcal{A} \cdot H}{2|H|} \leq \frac{(r-1)[c_1\mathcal{A} \cdot H]_+}{2|H|},$$

which proves the second part of the lemma. The first one can be proven in the same way. $\square$

Corollary 5.2.3. *Assume that $h^0(E, E \otimes \mathcal{A}) > h^0(\mathcal{A}^r)$.*

1. *If $c_1\mathcal{A} \cdot H < 0$ then E is not slope semistable.*
2. *If $c_1\mathcal{A} \cdot H \geq 0$ then E is not e-stable for all*

$$e \geq \frac{(r-1)c_1\mathcal{A} \cdot H}{2|H|}$$

Proof. Consider the (non-linear) map $\varphi : \operatorname{Hom}(E, E \otimes \mathcal{A}) \to H^0(\mathcal{A}^r)$ given by the determinant. Since $\varphi^{-1}(0) \neq \emptyset$, the fiber $\varphi^{-1}(0)$ is at least 1-dimensional. Hence there exists a non-zero $s \in \operatorname{Hom}(E, E \otimes \mathcal{A})$ with trivial determinant and we can apply the previous lemma. $\qquad\square$

The above corollary generalizes a well-known fact that a slope stable sheaf is simple (we get this for $\mathcal{A} = \mathcal{O}_X$).

If E is a torsion free slope H-semistable sheaf which is not e-stable for some $e \geq 0$ then we have a filtration

$$0 = E_0 \subset E_1 \subset \cdots \subset E_k = E$$

such that all quotients $F_i = E_i/E_{i-1}$ are slope semistable,

$$\mu(E_1) > \mu(E) - \frac{e|H|}{\operatorname{rk} E_1}$$

and

$$\mu(F_2) \geq \cdots \geq \mu(F_k).$$

Such a filtration can be constructed first by taking e-destabilizing subsheaf of E with torsion free quotient E/E_1 and then lifting the Harder–Narasimhan filtration of this quotient to E.

Let X be a smooth projective surface. In the following we always fix the determinant $\mathcal{L}$. Our aim is to bound the dimension of the locus $R(e)$ of e-unstable sheaves in $R_{\mathcal{L}}$ and the corresponding locus in $M(r, \mathcal{L}, c_2)$.

Note that for a point $[\rho : \mathcal{H} \to E] \in R(e)$ we have a filtration

$$0 \subset \mathcal{H}_0 \subset \mathcal{H}_1 \subset \cdots \subset \mathcal{H}_k = \mathcal{H}$$

such that $\mathcal{H}_0 = \ker \rho$ and $\mathcal{H}_i/\mathcal{H}_{i-1} = F_i$ for $i \geq 0$.

Set $P_i = P(F_i)$. Then $[\rho : \mathcal{H} \to E]$ lies in the image of the map

$$Y = \operatorname{Flag}(\mathcal{H}; P_0, P_1, \ldots, P_k) \to \operatorname{Quot}(\mathcal{H}; P)$$

obtained by forgetting all of the flag except for $\mathcal{H}_0 \to \mathcal{H}$. There exists only a finite number of such Flag-schemes which are non-empty. So in order to bound the dimension of $R(e)$ it is sufficient to bound the dimension of Y.

To do this one can use a version of Theorem 5.1.1 for Flag-schemes. The only thing we need to know about this is that the dimension of $\operatorname{Flag}(\mathcal{H}; P_0, P_1, \ldots, P_k)$ at $[\mathcal{H}_\bullet]$ is less or equal to the dimension of a certain Ext-group:

$$\dim \operatorname{Ext}^0_+(\mathcal{H}_\bullet, \mathcal{H}_\bullet) \leq h^0(\mathcal{H}om(\mathcal{H}, \mathcal{H})) - 1 + \sum_{i \leq j} \dim \operatorname{Ext}^1(F_j, F_i).$$

Now the dimension of $\mathrm{Ext}^1(F_j, F_i)$ can be computed using the Euler characteristic $\chi(F_j, F_i)$ (which can be computed by the Riemann–Roch formula; see Exercise 4.1.5) and the dimensions of $\mathrm{Hom}(F_j, F_i)$ and $\mathrm{Hom}(F_i, F_j \otimes \omega_X) \simeq (\mathrm{Ext}^2(F_j, F_i))^*$ (see Theorem 5.3.2). These dimensions can be bounded using Theorem 3.1.3. After some computation O'Grady used this idea to prove the following theorem.

Theorem 5.2.4 (O'Grady).

$$\dim R(e) \le \left(1 - \frac{1}{2r}\right)\Delta + d_1 e^2 + d_2 e + d_3 + (\dim V)^2,$$

where d_1, d_2 and d_3 are some explicit functions depending only on r, X and H.

Let $M(e)$ denote the locus of e-unstable sheaves in the moduli space $M(r, \mathcal{L}, c_2)$.

Corollary 5.2.5.

$$\dim M(e) \le \left(1 - \frac{1}{2r}\right)\Delta + d_1 e^2 + d_2 e + d_3 + r^2.$$

Proof. By the construction $\pi : R(e) \to M(e) = R(e)/\mathrm{GL}(V)$, so that

$$\dim M(e) \le \dim R(e) - \dim \pi^{-1}([E])$$

for some $[E] \in M(e)$. But the stabilizer of $\mathrm{GL}(V)$ action at $[\mathcal{H} \to E]$ is equal to $\mathrm{Aut}\, E$, so

$$\dim \pi^{-1}([E]) = \dim \mathrm{GL}(V) - \dim \mathrm{Aut}(E) \ge (h^0(\mathcal{H}))^2 - r^2$$

and the asserted inequality follows from the above theorem. $\square$

5.3. Examples of unobstructed moduli spaces.

Theorem 5.3.1 (Maruyama). *Let X be a smooth projective surface and let H be an ample divisor such that $HK_X < 0$. Then for any slope H-semistable torsion free sheaf E we have $\mathrm{Ext}^2(E, E) = 0$. In particular, any moduli space M_P^s is smooth.*

Proof. We will use the following useful version of the Serre duality theorem:

Theorem 5.3.2. *Let X be a smooth n-dimensional projective variety. Then for any two coherent sheaves F and G we have*

$$\mathrm{Ext}^p(F, G) \simeq (\mathrm{Ext}^{n-p}(G, F \otimes \omega_X))^*.$$

The above theorem implies that $\mathrm{Ext}^2(E, E) \simeq (\mathrm{Hom}(E, E \otimes \omega_X))^*$. But $h^0(\omega_X^{\otimes r}) = 0$ since $K_X H < 0$. Hence if $\mathrm{Ext}^2(E, E) \ne 0$ then $h^0(E, E \otimes \omega_X) > h^0(\omega_X^{\otimes r})$ and E is not slope semistable by Lemma 5.2.3, a contradiction. $\square$

In fact, we can also deal with another case:

Proposition 5.3.3. *Let X be a smooth projective surface with $\omega_X \simeq \mathcal{O}_X$. Then the moduli space of Gieseker stable sheaves $M^s(\mathcal{L})$ with fixed determinant $\mathcal{L}$ is smooth. In particular, if the characteristic of the base field does not divide r then the moduli space $M^s(\mathcal{L})$ has dimension $\Delta(E) - (r^2 - 1)\chi(\mathcal{O}_X)$. Moreover, if the Picard scheme of X is reduced then M_P^s is smooth.*

Proof. To prove smoothness it is sufficient to prove that for any Gieseker H-stable sheaf E the kernel of the trace map $\mathrm{Ext}^2(E, E) \to H^2(\mathcal{O}_X)$ is zero. Since the trace map is dual to the map $H^0(\omega_X) \to \mathrm{Hom}(E, E \otimes \omega_X)$ induced by the diagonal embedding, it suffices to show that the cokernel of this map is trivial. But this is obvious as Gieseker stable sheaves are simple.

The second part follows from the easy dimension count:

$$\dim \mathrm{Ext}^1(E, E)_0 = -\chi(E, E) + \chi(\mathcal{O}_X) = \Delta(E) - (r^2 - 1)\chi(\mathcal{O}_X). \qquad \square$$

Note that in positive characteristic there exist smooth projective surfaces with $\omega_X \simeq \mathcal{O}_X$, whose Picard scheme is non-reduced.

If X is a smooth projective surface with K_X numerically trivial, e.g., if X is an Enriques surface (i.e., such that $h^0(\omega_X) = 0$, but $\omega^2 \simeq \mathcal{O}_X$), then the moduli space $M^s(\mathcal{L})$ need not be smooth. Here we show when it happens referring to [Kim] for a more precise description of the moduli space in this case.

Proposition 5.3.4 ([Kim]). *Let X be an Enriques surface. Then the moduli space of slope stable rank r torsion free sheaves $M^\mu(\mathcal{L})$ with fixed determinant $\mathcal{L}$ is smooth if r is odd and it is singular precisely at points corresponding to E such that $E \simeq E \otimes \omega_X$ if r is even.*

Proof. Let E be a rank r slope stable sheaf on X. We have $\mathrm{Ext}^2(E, E)_0 = \mathrm{Ext}^2(E, E) \simeq (\mathrm{Hom}(E, E \otimes \omega_X))^*$. If there exists a non-zero $s \in \mathrm{Hom}(E, E \otimes \omega_X)$ then $\det s \neq 0$ by Lemma 5.2.2. Since K_X is numerically trivial, this implies that s gives rise to an isomorphism $E \simeq E \otimes \omega_X$ (E and $E \otimes \omega_X$ have equal Hilbert polynomials). In particular, comparing determinants we get $\omega_X^r \simeq \mathcal{O}_X$, so the rank r must be even. It is easy to see that at such points the moduli space is singular as the dimension of the tangent space is larger than the actual dimension. $\qquad \square$

There are many examples where the moduli space $M^s(\mathcal{L})$ is singular for an Enriques surface. We should note that in [Kim] the author's statement is different to ours and he distinguishes some cases where $E \simeq E \otimes \omega_X$ but $[E]$ is a smooth point of the moduli space. This happens because he forgets about the scheme structure and looks only at the underlying reduced scheme which is not a moduli space in the functorial sense of Definition 2.1.1.

Exercise 5.3.5. Assume that r and $c_1 \mathcal{L} \cdot H$ are relatively prime. Show that every rank r slope H-semistable sheaf with determinant $\mathcal{L}$ is slope H-stable. In particular, in this case the moduli spaces $M^\mu(\mathcal{L})$ and $M^s(\mathcal{L})$ are projective.

6. Lecture 6

- Spectral covers
- An algebraic interlude on local deformations
- Generic smoothness of moduli spaces of sheaves on surfaces

6.1. Spectral covers

Let N be a line bundle on a smooth projective variety X. Let $\pi : \mathbb{V}(N^*) \to X$ be the corresponding geometric bundle, i.e., $\mathbb{V}(N^*) = \operatorname{Spec} \bigoplus_{r \geq 0} (N^*)^r$.

Let us recall that the *F-theoretic support* of a sheaf supported on a divisor is the scheme corresponding to its 0th Fitting ideal. This should not be mistaken for a *scheme theoretic support* which is defined by the annihilator ideal sheaf.

A sheaf F on a regular scheme Z is called *Cohen–Macaulay* if at every point z in the scheme-theoretic support Y of F, the depth of F at z is equal to the codimension of z in Y. We will use the fact that if $\pi : Y \to X$ is a finite morphism onto a regular scheme X then F is Cohen–Macaulay if and only if $\pi_* F$ is locally free. A Cohen–Macaulay sheaf supported on a divisor in Z can be thought of as a pure sheaf of rank 1 on its F-theoretic support.

Proposition 6.1.1. *There exists a canonical bijection between*

1. *Isomorphism classes of Cohen–Macaulay (or pure) sheaves L on $\mathbb{V}(N^*)$, whose F-theoretic support $Y \subset \mathbb{V}(N^*)$ is a proper degree r cover of X,*
2. *Isomorphism classes of pairs (E, s), where E is a rank r locally free (torsion free, respectively) sheaf on X and $s : E \to E \otimes N$ is a homomorphism.*

The cover $Y \to X$ corresponding to (E, s) is called a spectral cover.

Proof. If we are given L as in 1, then we set $E = \pi_* L$ and we recover s from the $\pi_* \mathcal{O}_{\mathbb{V}(N^*)}$-module structure on $\pi_* L$:

$$\pi_* \mathcal{O}_{\mathbb{V}(N^*)} \otimes \pi_* L \to \pi_* L.$$

To recover L from (E, s) we need to define a section $x \in H^0(\mathbb{V}(N^*), \pi^* N) = H^0(X, \pi_* \mathcal{O}_{\mathbb{V}(N^*)} \otimes N)$, as the section corresponding to the constant section 1 of $\mathcal{O}_X$ in $\pi_* \mathcal{O}_{\mathbb{V}(N^*)} \otimes N = N \oplus \mathcal{O}_X \oplus N^{-1} \oplus \cdots$.

Then we define L by the following exact sequence

$$\pi^* E \xrightarrow{x \operatorname{Id}_{\pi^* E} - \pi^* s} \pi^* E \otimes \pi^* N \to L \otimes \pi^* N \to 0,$$

i.e., $L = coker(x \operatorname{Id}_{\pi^* E} - \pi^* s) \otimes \pi^*(N^*)$. By definition the F-theoretic support Y is given by vanishing of the determinant $\det(x \operatorname{Id}_{\pi^* E} - \pi^* s)$. $\qquad\square$

The above proposition is a generalization of [BNR, Proposition 3.6] to non-integral spectral covers in higher dimension.

Let (E, s) be a pair consisting of a rank r torsion free sheaf E and a homomorphism $s : E \to E \otimes N$. Let $a_i \in H^0(X, N^i)$, $i = 0, 1, \ldots, r$ be elementary symmetric function in eigenvalues of s (they are easily defined on the open subset where E is locally free and then uniquely extended to the whole X). Then by the Cayley–Hamilton theorem we have

$$\sum_{i=0}^{r} a_i s^{r-i} = 0$$

as a homomorphism $E \to E \otimes N^r$, whereas the spectral cover is given by

$$\sum_{i=0}^{r} a_i x^i s^{r-i} = 0.$$

6.2. An algebraic interlude on local deformations

Let $S = k[[x_1, \ldots, x_l]]$. Let us choose $0 \neq f \in S$ and set $R = S/(f)$. Let M be a Cohen–Macaulay R-module. As an S-module it has for some integer n a resolution of the following form:

$$0 \to S^n \xrightarrow{\varphi} S^n \to M \to 0.$$

Then there exists $\psi : S^n \to S^n$ such that $\varphi\psi = \psi\varphi = f \cdot \mathrm{Id}_{S^n}$. This is called a *matrix factorization* of f. After reducing modulo ideal (f) we get an infinite periodic free resolution of M:

$$\cdots \to R^n \xrightarrow{\overline{\psi}} R^n \xrightarrow{\overline{\varphi}} R^n \xrightarrow{\overline{\psi}} R^n \xrightarrow{\overline{\varphi}} R^n \to M \to 0.$$

Let N denote the image of $\overline{\varphi}$. Then we have two short exact sequences:

$$0 \to N \to R^n \to M \to 0$$

and

$$0 \to M \to R^n \to N \to 0.$$

Using the second sequence we see that there is an exact sequence

$$\mathrm{Hom}_R(M, N) \to \mathrm{Ext}^1_R(M, M) \to \mathrm{Ext}^1_R(M, R^n) = 0.$$

Lemma 6.2.1. *Let $J = (\frac{\partial f}{\partial x_1}, \ldots, \frac{\partial f}{\partial x_l})$. Then $J \mathrm{Ext}^1_R(M, M) = 0$.*

Proof. Then

$$\frac{\partial \varphi}{\partial x_i}\psi + \varphi\frac{\partial \psi}{\partial x_i} = \frac{\partial f}{\partial x_i}\mathrm{Id}_{S^n},$$

so multiplication of the resolution of M by $\frac{\partial f}{\partial x_i}$ is homotopical to the zero map. This easily implies the required assertion. $\qquad\square$

Using the above lemma we see that there exists a surjection

$$\mathrm{Hom}_R(M, N)/J\mathrm{Hom}_R(M, N) \to \mathrm{Ext}^1_R(M, M) \to 0.$$

Assume that R is an isolated singularity. Then $\mathrm{Ext}^1_R(M, M)$ is a module of finite length bounded from the above by

$$l(\mathrm{Hom}_R(M, N)/J\mathrm{Hom}_R(M, N)),$$

where $l(S)$ denotes the length of S. Let us also assume that the number of variables $l = 3$ and $f = x_3^2 - g(x_1, x_2)$. Then $J' = (\frac{\partial f}{\partial x_1}, \frac{\partial f}{\partial x_2})$ is an ideal generated by a system of parameters of R. Since $\mathrm{Hom}_R(M,N)$ is reflexive (see Exercise 6.2.3) over a normal surface singularity, it is Cohen–Macaulay. Therefore we can apply [BH, Corollary 4.6.11] and we get

$$l(\mathrm{Hom}_R(M,N)/J\mathrm{Hom}_R(M,N)) \le l(\mathrm{Hom}_R(M,N)/J'\mathrm{Hom}_R(M,N))$$
$$= \mathrm{rk}\,M \cdot \mathrm{rk}\,N \cdot l(R/J').$$

But $R/J' \simeq (k[[x_1,x_2]]/(\frac{\partial g}{\partial x_1}, \frac{\partial g}{\partial x_2}))[x_3]/(x_3^2 - g)$, so

$$l(\mathrm{Ext}_R^1(M,M)) \le 2\,\mathrm{rk}\,M \cdot (n - \mathrm{rk}\,M) \cdot \mu_0,$$

where μ_0 is the Milnor number of the plane singularity $g(x_1,x_2) = 0$ (at 0).

The above arguments prove in particular the following proposition:

Proposition 6.2.2. *Let $(E, s : E \to E \otimes N)$ be as in Proposition 6.1.1. Assume that X is a surface, Y is normal and $\mathrm{rk}\,E = 2$. Then L from Proposition 6.1.1 is a rank 1 reflexive sheaf on Y and at a point $y \in Y$*

$$h^0(Y, \mathcal{E}xt^1(L,L)_y) \le 2\mu_x,$$

where μ_x is the Milnor number of the branching curve $(\det s = 0)$ at $x = \pi(y)$.

Exercise 6.2.3. Assume that R is a normal ring and let A and B be R-modules. Prove that if B is reflexive then $\mathrm{Hom}_R(A,B)$ is reflexive. (Hint: prove that one can extend sections of $\mathrm{Hom}_R(A,B)$ defined outside of codimension 2 to the whole ring).

Exercise 6.2.4. Assume that $\mathrm{char}\,k \ne 2$. Prove that if $C := (xy = 0) \subset \mathbb{A}_k^2$ is a nodal curve then on the double cover of $\mathbb{A}^2$ branched along C there exist precisely one rank 1 reflexive sheaf L which is not locally free (cf. [Ha, II, Example 6.5.2]). Prove that $\mathrm{Ext}^1(L,L) = k$.

6.3. Generic smoothness of moduli spaces of sheaves on surfaces

Let $M = M_{\mathrm{lf}}^\mu(2, L, c_2)$ be the moduli space of slope stable locally free rank 2 sheaves with determinant L and second Chern class c_2.

The following theorem was proven (over $\mathbb{C}$) by S. Donaldson in the rank 2 case and then generalized to higher rank (and strengthened) by D. Gieseker, J. Li and K. O'Grady (all in $\mathrm{char}\,k = 0$). The author generalized this to higher characteristic (also strengthening a part of the theorem in $\mathrm{char}\,k = 0$). We will try to imitate the original Donaldson's proof. Although it gives weaker results, its idea is much simpler and more straightforward. Assumption on the characteristic is added mostly for our convenience (otherwise some covers become non-separable and the dimension of the moduli space is not so easy to compute).

Theorem 6.3.1. *Assume $\mathrm{char}\,k \ne 2$. If $c_2 \gg 0$ then M is generically smooth, i.e., smooth at a generic point of each irreducible component. Moreover, the dimension of each irreducible component is equal to $\Delta - 3\chi(\mathcal{O}_X)$ (and for even larger c_2 there is only one irreducible component).*

Proof. If E is a Gieseker stable locally free sheaf then by Theorem 5.1.3 we see that

$$T_{[E]}M \simeq \mathrm{Ext}^1(E, E)_0 \simeq H^1(\mathrm{End}_0 E)$$

and obstructions for M to be smooth lie in

$$\mathrm{Ext}^2(E, E)_0 \simeq H^2(\mathrm{End}_0 E) \simeq (H^0(\mathrm{End}_0 E \otimes \omega_X))^*.$$

Here we use the fact that the trace map splits as the characteristic is not equal to 2. So by the Riemann–Roch theorem M has at $[E]$ dimension at least

$$h^1(\mathrm{End}_0 E) - h^2(\mathrm{End}_0 E) = \Delta - 3\chi(\mathcal{O}_X).$$

To prove the theorem (except for the last part in the brackets which will be skipped), it is sufficient to show that for $c_2 \gg 0$ the closed subset $S = \{[E] \in M : H^0(\mathrm{End}_0 E \otimes \omega_X) \neq 0\}$ of M has dimension smaller than $\Delta - 3\chi(\mathcal{O}_X)$. This follows from Theorem 6.3.2. $\qquad\square$

Theorem 6.3.2. *Let $\mathcal{A}$ be a line bundle on X and let*

$$S_{\mathcal{A}} = \{[E] \in M : H^0(\mathrm{End}_0 E \otimes \mathcal{A}) \neq 0\}.$$

If $c_1\mathcal{A} \cdot H < 0$ then $S_{\mathcal{A}} = \emptyset$. If $c_1\mathcal{A} \cdot H \geq 0$ and $c_2 \gg 0$ then

$$\dim S_{\mathcal{A}} < \Delta - 3\chi(\mathcal{O}_X).$$

Proof. Since the first part is clear (see Lemma 5.2.2) we can assume that $c_1\mathcal{A} \cdot H \geq 0$.

Let $|\mathcal{A}| \to |\mathcal{A}^2|$ be a map sending divisor D to $2D$. We denote the image by $2|\mathcal{A}|$. We will prove the theorem under the simplifying assumption that all divisors in $|\mathcal{A}^2| - 2|\mathcal{A}|$ are reduced (this is satisfied, e.g., if $\mathrm{Pic}\, X$ is generated by $\mathcal{A}$).

We decompose $S_{\mathcal{A}}$ into two locally closed subsets:

$$S'_{\mathcal{A}} = \{[E] \in M : \text{there exists a non-zero } s \in H^0(\mathrm{End}_0 E \otimes \mathcal{A})$$
$$\text{such that } \det s = 0 \text{ or } \det s \in 2|\mathcal{A}|\}$$

and

$$S''_{\mathcal{A}} = \{[E] \in M : \text{for every non-zero } s \in H^0(\mathrm{End}_0 E \otimes \mathcal{A})$$
$$\text{we have } \det s \in |\mathcal{A}^2| - 2|\mathcal{A}|\}.$$

First let us note that $S'_{\mathcal{A}} \subset M(\frac{c_1 \mathcal{A} \cdot H}{2|H|})$. This follows by Lemma 5.2.2 since if $E \in S'_{\mathcal{A}}$ then there exists $s \in H^0(\mathrm{End}_0 E \otimes \mathcal{A})$ such that either $\det s = 0$ or there exists $t \in H^0(\mathcal{A})$ such that $\det s + t^2 = 0$ and then for $s' = s - t \cdot \mathrm{Id}_E$ (which is non-zero as $\mathrm{Tr}\, s = 0$) we have $\det s' = 0$. Hence $\dim S'_{\mathcal{A}}$ can be bounded using Corollary 5.2.5.

To bound the dimension of $S''_{\mathcal{A}}$ we will use the moduli space M' of pairs (E, s) such that $E \in S''_{\mathcal{A}}$ and $0 \neq s \in H^0(\mathrm{End}_0 E \otimes \mathcal{A})$. This moduli space exists and it can be constructed as a locally closed subscheme in the moduli space of Gieseker stable pure sheaves on a compactification $\mathbb{P}(\mathcal{O} \oplus \mathcal{A}^*)$ of $\mathbb{V}(\mathcal{A}^*)$ (with respect to polarization $\pi^* H + aZ$, where Z is the complement on $\mathbb{V}(\mathcal{A}^*)$ and a is a small positive rational number). This is a special case of the moduli space of (generalized) Higgs bundles on X.

Note that we have a surjection $M' \to S''_{\mathcal{A}}$ so it suffices to bound the dimension of M'. There exists a partial Hitchin's map $M' \to |\mathcal{A}^2|$ sending (E, s) to the determinant of s. By Proposition 6.1.1 the fiber over $B \in |\mathcal{A}^2|$ is a subset of the moduli space M''_B of reflexive rank 1 sheaves on the double cover Y of X branched along B. The cover $p : Y \to X$ is constructed as $\mathrm{Spec}(\mathcal{O}_X \oplus \mathcal{A}^*)$, where the $\mathcal{O}_X$-algebra structure on $\mathcal{O}_X \oplus \mathcal{A}^*$ is given by $\mathcal{A}^* \otimes \mathcal{A}^* \to \mathcal{O}_X$ coming from multiplication by the chosen equation of B. By assumption B is reduced, so the surface Y is normal (it is S_2 as a hypersurface in a smooth 3-fold and it has only isolated singular points lying over singular points of B).

The dimension of M''_B at a rank 1 reflexive sheaf L is bounded from the above by $\dim \mathrm{Ext}^1(L, L)$. Using the local to global Ext spectral sequence we get

$$\dim \mathrm{Ext}^1(L, L) \leq h^0(\mathcal{E}xt^1_Y(L, L)) + h^1(\mathcal{H}om_Y(L, L)).$$

Note that $\mathcal{H}om_Y(L, L) \simeq \mathcal{O}_Y$, so $h^1(\mathcal{H}om_Y(L, L)) = h^1(p_*\mathcal{O}_Y) = h^1(\mathcal{O}_X) + h^1(\mathcal{A}^*)$ and we need to bound only local contributions. In our case this can be done using Proposition 6.2.2, which implies that

$$h^0(\mathcal{E}xt^1_Y(L, L)) \leq 2 \sum_{x \in \mathrm{Sing}\, B} \mu_x = 2(K_X + B)B + 2e_{\mathrm{top}}(B),$$

where $\mathrm{Sing}\, B$ denotes the set of singular points of B and $e_{\mathrm{top}}(B)$ is the "topological" Euler characteristic of B (it also makes sense in positive characteristic and it is bounded from the above by twice the number of irreducible components of B). Summing up, we get

$$\dim S''_{\mathcal{A}} < h^0(\mathcal{A}^2) + h^1(\mathcal{O}_X) + h^1(\mathcal{A}^*) + 2(K_X + B)B + 2e_{\mathrm{top}}(B),$$

where the right-hand side does not depend on c_2. $\qquad\square$

One should note that K. Zuo also used Donaldson's idea to give a proof of Donaldson's Theorem 6.3.2 (in the rank 2 case) using some results of R. Friedman. However, his proof is incomplete as he "forgets" to consider the case when there exists $s \in H^0(\mathrm{End}_0 E \otimes \mathcal{A})$ such that $\det s \notin 2|\mathcal{A}|$ but the corresponding spectral cover is reducible. This happens if the curve $\det s = 0$ decomposes into a sum $D_+ + D_-$ for some divisors $D_\pm \in |\mathcal{A} \pm \tau|$, where τ is a 2-torsion. Nowadays, Theorem 6.3.2 is usually proven using O'Grady's approach which is based on a completely different idea than spectral covers.

Acknowledgements

The author would like to thank Piotr Pragacz without whose persistence these notes would have not been written. The author would also like to thank Halszka Tutaj-Gasińska. A part of these notes (part of Lecture 5 and Lecture 6) was written while the author's stay in the Max Planck Institute für Mathematik.

References

[BP]   C. Bănică, M. Putinar, On the classification of complex vector bundles of stable rank, *Proc. Indian Acad. Sci. Math. Sci.* **116** (2006), 271–291.

[BNR] A. Beauville, M.S. Narasimhan, S. Ramanan, Spectral curves and the generalised theta divisor, *J. reine angew. Math.* **398** (1989), 169–179.

[Be]   P. Belkale, The strange duality conjecture for generic curves, to appear in *J. Amer. Math. Soc.*

[BH]   W. Bruns, J. Herzog, Cohen-Macaulay rings, Cambridge Studies in Advanced Mathematics **39**, Cambridge University Press, Cambridge, 1993.

[Da]   G. Danila, Sections du fibré déterminant sur l'espace de modules des faisceaux semi-stables de rang 2 sur le plan projectif, *Ann. Inst. Fourier (Grenoble)* **50** (2000), 1323–1374.

[Ha]   R. Hartshorne, Algebraic geometry, Graduate Texts in Mathematics **52**, 1977.

[HL]   D. Huybrechts, M. Lehn, The geometry of moduli spaces of sheaves, Aspects of Mathematics **31**, 1997.

[Kim] H. Kim, Moduli spaces of stable vector bundles on Enriques surfaces, *Nagoya Math. J.* **150** (1998), 85–94.

[La1] A. Langer, Semistable sheaves in positive characteristic, *Ann. of Math.* **159** (2004), 251–276.

[La2] A. Langer, Moduli spaces of sheaves in mixed characteristic, *Duke Math. J.* **124** (2004), 571–586.

[La3] A. Langer, Moduli spaces and Castelnuovo-Mumford regularity of sheaves on surfaces, *Amer. J. Math.* **128** (2006), 373–417.

[LP1] J. Le Potier, Lectures on vector bundles, Cambridge Studies in Mathematics **54**, 1997.

[LP2] J. Le Potier, Dualité étrange sur les surfaces, unpublished manuscript, 2001.

[MO]   A. Marian, D. Oprea, The level-rank duality for non-abelian theta functions, *Invent. Math.* **168** (2007), 225–247.

[Mi]   Y. Miyaoka, The Chern classes and Kodaira dimension of a minimal variety, Algebraic Geometry, Sendai, 1985, *Adv. Stud. Pure Math.* **10**, 1987, 449–476.

[O]   K. G. O'Grady, Involutions and linear systems on holomorphic symplectic manifolds, *Geom. Funct. Anal.* **15** (2005), 1223–1274.

Adrian Langer
Institute of Mathematics, Warsaw University
ul. Banacha 2
PL-02-097 Warszawa, Poland

and

Institute of Mathematics, Polish Academy of Sciences
ul. Śniadeckich 8
PL-00-956 Warszawa, Poland
e-mail: `alan@mimuw.edu.pl`

Algebraic Cycles, Sheaves, Shtukas, and Moduli
Trends in Mathematics, 105–116
© 2007 Birkhäuser Verlag Basel/Switzerland

Miscellany on the Zero Schemes of Sections of Vector Bundles

Piotr Pragacz

Notes by Ozer Ozturk

Abstract. This purely expository article is a summary of the author's lectures on topological, algebraic, and geometric properties of the zero schemes of sections of vector bundles. These lectures were delivered at the seminar *Impanga* at the Banach Center in Warsaw (2006), and at the METU in Ankara (December 11–16. 2006). A special emphasis is put on the connectedness of zero schemes of sections, and the "point" and "diagonal" properties in algebraic geometry and topology. An overview of recent results by V. Srinivas, V. Pati, and the author on these properties is given.

Mathematics Subject Classification (2000). 14F05, 14F45, 57R22.

Keywords. Vector bundle, section, zero scheme of a section, connectedness, diagonal, cohomologically trivial line bundle.

1. The role of global equations in algebraic geometry and topology

Algebraic objects like *polynomials* enable us to present geometric objects like varieties via equations. However, when we consider projective (compact) varieties there is a problem: every global polynomial function is constant. To overcome this problem, we can "glue" local polynomial equations with the help of some global objects: vector bundles, which are families of vector spaces over a base variety, with transition functions from full linear groups.

For a motivating example, consider the complex projective n-space $\mathbb{P}^n = \mathbb{P}^n(V)$ – the set of lines through zero in the $(n+1)$-dimensional complex vector space $V = \bigoplus_{i=0}^n \mathbb{C}e_i$. Over this variety we have a subbundle of the trivial vector bundle:

$$\mathcal{O}(-1) = \{(x, l) : x \in \mathbb{P}^n, x \in l\}.$$

Research supported by TÜBİTAK (during the stay at the METU in Ankara).

This bundle is called the *tautological line bundle* or the *Hopf bundle*. The dual bundle $\mathcal{O}(1) = \mathcal{O}_{\mathbb{P}^n}(1) = \mathcal{O}(-1)^*$ is called the *Grothendieck bundle*. Note that the space of global sections of this bundle, $\Gamma(\mathbb{P}^n(V), \mathcal{O}(1))$, is isomorphic to $S^\bullet(V^*)$. So homogeneous polynomials in the dual coordinates $x_i = e_i^*$ can be identified with the global sections of $\mathcal{O}(1)$. This – most classical example of "global equations" – admits a natural generalization to *sections* of any vector bundle.

Suppose that s is a section of a vector bundle $\mathcal{E} \to X$. Consider the *zero scheme* of s

$$Z(s) = \{x \in X : s(x) = 0\}.$$

We wish to discuss the following question:

Which properties of $Z(s)$ can be deduced from those of $\mathcal{E}$?

Apart from this question (and also in connection with it), we shall study the following two properties of varieties: We will say that a variety X has the

- *Weak point property* if for a point $x \in X$ there exist a vector bundle $\mathcal{E}$ on X of rank $\dim X$, and a section s of $\mathcal{E}$ such that $\{x\} = Z(s)$.

- *Diagonal property* if there exist a vector bundle $\mathcal{E}$ on $X \times X$ of rank $\dim X$, and a section s of $\mathcal{E}$ such that $\Delta = Z(s)$, where Δ denotes the diagonal in $X \times X$.

We shall write (D) for the diagonal property and (P) for the week point property. Notice that (D) implies (P) – in fact, for any point $x \in X$ – via restriction from $X \times X$ to $X \times \{x\}$.

To the best of our knowledge, (P) was a popular topic neither in algebraic geometry nor topology. It appears that a stronger variant[1] of (P) was studied in algebra and arithmetics. Let A be a finitely generated reduced algebra over an algebraically closed field k with Krull dimension d. Recall that a point x of $X = \operatorname{Spec} A$ is a *complete intersection* if the corresponding maximal ideal has height d, and is generated by d elements of A. In this case x is a regular point, but not conversely. We record the following problem:

Characterize reduced affine k-varieties such that all smooth points are complete intersections.

This problem is discussed in detail by V. Srinivas in his paper [10] in the present volume. For instance, we have the following "Affine Bloch–Belinson conjecture":

Let $k = \mathbb{Q}$. Then for any finitely generated smooth k-algebra of dimension greater than 1 every maximal ideal is a complete intersection.

(Note that this conjecture is confirmed yet by *no* nontrivial example (!).)

It appears that also (D) was not studied systematically before. First examples of varieties with (D) are curves. When X is a smooth curve, the diagonal Δ

---

[1] The word "stronger" means here that we want (P) for any point $x \in X$, and the bundle involved in (P) should be trivial.

is a Cartier divisor in $X \times X$, so (D) holds. If (D) holds for varieties X_1 and X_2 then (D) holds for $X_1 \times X_2$ too. Other known examples having (D) are $\mathbb{P}^n$, Grassmannians and, in general, flag varieties of the type SL_n/P, where P is any parabolic subgroup of SL_n. Though this is well known to experts, we sketch a simple argument, since we could not find it in the literature. The argument in the Grassmannian case, given below, is well known. Also, (D) for the variety of *complete* flags was the starting point for the theory of *Schubert polynomials* of Lascoux and Schützenberger [6]. In [2], this property was proved and used to compute the fundamental classes of flag degeneracy loci. In fact, the argument for an arbitrary flag variety, follows closely that given in [2].

Let V be an n-dimensional vector space. Fix an increasing sequence of integers

$$d_\bullet : 0 < d_1 < d_2 < \cdots < d_{k-1} < d_k = n\,.$$

Then by a $d_\bullet$-*flag* we mean an increasing sequence of linear subspaces

$$V_1 \subset V_2 \subset \cdots \subset V_{k-1} \subset V_k = V$$

of V such that $\dim V_i = d_i$ for $i = 1, \ldots, k$. The set of all $d_\bullet$-flags forms the flag variety $F\ell^{d_\bullet}$. For example, the sequence

$$d_1 = r < d_2 = n$$

gives rise to the Grassmannian $G_r(V)$ parametrizing r-dimensional linear subspaces of V.

In analogy to the Hopf bundle $\mathcal{O}(-1)$ on the projective space, we have a rank r vector bundle on $G_r(V)$. Consider the subbundle $\mathcal{S}$ of the trivial vector bundle $\mathcal{V}_{G_r(V)}$ of rank r for which the fiber $\mathcal{S}_g$ over $g \in G_r(V)$ is just the r-dimensional subspace corresponding to g. This bundle is called the rank r *tautological subbundle* and is denoted by $\mathcal{S}$. We have the tautological vector bundle sequence over $G_r(V)$:

$$0 \to \mathcal{S} \to \mathcal{V}_{G_r(V)} \to \mathcal{Q} \to 0\,,$$

where $\mathcal{Q}$ is the rank $(n - r)$ *tautological quotient*. Let G_1 and G_2 be two copies of $G_r(V)$. Let $\mathcal{S}_1$ be the tautological subbundle on G_1, and let $\mathcal{Q}_2$ be the tautological quotient bundle on G_2. Moreover, let

$$p_i : G_1 \times G_2 \to G_i$$

denote the projection for $i = 1, 2$. Then the composition

$$p_1^* \mathcal{S}_1 \hookrightarrow p_1^*(\mathcal{V}_{G_1}) = p_2^*(\mathcal{V}_{G_2}) \to p_2^* \mathcal{Q}_2$$

gives rise to the section of the vector bundle

$$\underline{\mathrm{Hom}}(p_1^* \mathcal{S}_1, p_2^* \mathcal{Q}_2)$$

over $G_1 \times G_2$ which vanishes precisely on the diagonal. We conclude that $G_r(V)$ has (D).

In the case of arbitrary $d_\bullet$-flags, the tautological sequence takes the form:

$$\mathcal{S}_1 \hookrightarrow \mathcal{S}_2 \hookrightarrow \cdots \hookrightarrow \mathcal{S}_{k-1} \hookrightarrow \mathcal{S}_k = \mathcal{V} \xrightarrow{q_1} \mathcal{Q}_1 \xrightarrow{q_2} \mathcal{Q}_2 \xrightarrow{q_3} \cdots \xrightarrow{q_k} \mathcal{Q}_k\,,$$

where $\mathrm{rank}(\mathcal{S}_i) = d_i$ for $i = 1, \ldots, k$, and $\mathcal{Q}_i$ is the quotient of $\mathcal{V}$ by $\mathcal{S}_i$, so that $\mathrm{rank}(\mathcal{Q}_i) = n - d_i$.

Let F_1 and F_2 be two copies of $F\ell^{d_\bullet}$ and

$$p_i : F_1 \times F_2 \to F_i$$

denote the projection for $i = 1, 2$. Consider the map

$$\varphi : \bigoplus_{i=1}^{k-1} \mathrm{Hom}(p_1^* \mathcal{S}_i, p_2^* \mathcal{Q}_i) \to \bigoplus_{i=1}^{k-2} \mathrm{Hom}(p_1^* \mathcal{S}_i, p_2^* \mathcal{Q}_{i+1})$$

defined by

$$\varphi\Big(\sum_{i=1}^{k-1} f_i\Big) = \Big(\sum_{i=1}^{k-2} (f_{i+1}|_{\mathcal{S}_i} - q_i \circ f_i)\Big).$$

One checks that φ is surjective. Set $\mathcal{K} = \mathrm{Ker}\,\varphi$. Then the compositions

$$p_1^* \mathcal{S}_i \hookrightarrow p_1^*(\mathcal{V}_{F_1}) = p_2^*(\mathcal{V}_{F_2}) \to p_2^* \mathcal{Q}_i$$

for $i = 1, \ldots, k$, give rise to a section s of $\mathcal{K}$ such that $\Delta = Z(s)$.

Note that

$$\mathrm{rank}\,\mathcal{K} = \sum_{i=1}^{k-1} (n - d_i)(d_i - d_{i-1}) = \dim F\ell^{d_\bullet}.$$

Summing up, we conclude that the flag variety $F\ell^{d_\bullet}$ has (D).

2. Connectedness of the zero schemes of sections of vector bundles

In this section, we shall discuss the connectedness properties of the zero schemes. A prototype of all results here is

Theorem 1 (Lefschetz). *The hypersurface defined by a single homogeneous polynomial equation in $\mathbb{P}^n$ is connected provided $n \geq 2$.*

One should be careful with generalizations of this simple result: one cannot – in general – replace $\mathbb{P}^n$ by $\mathbb{C}^n$, a hypersurface in $\mathbb{P}^n$ by a hypersurface in another smooth projective variety, and single equation by several equations (these issues are discussed in detail in [11]).

Recall that a line bundle $\mathcal{L} \to X$ is called *very ample* if $\mathcal{L} \cong \mathcal{O}_{\mathbb{P}^n}(1)|_X$ for some embedding of X into $\mathbb{P}^n$. A line bundle $\mathcal{L}$ is called *ample* if there exists $m \geq 0$ such that $\mathcal{L}^{\otimes m}$ is very ample.

Theorem 2 (Lefschetz). *Let X be a smooth projective irreducible variety over $\mathbb{C}$. Let $\mathcal{L}$ be an ample line bundle over X and s be a section of $\mathcal{L}$. Then $H_q(Z(s), \mathbb{Z}) \to H_q(X, \mathbb{Z})$ is an isomorphism for $q < \dim X - 1$ and is a surjection when $q = \dim X - 1$.*

This theorem is called the *Lefschetz hyperplane theorem*. Let us record its simple consequence.

Corollary 3. *Under the assumptions of the theorem, if* $\dim X \geq 2$, *then* $Z(s)$ *is connected.*

A vector bundle $\mathcal{E} \to X$ is called *ample* if the vector bundle $\mathcal{O}(1)$ on $\mathbb{P}(\mathcal{E}^*)$ is ample.

Proposition 4 (Sommese). *If $\mathcal{E}$ is a rank e vector bundle on X and $s \in \Gamma(X, \mathcal{E})$, then $\mathbb{P}(\mathcal{E}^*) \setminus Z(s^*)$ is an affine-space bundle with fiber $\mathbb{C}^{e-1}$ over $X \setminus Z(s)$. So $H_0(\mathbb{P}(\mathcal{E}^*) \setminus Z(s^*), \mathbb{Z}) = H_0(X \setminus Z(s), \mathbb{Z})$.*

Indeed, if $x \in Z(s)$ then $s(x)^*$ vanishes on entire $\mathcal{E}_x^*$. If $x \notin Z(s)$ then $s(x)^*$ vanishes on a hyperplane in $\mathcal{E}_x^*$. Therefore, the fiber at x of

$$\mathbb{P}(\mathcal{E}^*) \setminus Z(s^*) \to X \setminus Z(s)$$

is

$$\mathbb{P}(\mathcal{E}^*) \setminus H = \mathbb{C}^{e-1},$$

where H is a hyperplane (cf. [9]).

Theorem 5 (Griffiths, Sommese). *Let X be an irreducible smooth projective variety over $\mathbb{C}$. Let $\mathcal{E}$ be a rank e vector bundle over X, and s a section of $\mathcal{E}$. Then $H_q(Z(s), \mathbb{Z}) \to H_q(X, \mathbb{Z})$ is an isomorphism for $q < \dim X - e$, and is a surjection when $q = \dim X - e$.*

(Cf. [4], [9].)

Corollary 6. *Under the assumptions of the last theorem, if* $\dim X \geq e + 1$, *then* $Z(s)$ *is connected.*

For a more detailed account to this theory, we refer the reader to the Tu's article [11]. In this article, the author also discusses the connectedness of *degeneracy loci* which provide natural generalizations of the zero schemes of sections of vector bundles.

3. Cohomologically trivial line bundles

We start with two simple consequences of (D). Consider the ideal sheaf $\mathcal{J}_\Delta \subset \mathcal{O}_{X \times X}$. The diagonal property (D) implies that

$$\mathcal{E}^*_{|\Delta} = \mathcal{J}_\Delta / \mathcal{J}_\Delta^2 \cong \Omega_X^1$$

(the last isomorphism uses the isomorphism $\Delta \cong X$). Therefore, Ω_X^1 is locally free of rank equal to dimension of X. Hence X is smooth. Also, by the Grothendieck formula [5], we obtain the following expression for the fundamental class of Δ:

$$[\Delta] = c_{\dim X}(\mathcal{E}).$$

From now on – unless otherwise is explicitly stated – all results and conjectures surveyed here come from the paper [8], written by the author, V. Srinivas, and V. Pati.

Definition 7. *A line bundle $\mathcal{L}$ over X is called cohomologically trivial (we shall write "c.t.") if $H^i(X, \mathcal{L}) = 0$ for all $i \geq 0$.*

Example 8. Any smooth projective curve supports a c.t. line bundle (this should be well known to experts – for a written account, cf. [8]). Any abelian variety supports a c.t. line bundle [7].

Theorem 9.

   (i) *Let $p_1, p_2 : X \times X \to X$ be the two projections. If X has (D), and moreover, the following isomorphism holds:*

$$\operatorname{Pic}(X \times X) \cong p_1^* \operatorname{Pic}(X) \oplus p_2^* \operatorname{Pic}(X), \tag{1}$$

     then there exists a c.t. line bundle $\mathcal{L}$ over X such that

$$\det(\mathcal{E}) = p_1^* \mathcal{L}^{-1} \otimes p_2^*(\mathcal{L} \otimes \omega_X^{-1}).$$

  (ii) *If $\dim X = 2$ and there exists a c.t. line bundle on X, then X has (D).*

This theorem was proved in [8]. We give now a sketch of this proof.

Apply to the exact sequence

$$0 \to \mathcal{I}_\Delta \to \mathcal{O}_{X \times X} \to \mathcal{O}_\Delta \to 0.$$

and any line bundle $\mathcal{L}$ on $X \times X$, the functor $\operatorname{Hom}(-, \mathcal{L})$ (and its derived functors) to get the sequence of global $\operatorname{Ext}_{X \times X}$'s :

$$\operatorname{Ext}^{n-1}(\mathcal{I}_\Delta, \mathcal{L}) \to \operatorname{Ext}^n(\mathcal{O}_\Delta, \mathcal{L}) \overset{\alpha}{\to} H^n(X \times X, \mathcal{L}). \tag{2}$$

We shall need the following cohomological result (cf. [8] and the references there):

Proposition 10. *Let $\mathcal{L}$ be a line bundle on $X \times X$ whose restriction to Δ is isomorphic to ω_Δ. Assume that there exist a rank n vector bundle $\mathcal{E}$ on $X \times X$ with $\det(\mathcal{E}^*) = \mathcal{L}$, and $s \in \Gamma(X \times X, \mathcal{E})$ satisfying $\operatorname{Im}(s^*) = \mathcal{I}_\Delta$. Then α in the exact sequence (2) vanishes. The converse holds if $n = \dim(X) = 2$.*

Suppose that there exists a vector bundle $\mathcal{E}$ on $X \times X$ of rank n such that the diagonal is the zero scheme of its section s. Let $\mathcal{L} = \det(\mathcal{E}^*)$, and form the corresponding exact sequence (2). From the proposition, we have $\alpha = 0$. Consider the dual linear map to α:

$$\alpha^* : H^n(X \times X, \mathcal{L})^* \to \operatorname{Ext}^n(\mathcal{O}_\Delta, \mathcal{L})^* \cong H^n(\Delta, \omega_\Delta) = k.$$

Using (1) choose $\mathcal{M} \in \operatorname{Pic}(X)$ such that

$$\mathcal{L} = \det(\mathcal{E}^*) \cong p_1^*(\mathcal{M}) \otimes p_2^*(\mathcal{M}^{-1} \otimes \omega_X).$$

By Serre duality on $X \times X$, we get that

$$H^n(X \times X, \mathcal{L})^* \cong H^n(X \times X, \mathcal{L}^{-1} \otimes \omega_{X \times X}) \cong H^n(X \times X, p_1^*(\mathcal{M}^{-1} \otimes \omega_X) \otimes p_2^* \mathcal{M}).$$

From the Künneth formula, we have

$$H^n(X \times X, p_1^*(\mathcal{M}^{-1} \otimes \omega_X) \otimes p_2^*(\mathcal{M})) = \bigoplus_{i=0}^{n} H^i(X, \mathcal{M}^{-1} \otimes \omega_X) \otimes H^{n-i}(X, \mathcal{M}). \tag{3}$$

Further, on any summand on the right, the induced map

$$H^i(X, \mathcal{M}^{-1} \otimes \omega_X) \otimes H^{n-i}(X, \mathcal{M}) \hookrightarrow$$

$$H^n(X \times X, p_1^*(\mathcal{M}^{-1} \otimes \omega_X) \otimes p_2^*(\mathcal{M})) \xrightarrow{\alpha^*} H^n(\Delta, \omega_\Delta) = k$$

coincides with the *Serre duality pairing* on cohomology of X, and is hence a non-degenerate bilinear form, for each $0 \leq i \leq n$. Thus, α^* vanishes if and only if all the summands on the RHS of (3) vanish, which says that $\mathcal{M}$ is c.t.

Conversely, if $\mathcal{M}$ is c.t., then in the exact sequence (2) determined by

$$\mathcal{L} = p_1^* \mathcal{M} \otimes p_2^*(\mathcal{M}^{-1} \otimes \omega_X),$$

the map α is the zero map, by reversing the above argument. Hence, if $n = 2$, we deduce that $X \times X$ supports a vector bundle $\mathcal{E}$ of rank 2 and a section s with zero scheme Δ, by the $n = 2$ case of the proposition.

This ends our sketch of the proof from [8].

Corollary 11. *Suppose that the isomorphism* (1) *holds for X, and X supports no c.t. bundle. Then X has not* (D).

Let X be a smooth proper variety over an algebraic closed field. The isomorphism (1) holds for X if and only if $\operatorname{Pic} X$ is a finitely generated abelian group. Also, if $H^1(X, \mathcal{O}_X) = 0$ then (1) holds for X.

4. When a smooth projective surface has (D)?

We remind that for a surface to have (D) is almost equivalent to the existence of a c.t. line bundle on it. In [8], the following results were proved for a smooth projective surface X over an algebraically closed field:

1. There exists a surface Y having (D), and a birational proper map $f : Y \to X$.
2. If $f : Y \to X$ is a birational map, X has (D) and $\operatorname{Pic} X$ is finitely generated then Y has (D).
3. If X is birational to one of the following: a ruled or an abelian surface or a K3 surface with 2 disjoint smooth rational curves or an elliptic surface with a section or a complex Enriques or hyperelliptic surface, then X has (D).
4. If $\operatorname{Pic} X = \mathbb{Z}$, $\Gamma(X, \mathcal{O}_X(1)) \neq 0$ and X has (D), then $X = \mathbb{P}^2$.

More precisely, the first item says that any surface – after blowing up sufficiently many points – becomes a surface having (D).

The last item implies that (D) fails for general K3 surfaces or general hypersurfaces in $\mathbb{P}^3$ of degree greater than 3.

5. When a higher-dimensional variety has (D)?

In this section, we consider varieties of dimension ≥ 3. We first consider the varieties with Picard group $\mathbb{Z}$.

Proposition 12. *Let X be a smooth projective variety of dimension $d \geq 3$ over a field with $\operatorname{Pic} X = \mathbb{Z}$. If X has (D) and $H^0(X, \mathcal{O}_X(1)) \neq 0$ then X is a Fano variety and $\omega_X \cong \mathcal{O}_X(-n)$ for some $n \geq 2$.*

This result has two useful consequences.

Let $X \subset \mathbb{P}^n$ be a smooth complete intersection of multidegree $(d_1, \ldots, d_r)$ such that $r < n-3$ and $\sum d_i \geq n$. Then X has not (D). In particular, the non-Fano hypersurfaces in the projective spaces have not (D).

Also, let X be a smooth projective Fano variety such that $b_2(X) = 1$ and $\omega_X = \mathcal{O}_X(-1)$ (i.e., X is of index 1). Then X has not (D).

Let X be a scheme and $\mathcal{L}$ be a line bundle over X. We say that X has the

- $\mathcal{L}$-*point property* if for every point $x \in X$ there exists a vector bundle $\mathcal{F}$ over X such that $d = \operatorname{rank} \mathcal{F} = \dim X$, $\det(\mathcal{F}) = \mathcal{L}$, and there exists a section s of $\mathcal{F}$ such that $\{x\} = Z(s)$.

Note that in this case $c_1(\mathcal{F}) = c_1(\mathcal{L})$ and $c_d(\mathcal{F}) = [x]$.

Theorem 13. *Let X be a smooth proper variety over an algebraically closed field. If $\operatorname{Pic} X$ is finitely generated and X has (D), then there exists a c.t. line bundle $\mathcal{L}$ on X such that*

(i) *X has the $\mathcal{L}^{-1}$-point property, and*

(ii) *X has the $\mathcal{L} \otimes \omega_X^{-1}$-point property.*

Corollary 14. *Let X be a scheme with finitely generated $\operatorname{Pic} X$. If for any c.t. line bundle $\mathcal{L}$, either $\mathcal{L}^{-1}$-point property fails or $\mathcal{L} \otimes \omega_X^{-1}$-point property fails then X has not (D).*

For example, if X is a smooth complex projective quadric of dimension 3, then $\mathcal{O}_X(-1)$ and $\mathcal{O}_X(-2)$ are the unique c.t. line bundles on X. One checks – with the help of the corollary – that X has not (D).

Sometimes, (D) boils down to (P). For instance, we have

Proposition 15. *Let X be a group variety over an algebraically closed field. Then X has (D) if and only if X has (P).*

Indeed, assume that X has (P). Let $\mathcal{E}$ be a vector bundle over X such that $\operatorname{rank} \mathcal{E} = \dim X$, and let $s \in \Gamma(X, \mathcal{E})$ be such that $Z(s) = \{x\}$ for some $x \in X$. Since X is a group variety we have the morphisms $\mu : X \times X \to X$ of multiplication and $i : X \to X$ of inverse. Consider the morphism

$$f : X \times X \to X$$

defined by

$$f(u, v) = \mu(\mu(u, i(v)), x).$$

Since $f^{-1}(x) = \Delta$, the vector bundle $f^*\mathcal{E}$, together with section f^*s, implies (D) for X.

In particular, an abelian variety has (D) if and only if it has (P). Recently O. Debarre [1] has proved that the Jacobian of a smooth projective connected curve has (P), and that there exist non-principally polarized abelian varieties in dimension greater than 2, which fail to have (P). Moreover, he suggests that (P) may characterize Jacobians among all principally polarized abelian varieties.

6. Affine case

Let k be an algebraically closed field and A be a finitely generated k-algebra. Let $X = \operatorname{Spec} A$. If $\dim X = 2$, then X has (D) by Serre's construction. What about higher dimensions?

Theorem 16. *An affine algebraic group over an algebraically closed field has* (D).

Indeed, M. Kumar and P. Murthy proved that an affine algebraic group over an algebraically closed field has (P). It suffices then to invoke Proposition 15.

Conjecture 17. *There exists smooth complex varieties of any dimension greater than 2 for which* (D) *fails.*

This conjecture leads to the following question:

Let k be an algebraically closed field and A be a regular k-algebra. Let K be an extension field of k (not necessarily algebraically closed), $A_K := A \otimes_k K$ and let $M \subset A_K$ be a maximal ideal with residue field K. Does there exist a projective A_K-module P of rank $n = \dim A$ such that there is a surjection $P \twoheadrightarrow M$?

The question has a positive answer by P. Murthy when K is an algebraically closed extension of k. If $X = \operatorname{Spec} A$ has (D), then the question has a positive answer for any field extension K. So a negative answer to the question would produce counterexamples to (D).

7. Diagonal property in topology

In this section, we use mostly the notation used by topologists. Let M be a compact connected oriented smooth manifold of real dimension n, and Δ be the diagonal submanifold of $M \times M$. We say that M has property (D_r) if there exist a smooth *real* vector bundle $\mathcal{E}$ over $M \times M$ with $\operatorname{rank}(\mathcal{E}) = n$ and a smooth section s of $\mathcal{E}$ such that s is transverse to the zero section $0_{\mathcal{E}}$ of $\mathcal{E}$ and $s^{-1}(0_{\mathcal{E}}) = \Delta$. If the vector bundle $\mathcal{E}$ is *orientable* then we say that M has the property (D_o). If $\dim_{\mathbb{R}} M = 2m$ and the vector bundle $\mathcal{E}$ is a smooth *complex* vector bundle of $\operatorname{rank}_{\mathbb{C}}(\mathcal{E}) = m$ then we say that M has the property (D_c). We have the following relation between these properties:

$$(D) \Rightarrow (D_c) \Rightarrow (D_o) \Rightarrow (D_r).$$

Take a Riemannian metric on M. It induces a Riemannian metric on $M \times M$, on the tangent bundle τ_M, and on all its subbundles. Let U be a closed ϵ-tubular neighborhood of Δ in $M \times M$. Then, by the tubular neighborhood theorem, we have a diffeomorphism

$$\phi : (U, \partial U) \xrightarrow{\sim} (D(\nu), S(\nu)),$$

where $D(\nu)$ is the ϵ-disc bundle of the normal bundle $\rho : \nu \to \Delta$ of Δ in $M \times M$ and $S(\nu)$ is the ϵ-sphere bundle. Then $r := \rho \circ \phi : U \to \Delta$ is a strong deformation retraction of U to Δ. So we have the following bundle diagram:

$$
\begin{array}{ccc}
r^*(\nu) & \longrightarrow & \rho^*(\nu) \\
\rho \downarrow & & \downarrow \rho \\
U & \xrightarrow{\ \phi\ } & D(\nu).
\end{array}
$$

The bundle $\rho^*(\nu)_{|S(\nu)} \to S(\nu)$ has a tautological section s given by $v \mapsto v$, which satisfies $\|s(v)\| = \epsilon$ for all $v \in S(\nu)$. So

$$\rho^*(\nu)_{|S(\nu)} \to S(\nu) = \xi \oplus \mathcal{L},$$

where $\mathcal{L}$ is the trivial line subbundle spanned by s and ξ is its orthogonal complement. Under the identification $\nu \to \Delta$ is isomorphic to $\tau \to M$. Since M is orientable, so is ξ, and thus it is isomorphic to the quotient bundle

$$\rho^*(\tau_M)/\mathcal{L} \to S(\tau_M).$$

Let $\mathcal{F} := \phi^*(\xi)$. $\mathcal{F}$ is a rank $(n-1)$ subbundle of $r^*(\nu)_{|\partial U}$, and is isomorphic to $\rho^*(\tau_M)/\mathcal{L} \to S(\tau_M)$. Moreover, $\mathcal{F}_{|\partial U}$ is orientable.

Note that the restriction of ξ to each fiber $S(\nu_x)$ of $\rho : S(\nu) \to M$ is the tangent bundle τ_{n-1} of the sphere $S(\nu_x)$. Consequently, the bundle $\mathcal{F}$ when restricted to the fiber $r^{-1}(x)$ of the bundle $r : \partial U \to \Delta$ is isomorphic to τ_{n-1}. The following result is *key* tool in analyzing the topological diagonal properties:

Lemma 18. *Let M, Δ and U be as above. Set $X := M \times M \setminus Int(U)$. Then M has* (D_r) *if and only if the rank $(n-1)$ bundle $\mathcal{F} \to \partial U$ is isomorphic to the restriction to $\partial U = \partial X$ of a smooth rank $(n-1)$ bundle $\mathcal{G}$ on X. Moreover, M has (D_o) if and only if the bundle G can be chosen to be orientable.*

We list here some results from [8]:

1. S^n has (D_r) if and only if $n = 1, 2, 4$ or 8 (all except the first have (D_o)).
2. Let M be an almost complex manifold of $\dim_{\mathbb{C}} M = 2$. Then M has (D_c). (This is in contrast with the algebraic situation.)
3. Let M be an almost complex manifold of $\dim_{\mathbb{C}} M = 3$. Assume that $H^1(M, \mathbb{Z}) = 0$ and $H^2(M, \mathbb{Z}) = \mathbb{Z}$. Then if M satisfies (D_c), the second Stiefel-Whitney class $w_2(M)$ vanishes (i.e., M is spin).
4. Let $M \subset \mathbb{CP}^N$ be a smooth projective variety of $\dim_{\mathbb{C}} M = 3$. Assume that M is a strict complete intersection or a set-theoretically complete intersection with $H_1(M, \mathbb{Z}) = 0$. Then M has (D_c) if and only if M is spin.

5. Let M be a smooth strict complete intersection of $\dim_{\mathbb{C}} M = 3$ in $\mathbb{CP}^n$ with $M = X_1 \cap \cdots \cap X_{n-3}$ where X_i is a smooth hypersurfaces of degree d_i. Then M has (D_c) only if $(n + 1 - \sum d_i)$ is even. In particular, a smooth hypersurface M in $\mathbb{CP}^4$ has (D_c) only if it is of odd degree.

6. Let $m \geq 2$. Then a smooth quadric hypersurface $Q_{2m-1} \subset \mathbb{P}^{2m}$ has not (D_c).

Note that the converse of the third item is not true. For example, S^6 is an almost complex manifold which has a spin structure but fails to have (D_c) by the first item.

From the last item one can deduce that a smooth quadric hypersurface $Q_{2m-1} \subset \mathbb{P}^{2m}$ over any algebraically closed field, has not (D).

Conjecture 19. *Let Q_n be a smooth quadric hypersurface in $\mathbb{P}^{n+1}$ (over an algebraically closed field). Then Q_n has (D) if and only if $n = 1, 2$ or 4.*

8. Three other conjectures

1. From the results on surfaces, it follows that any smooth projective toric surface has (D). We conjecture that any smooth toric variety has (D).

2. The Grassmannian of Lagrangian 2-planes in $\mathbb{C}^4$ is identified with the quadric Q_3, which has not (D). We conjecture that, in general, the homogeneous spaces Sp_n/P and $SO(n)/P$ (P being any parabolic subgroup) have not (D).[2]

3. We conjecture that (D) fails for a general cubic threefold X, though X has (P) and even the $\mathcal{O}_X(1)$-point property ($\mathcal{O}_X(-1)$ is the unique c.t. line bundle on X) – cf. [8] for details.

Acknowledgments

The present article is basically a report on the joint work [8] with Vasudevan Srinivas and Vishwambhar Pati. It is our pleasure to acknowledge cordially this cooperation.

We thank the audience of our lectures both at the Banach Center in Warsaw, and METU in Ankara, for useful comments.

Finally, we are grateful to O. Ozturk for writing up these notes.

---

[2] The question: "Do the flag varieties for the symplectic and orthogonal groups have (D)?" arose in discussions of the author with W. Fulton while writing up [3] at the University of Chicago in 1996.

116   P. Pragacz

References

[1] O. Debarre, *Letter to the authors of* [8], 20.12.2006.

[2] W. Fulton, *Flags, Schubert polynomials, degeneracy loci, and determinantal formulas,* Duke. Math. J. **65** (1992), 381–420.

[3] W. Fulton, P. Pragacz, *Schubert varieties and degeneracy loci,* Lecture Notes in Math. **1689**, Springer-Verlag, Berlin, 1998.

[4] P.A. Griffiths, *Hermitian differential geometry, Chern classes, and positive vector bundles,* in: "Global Analysis" (D. Spencer and S. Iyanaga eds.), Princeton Math. Ser. **29**, Tokyo 1969, 185–251.

[5] A. Grothendieck, *La théorie des classes de Chern,* Bull. Soc. Math. France **86** (1958), 137–154.

[6] A. Lascoux, M.-P. Schützenberger, *Polynômes de Schubert,* C. R. Acad. Sci. Paris **294** (1982), 447–450.

[7] D. Mumford, *Abelian varieties,* Tata Inst. Studies in Math. **5**, Oxford Univ. Press, 1970.

[8] P. Pragacz, V. Srinivas, V. Pati, *Diagonal subschemes and vector bundles,* math.AG/ 0609381, to appear in the special volume of Quart. Pure Appl. Math., dedicated to Jean-Pierre Serre on his 80th Birthday.

[9] A. Sommese, *Submanifolds of abelian varieties,* Math. Ann. **233** (1978), 229–256.

[10] V. Srinivas, *Some applications of algebraic cycles to affine algebraic geometry,* this volume.

[11] L. Tu, *The connectedness of degeneracy loci,* in: "Topics in algebra" (S. Balcerzyk et al. eds.), Banach Center Publications **26(2)**, Warszawa 1990, 235–248.

Piotr Pragacz
Institute of Mathematics of Polish Academy of Sciences
Śniadeckich 8
PL-00-956 Warszawa, Poland
e-mail: P.Pragacz@impan.gov.pl

Algebraic Cycles, Sheaves, Shtukas, and Moduli
Trends in Mathematics, 117–129
© 2007 Birkhäuser Verlag Basel/Switzerland

Thom Polynomials of Invariant Cones, Schur Functions and Positivity

Piotr Pragacz and Andrzej Weber

Abstract. We generalize the notion of Thom polynomials from singularities of maps between two complex manifolds to invariant cones in representations, and collections of vector bundles. We prove that the generalized Thom polynomials, expanded in the products of Schur functions of the bundles, have nonnegative coefficients. For classical Thom polynomials associated with maps of complex manifolds, this gives an extension of our former result for stable singularities to nonnecessary stable ones. We also discuss some related aspects of Thom polynomials, which makes the article expository to some extent.

Mathematics Subject Classification (2000). 05E05, 14C17, 14N15, 55R40, 57R45.

Keywords. Thom polynomials, representations, orbits, invariant cones, Schur functions, globally generated and ample vector bundles, numerical positivity.

1. Introduction

The present paper is both of the research and expository character. It concerns global invariants for singularities. Our main new result here is Theorem 5 (see also Corollary 6 and 7).

To start with, we recall that the global behavior of singularities is governed by their *Thom polynomials* (cf. [33], [1], [17], and [32]). By a *singularity*, we shall mean in the paper a class of germs

$$(\mathbf{C}^m, 0) \to (\mathbf{C}^n, 0),$$

where $m, n \in \mathbf{N}$, which is closed under the right-left equivalence (i.e., analytic reparametrizations of the source and target).

Suppose that $f : M \to N$ is a map between complex manifolds, where $\dim(M) = m$ and $\dim(N) = n$. Let $V^\eta(f)$ be the cycle carried by the *closure* of the set

$$\{x \in M : \text{the singularity of } f \text{ at } x \text{ is } \eta\}. \tag{1}$$

Research of A. Weber supported by the KBN grant 1 P03A 005 26.

We recall that the *Thom polynomial* $\mathcal{T}^\eta$ of a singularity η is a polynomial in the formal variables

$$c_1, c_2, \ldots, c_m; \; c_1', c_2', \ldots, c_n',$$

such that after the substitution

$$c_i = c_i(TM), \quad c_j' = c_j(f^*TN), \tag{2}$$

$(i = 1, \ldots, m, \; j = 1, \ldots, n)$ for a general map $f : M \to N$ between complex manifolds, it evaluates the Poincaré dual[1] of $[V^\eta(f)]$. This is the content of the Thom theorem [33]. For a detailed discussion of the *existence* of Thom polynomials, see, e.g., [1]. Thom polynomials associated with group actions were studied by Kazarian in [17].

Recall that – historically – the first "Thom polynomial" appeared in the so-called "Riemann-Hurwitz formula". Let $f : M \to N$ be a general holomorphic map of compact Riemann surfaces. This means that f is a simple covering, that is, the critical points are nondegenerate and at most one appears in each fiber. Denoting by e_x the *ramification index* of f at $x \in M$ (i.e., the number of sheets of f meeting at x), the Riemann-Hurwitz formula asserts that

$$\sum_{x \in M} (e_x - 1) = 2g(M) - 2 - \deg(f)\big(2g(N) - 2\big). \tag{3}$$

Denoting by A_1 the singularity of $z \mapsto z^2$ at 0, this is equivalent to saying that the fundamental class of the *ramification divisor* of f,

$$\sum_x (e_x - 1)[x] = [V^{A_1}(f)],$$

where x runs over the set of critical points of f, is given by the following expression in the first Chern classes:

$$c_1(f^*TN) - c_1(TM) = c_1(f^*TN - TM). \tag{4}$$

In other words, the Riemann-Hurwitz formula says:

$$\mathcal{T}^{A_1} = c_1' - c_1. \tag{5}$$

For a wider discussion of the Riemann-Hurwitz formula and early history of Thom polynomials, we refer to Kleiman's survey article [18]. The Riemann-Hurwitz formula is true also in positive characteristic for finite separable morphisms of algebraic curves (cf. [15], Chap. IV, Sect. 2).

Thom [33] generalized the Riemann-Hurwitz formula to general maps $f : M \to N$ of complex manifolds with $n - m > 0$, the singularity being always A_1:

$$[V^{A_1}(f)] = c_{n-m+1}(f^*TN - TM) = \sum_{i=0}^{n-m+1} S_{n-m+1-i}(TM^*)c_i(f^*TN), \tag{6}$$

where S_j denotes the jth Segre class.

---

[1] In the following, we shall often omit the expression: "the Poincaré dual of".

Though for the singularity A_1 the Thom polynomials are rather simple, they start to be quite complicated even for simplest singularities coming "just after A_1", say (cf. the tables in [32]). For example, the Thom polynomial for the singularities A_4 is known only for small values of k (cf. [23]). Therefore, it is important to study the *structure* of Thom polynomials. It appears that a good tool for this task is provided by *Schur functions* [3], [26], [27], [28], [29]. Let us quote some results related to Schur function expansions of Thom polynomials of stable[2] singularities.

First, it was shown in [27] that if a representative of a stable singularity

$$\eta : (\mathbf{C}^m, 0) \to (\mathbf{C}^n, 0)$$

is of Thom-Boardman type Σ^i, then all summands in the Schur function expansion of $\mathcal{T}^\eta$ are indexed by partitions containing[3] the rectangle partition

$$(n - m + i, \ldots, n - m + i) \quad (i \text{ times}).$$

This is a consequence of the structure of the $\mathcal{P}$-*ideals of the singularities* Σ^i, which were introduced and investigated in [24]. Second, in [31], the authors proved that for any partition I the coefficient α_I in the Schur function expansion of the Thom polynomial

$$\mathcal{T}^\eta = \sum_I \alpha_I S_I(TM^* - f^*TN^*)$$

is nonnegative. This result was conjectured before in [3] and independently in [26]. It appears to be a consequence of the Fulton-Lazarsfeld theory of numerical positivity of cones in ample vector bundles [8] (cf. also [21, §8]), combined with a functorial version of the bundles of jets, appearing in the approach to Thom polynomials via classifying spaces of singularities [17].

In the present paper, we shall prove a more general result that – we believe – will better explain a reason of the positivity in the above classical case, as well as in many other situations. To this end, we extend the definition of Thom polynomials from the singularities of maps $f : M \to N$ of complex manifolds [33] to the invariant cones in representations of the product of general linear groups

$$\prod_{i=1}^{p} GL_{n_i} .$$

Such Thom polynomials are naturally defined on p-tuples of vector bundles of ranks n_i. It is convenient to pass to *topological homotopy category*, where each p-tuple of bundles can be pulled back from the universal p-tuple of bundles on the

---

[2]By a *stable* singularity we mean an equivalence class of stable germs $(\mathbf{C}^\bullet, 0) \to (\mathbf{C}^{\bullet+k}, 0)$, where $\bullet \in \mathbf{N}$, under the equivalence generated by right-left equivalence and suspension (by suspension of a germ κ we mean its trivial unfolding: $(x, v) \mapsto (\kappa(x), v)$). For a stable singularity, its Thom polynomial is of the form $\sum_I \alpha_I S_I(TM^* - f^*TN^*)$, where S_I denotes a Schur function, cf. Sections 3 and 5.

[3]We say that one partition *is contained* in another if this holds for their Young diagrams.

product of p classifying spaces

$$\prod_{i=1}^{p} BGL_{n_i}.$$

Suppose that the functor associated with such a representation preserves global generateness. Our main result – Theorem 5 – then asserts that the Thom polynomial for a p-tuple of vector bundles $(E_1, E_2, \ldots, E_p)$, when expanded in the basis

$$\{S_{I_1}(E_1) \cdot S_{I_2}(E_2) \; \cdots \; S_{I_p}(E_p)\}$$

of products of Schur functions applied to the successive bundles, has *nonnegative* coefficients. The key tool is *positivity of cone classes* for globally generated vector bundles combined with the *Giambelli formula*. For a polynomial representation of $\prod_{i=1}^{p} GL_{n_i}$ of positive degree, we get, in addition, that the sum of the coefficients is *positive* (cf. Corollary 6).

Theorem 5, in the classical situation of singularities of maps $f : M \to N$ between complex manifolds, implies that for a given singularity its Thom polynomial, when expanded in the basis

$$S_I(TM^*) \cdot S_J(f^*TN),$$

has nonnegative coefficients (cf. Corollary 7).

We also note that Theorem 5 implies the main result of our former paper [31] for Thom polynomials of *stable* of singularities of maps between complex manifolds, where, however, the Schur functions *in difference of bundles* were used (cf. Theorem 8).

2. Thom polynomials of invariant cones

In this section, we define "generalized Thom polynomials". Our construction is modeled on that used to the construction of classical Thom polynomials with the help of the "classifying spaces of singularities" (cf., e.g., [17]).

Suppose that $(n_1, n_2, \ldots, n_p) \in \mathbf{N}^p$ and that V is a representation of

$$G = \prod_{i=1}^{p} GL_{n_i}. \tag{7}$$

The representation V gives rise to a *functor* ϕ defined for a collection of bundles on a variety X:

$$E_1, E_2, \ldots, E_p \mapsto \phi(E_1, E_2, \ldots, E_p),$$

with $\dim E_i = n_i$, $i = 1, \ldots, p$. By passing to the dual bundles, we may assume that the functor ϕ is covariant in each variable.

Let

$$P(E_\bullet) = P(E_1, E_2, \ldots, E_p) \tag{8}$$

be the principal G-bundle associated with the bundles $E_1, E_2, \ldots, E_p$. We define a new vector bundle:

$$V(E_\bullet) = V(E_1, E_2, \ldots, E_p) := P(E_\bullet) \times_G V . \tag{9}$$

Suppose now that a G-invariant cone $\Sigma \subset V$ is given. We set

$$\Sigma(E_\bullet) = \Sigma(E_1, E_2, \ldots, E_p) := P(E_\bullet) \times_G \Sigma \subset V(E_\bullet) . \tag{10}$$

We define the "Thom polynomial" $\mathcal{T}^\Sigma$ to be the dual class[4] of

$$[\Sigma(R^{(1)}, \ldots, R^{(p)})] \in H^*\big(V(R^{(1)}, \ldots, R^{(p)}), \mathbf{Z}\big) = H^*(BG, \mathbf{Z}) ,$$

where $R^{(i)}$, $i = 1, \ldots, p$, is the pullback of the tautological vector bundle from BGL_{n_i} to

$$BG = \prod_{i=1}^{p} BGL_{n_i} .$$

Then, the so-defined Thom polynomial

$$\mathcal{T}^\Sigma \in H^*(BG, \mathbf{Z})$$

depends on the Chern classes of the $R^{(i)}$'s.

We shall write $\mathcal{T}^\Sigma(E_1, \ldots, E_p)$ for the Thom polynomial $\mathcal{T}^\Sigma$, with $c_j(R^{(i)})$ replaced by $c_j(E_i)$ for $i = 1, \ldots, p$.

Lemma 1. *For any vector bundles $E_1, E_2, \ldots, E_p$ on a variety X, the dual class[5] of $[\Sigma(E_\bullet)]$ in*

$$H^{2\operatorname{codim}(\Sigma)}(V(E_\bullet), \mathbf{Z}) = H^{2\operatorname{codim}(\Sigma)}(X, \mathbf{Z})$$

is equal to $\mathcal{T}^\Sigma(E_1, \ldots, E_p)$.

Proof. Each p-tuple of bundles can be pulled back from the universal p-tuple $(R^{(1)}, R^{(2)}, \ldots, R^{(p)})$ of bundles on BG using a C^∞-map. It is possible to work entirely with the algebraic varieties and maps. One can use the Totaro construction and representability for affine varieties ([34, proof of Theorem 1.3]).

Remark 2. In the situation of classical Thom polynomials [33], the functor ϕ is the functor of k-jets :

$$(E, F) \mapsto \mathcal{J}^k(E, F) = \left(\bigoplus_{i=1}^{k} \operatorname{Sym}^i E^* \right) \otimes F ,$$

where k is large enough, adapted to the investigated class of singularities – cf. [31] for details and applications. (We note that in this situation an invariant closed subset Σ, called in [31] a "class of singularities", is automatically a cone.)

---

[4] Compare the footnote 4 in [31].
[5] The meaning of the "dual class" for singular X is explained in [31], Note 6.

3. Schur functions and the Giambelli formula

In this section, we recall the notion of *Schur functions*. We also recall a geometric interpretation of them, namely the classical *Giambelli formula*.

Given a partition $I = (i_1, i_2, \ldots, i_l) \in \mathbf{N}^l$, where

$$i_1 \geq i_2 \geq \cdots \geq i_l \geq 0 \ ^6,$$

and vector bundles E and F on a variety X, the *Schur function*[7] $S_I(E - F)$ is defined by the following determinant:

$$S_I(E - F) = \left| S_{i_p - p + q}(E - F) \right|_{1 \leq p, q \leq l}, \tag{11}$$

where the entries are defined by the expression

$$\sum S_i(E - F) = \prod_b (1 - b) / \prod_a (1 - a). \tag{12}$$

Here, the a's and b's are the Chern roots of E and F and the LHS of Eq. (12) is the *Segre class* of the virtual bundle $E - F$. So the Schur functions $S_I(E - F)$ lie in a ring containing the Chern classes of E and F; e.g., we can take the cohomology ring $H^*(X, \mathbf{Z})$ or the Chow ring $A^*(X)$.

Given a vector bundle E and a partition I, we shall write $S_I(E)$ for $S_I(E - 0)$, where 0 is the zero vector bundle.

We refer to [20], [22], and [30] for the theory of Schur functions $S_I(E)$ and $S_I(E - F)$.

Given a smooth variety X, we shall identify its cohomology $H^*(X, \mathbf{Z})$ with its homology $H_*(X, \mathbf{Z})$, as is customary. More precisely, this identification is realized via capping the cohomology classes with the fundamental class $[X]$ of X, using the standard map:

$$\cap : H^*(X, \mathbf{Z}) \otimes H_*(X, \mathbf{Z}) \to H_*(X, \mathbf{Z}).$$

Let V be a complex vector space of dimension N, and let $G_m(V)$ be the *Grassmannian parametrizing* m-dimensional subspaces of V. On knows that $G_m(V)$ is a smooth projective variety of dimension mn, where $n = N - m$. We shall also use the notation $G^n(V)$ for this Grassmannian. The Grassmannian $G_m(V)$ is stratified by Schubert cells; the closures of these cells are Schubert varieties $\Omega_I(V_\bullet)$, where

$$I = (n \geq i_1 \geq i_2 \geq \cdots \geq i_m \geq 0)$$

is a partition, and

$$V_\bullet : 0 = V_0 \subset V_1 \subset \cdots \subset V_N = V$$

is a complete flag of subspaces of V, with $\dim V_j = j$ for $j = 0, 1, \ldots, N$.

---

[6] Since the most common references to Schubert Calculus use weakly decreasing partitions, we follow this convention in the present paper.

[7] Usually this family of functions is called "super Schur functions" or "Schur functions in difference of bundles".

The precise definition of $\Omega_I(V_\bullet)$ is

$$\Omega_I(V_\bullet) = \{\Lambda \in G_m(V) : \dim(\Lambda \cap V_{n+j-i_j}) \geq j,\ j = 1,\ldots,m\}. \qquad (13)$$

This is a subvariety of codimension $|I| = i_1 + i_2 + \cdots + i_m$ in $G_m(V)$. The cohomology class $[\Omega_I(V_\bullet)]$ does not depend on a flag $V_\bullet$. We denote it by σ_I and call a *Schubert class*.

Let Q denote the tautological quotient bundle on $G_m(V)$. Then

$$\sigma_{(i)} = c_i(Q) = S_{(1,\ldots,1)}(Q),$$

where 1 appears i times, and – more generally – the following *Giambelli formula* [10] holds:

Proposition 3. *In the cohomology ring of $G_m(V)$, we have*

$$\sigma_I = \left| c_{i_p - p + q}(Q) \right|_{1 \leq p,q \leq m} = S_{I^\sim}(Q), \qquad (14)$$

where $I^\sim$ is the conjugate partition of I (i.e., the consecutive rows of the diagram of $I^\sim$ are the transposed consecutive columns of the diagram of I).

(Cf. [12, Chap. 1, Sect. 5], [6, §9.4]).

Given a partition I, consider the partition

$$J = (n - i_m, n - i_{m-1}, \ldots, n - i_1).$$

Then (*loc. cit.*) σ_J is the unique Schubert class of complementary codimension to σ_I whose intersection with σ_I is nonzero, and in fact

$$\int_{G_m(V)} \sigma_I \cdot \sigma_J = 1. \qquad (15)$$

The class σ_J is called the *complementary class* to σ_I.

4. Cone classes for globally generated and ample vector bundles

In the proof of our main result, we shall use the following results of Fulton and Lazarsfeld from [7], [8] (cf. also [5, Chap. 12], [21, §8]). Recall first some classical definitions and facts from [5] (we shall also follow the notation from this book). Let E be a vector bundle of rank e on X. By a *cone* in E we mean a subvariety of E which is stable under the natural $\mathbb{G}_m$-action on E. If $C \subset E$ is a cone of pure dimension d, then one may intersect its cycle $[C]$ with the zero-section of the vector bundle:

$$z(C, E) := s_E^*([C]) \in A_{d-e}(X), \qquad (16)$$

where $s_E^* : A_d(E) \to A_{d-e}(X)$ is the Gysin map determined by the zero-section $s_E : X \to E$. For a projective variety X, there is a well-defined *degree* map $\int_X : A_0(X) \to \mathbf{Z}$.

The following results stem from [7, Theorem 1 (A)] and [8, Theorem 2.1].

Theorem 4. *Suppose that E is a vector bundle of rank e on a projective variety X, and let $C \subset E$ be a cone of pure dimension e.*

(1) *If some symmetric power of E is globally generated, then*

$$\int_X z(C, E) \geq 0.$$

(2) *If E is ample, then*

$$\int_X z(C, E) > 0.$$

Under the assumptions of the theorem, we also have in $H_0(X, \mathbf{Z})$ the homology analog of $z(C, E)$, denoted by the same symbol, and the homology degree map $\deg_X : H_0(X, \mathbf{Z}) \to \mathbf{Z}$. They are compatible with their Chow group counterparts via the cycle map: $A_0(X) \to H_0(X, \mathbf{Z})$ (cf. [5, Chap. 19]). We thus have the same inequalities for $\deg_X\big(z(C, E)\big)$.

5. Schur function expansions of Thom polynomials

We follow the setting from Section 2. Since the Schur functions form an additive basis of the ring of symmetric functions, the Thom polynomial $\mathcal{T}^\Sigma$ is uniquely written in the following form:

$$\mathcal{T}^\Sigma = \sum \alpha_{I_1,\ldots,I_p} \, S_{I_1}(R^{(1)}) \, S_{I_2}(R^{(2)}) \, \cdots \, S_{I_p}(R^{(p)}), \tag{17}$$

where $\alpha_{I_1,\ldots,I_p}$ are integer coefficients.

We say that the functor ϕ, associated with a representation V, *preserves global generateness* if for a collection of globally generated vector bundles $E_1, E_2, \ldots, E_p$, the bundle

$$\phi(E_1, E_2, \ldots, E_p)$$

is globally generated.

Examples of functors preserving global generateness over fields of characteristic zero are *polynomial functors*. They are, at the same time, quotient functors and subfunctors of the tensor power functors (cf. [14]). On the other hand, the functors: $\mathrm{Hom}(-, E)$ with fixed E, or $\mathrm{Hom}(-, -)$, do not preserve global generateness.

The main result of the present paper is

Theorem 5. *Suppose that the functor ϕ preserves global generateness. Then the coefficients $\alpha_{I_1,\ldots,I_p}$ in Eq. (17) are nonnegative. Assume additionally that there exists a projective variety X [8] of dimension greater than or equal to $\mathrm{codim}(\Sigma)$, and there exist vector bundles $E_1, \ldots, E_p$ on X such that the bundle $\phi(E_1, E_2, \ldots, E_p)$ is ample. Then at least one of the coefficients $\alpha_{I_1,\ldots,I_p}$ is positive.*

---

[8]The variety X can be singular.

Proof. We assume for simplicity that $p = 2$ (the reasoning in general case goes in the same way). We want to estimate the coefficients α_{IJ} in the universal expansion into products of Schur functions:

$$\mathcal{T}^\Sigma(E_1, E_2) = \sum_{I,J} \alpha_{IJ}\, S_I(E_1) \cdot S_J(E_2) \tag{18}$$

Let E_1 and E_2 be the pullbacks of the tautological quotient bundles from the Grassmannians $G^{n_1}(\mathbf{C}^{N_1})$ and $G^{n_2}(\mathbf{C}^{N_2})$ to

$$G^{n_1}(\mathbf{C}^{N_1}) \times G^{n_2}(\mathbf{C}^{N_2}),$$

where N_1 and N_2 are sufficiently large. It is enough to estimate the coefficients α_{IJ} for such E_1 and E_2. Let $\sigma_K \in H^*(G^{n_1}(\mathbf{C}^{N_1}), \mathbf{Z})$ be the complementary class to $\sigma_{I\sim}$ and $\sigma_L \in H^*(G^{n_2}(\mathbf{C}^{N_2}), \mathbf{Z})$ be the complementary class to $\sigma_{J\sim}$. By the Giambelli formula (Proposition 3) and properties of complementary Schubert classes (15), we have

$$\alpha_{IJ} = \int_{G^{n_1}(\mathbf{C}^{N_1}) \times G^{n_2}(\mathbf{C}^{N_2})} \mathcal{T}^\Sigma(E_1, E_2) \cdot (\sigma_K \times \sigma_L).$$

The vector bundles E_1 and E_2 are globally generated. Hence, by the assumption, the bundle $\phi(E_1, E_2)$ is globally generated. By Theorem 4(1), we thus have $\alpha_{IJ} \geq 0$.

Now, suppose that there exists a projective variety of dimension greater than or equal to $\mathrm{codim}(\Sigma)$, and there exist vector bundles E_1, E_2 on X such that the bundle $\phi(E_1, E_2)$ is ample. Let Y be a subvariety of X of dimension equal to $\mathrm{codim}(\Sigma)$. Then, by Theorem 4(2), we have

$$\int_Y \mathcal{T}^\Sigma(E_1, E_2) > 0.$$

Therefore, $\mathcal{T}^\Sigma \neq 0$, which implies that at least one of the coefficients α_{IJ} is positive.

Consider now the projective variety

$$X = \prod_{i=1}^p G^{n_i}(\mathbf{C}^N),$$

where N is sufficiently large. We denote by Q_i the pullback to X of the tautological quotient bundle on $G^{n_i}(\mathbf{C}^N)$, $i = 1, \ldots, p$. The bundle Q_i is not ample, but it is globally generated. Let L be an ample line bundle on X. Then each bundle

$$E_i = Q_i \otimes L$$

is ample (cf. [14]).

Observe the hypotheses of the theorem are satisfied by the variety X, vector bundles $E_1, \ldots, E_p$, and any polynomial functor ϕ of positive degree. We thus obtain

Corollary 6. *If ϕ is a polynomial functor of positive degree, then the coefficients $\alpha_{I_1,\ldots,I_p}$ in Eq. (17) are nonnegative, and their sum is positive.*

In the next corollary, we use the concept of a classical Thom polynomial associated with a map $f : M \to N$ of complex manifolds and a nontrivial class of singularities Σ (cf. [31]). We *do not*, however, assume now that Σ is stable.

By the theory of Schur functions, there exist universal coefficients $\beta_{IJ} \in \mathbf{Z}$ such that

$$\mathcal{T}^{\Sigma} = \sum_{I,J} \beta_{IJ} S_I(TM^*) \cdot S_J(f^*TN). \tag{19}$$

The following result follows from Theorem 5.

Corollary 7. *For any pair of partitions I, J, we have $\beta_{IJ} \geq 0$.*

We also give an alternative proof of the main result from [31]. Let Σ be a *stable* singularity. Then by the Thom-Damon theorem ([33], [2]),

$$\mathcal{T}^{\Sigma}(c_1(M), \dots, c_m(M), c_1(N), \dots, c_n(N))$$

is a universal polynomial in

$$c_i(TM - f^*TN) \quad \text{where} \quad i = 1, 2, \dots$$

(Cf. also [17, Theorem 2].)

Using the theory of supersymmetric functions (cf. [20], [22], [30]), the Thom-Damon theorem can be rephrased by saying that there exist coefficients $\alpha_I \in \mathbf{Z}$ such that

$$\mathcal{T}^{\Sigma} = \sum_{I} \alpha_I S_I(TM^* - f^*TN^*), \tag{20}$$

the sum is over partitions I with $|I| = \operatorname{codim}(\Sigma)$. The expression in Eq. (20) is unique (*loc. cit.*).

Theorem 8. *Let Σ be a stable singularity. Then for any partition I the coefficient α_I in the Schur function expansion of the Thom polynomial $\mathcal{T}^{\Sigma}$ (cf. Eq. (20)) is nonnegative.*

Proof. By the theory of Schur functions (*loc. cit.*), we have that the coefficient of $S_I(TM^* - f^*TN^*)$ in the RHS of (20) is equal to the coefficient of $S_I(TM^*)$ in the RHS of (19), that is, $\alpha_I = \beta_{I,\emptyset}$ for any partition I. The assertion now follows from Corollary 7.

Remark 9. Note that Theorem 5 overlaps various situations already studied in the literature. Consider, e.g., a *family of quadratic forms* on the tangent bundle of an m-fold M with values in a line bundle L, i.e., a section of

$$\operatorname{Hom}(\operatorname{Sym}^2(TM), L).$$

The singularities of such forms lead to Thom polynomials. The group which is relevant here is $GL_m \times GL_1$ with the natural representation in the vector space

$$\bigoplus_{i=0}^{r} \operatorname{Sym}^i(\mathbf{C}^m) \otimes \operatorname{Hom}(\operatorname{Sym}^2(\mathbf{C}^m), \mathbf{C}).$$

The singularity classes defined by the 0th jet are just invariant subsets of

$$\mathrm{Hom}(\mathrm{Sym}^2(\mathbf{C}^m), \mathbf{C}).$$

The corank of the quadratic form determines the singularity class.

We recover[9] the situation already described in the literature in the context of *degeneracy loci formulas* for morphisms with symmetries of rank m bundles:

$$E^* \to E \otimes L.$$

The degrees of projective symmetric varieties were computed in [11]. The Schur function formulas for a trivial L were given in [13], [16]. To give the formulas in full generality [25], we consider, for partitions $I = (i_1, \ldots, i_m)$ and $J = (j_1, \ldots, j_m)$, the following determinant studied in the paper [19] of Lascoux:

$$d_{I,J} = \left| \binom{i_a + m - a}{j_b + m - b} \right|_{1 \le a,b \le m}. \tag{21}$$

Then, the Thom polynomial associated with the locus of quadratic forms whose corank $\ge q$ is equal to

$$2^{-\binom{q}{2}} \sum_J 2^{|J|} \, d_{\rho_q, J} \, S_J(E) \cdot S_{\binom{q+1}{2} - |J|}(L),$$

where $J = (j_1, \ldots, j_q)$ runs over partitions contained in the partition

$$\rho_q = (q, q - 1, \ldots, 1).$$

(Cf. [25] for details. Similar formulas are valid for antisymmetric forms (*loc.cit.*).) In particular, we obtain the positivity of the $d_{\rho_q, J}$'s – a result known before by combinatorial methods (cf. [9]).

It seems to be interesting to apply Theorem 5 to other concrete situations, where Thom polynomials of invariant cones appear.

References

[1] V. Arnold, V. Vasilev, V. Goryunov, O. Lyashko: *Singularities. Local and global theory,* Enc. Math. Sci. Vol. **6** (Dynamical Systems VI), Springer, 1993.

[2] J. Damon, *Thom polynomials for contact singularities,* Ph.D. Thesis, Harvard, 1972.

[3] L. Feher, B. Komuves, *On second order Thom-Boardman singularities,* Fund. Math. **191** (2006), 249–264.

[4] L. Feher, R. Rimanyi, *Calculation of Thom polynomials and other cohomological obstructions for group actions,* in: "Real and complex singularities (São Carlos 2002)" (T. Gaffney and M. Ruas eds.), Contemporary Math. **354**, (2004), 69–93.

[5] W. Fulton, *Intersection theory,* Springer, 1984.

[6] W. Fulton, *Young tableaux,* Cambridge University Press, 1997.

[7] W. Fulton, R. Lazarsfeld, *Positivity and excess intersections,* in "Enumerative and classical geometry", Nice 1981, Progress in Math. **24**, Birkhäuser (1982), 97–105.

---

[9]See also [4].

[8] W. Fulton, R. Lazarsfeld, *Positive polynomials for ample vector bundles*, Ann. of Math. **118** (1983), 35–60.

[9] I. Gessel, X. Viennot, *Binomial determinants, paths and hook length formulae*, Adv. Math. **58** (1985), 300–321.

[10] G.Z. Giambelli, *Risoluzione del problema degli spazi secanti*, Mem. Accad. Sci. Torino (2) **52** (1902), 171–211.

[11] G.Z. Giambelli, *Sulle varietá rappresentata coll'annullare determinanti minori contenuti in un determinante simmetrico od emisimmetrico generico di forme*, Atti della R. Accad. delle Scienze di Torino **44** (1906), 102–125.

[12] P.A. Griffiths, J. Harris, *Principles of algebraic geometry*, Wiley & Sons Inc., 1978.

[13] J. Harris, L. Tu, *On symmetric and skew-symmetric determinantal varieties*, Topology **23** (1984), 71–84.

[14] R. Hartshorne, *Ample vector bundles,* Publ. Math. IHES **29** (1966), 63–94.

[15] R. Hartshorne, *Algebraic geometry*, Springer, 1977.

[16] T. Józefiak, A. Lascoux, P. Pragacz, *Classes of determinantal varieties associated with symmetric and skew-symmetric matrices*, Math. USSR Izv. **18** (1982), 575–586.

[17] M.E. Kazarian, *Classifying spaces of singularities and Thom polynomials*, in: "New developments in singularity theory", NATO Sci. Ser. II Math. Phys. Chem., **21**, Kluwer Acad. Publ. (2001), 117–134.

[18] S. Kleiman, *The enumerative theory of singularities*, in: "Real and complex singularities, Oslo 1976" (P. Holm ed.) Sijthoff&Noordhoff Int. Publ. (1978), 297–396.

[19] A. Lascoux, *Classes de Chern d'un produit tensoriel*, C.R. Acad. Sci. Paris **286** (1978), 385–387.

[20] A. Lascoux, *Symmetric functions and combinatorial operators on polynomials*, CBMS/AMS Lectures Notes **99**, Providence, 2003.

[21] R. Lazarsfeld, *Positivity in algebraic geometry*, Springer, 2004.

[22] I.G. Macdonald, *Symmetric functions and Hall polynomials*, 2nd edition, Oxford University Press, 1995.

[23] O. Ozturk, *On Thom polynomials for $A_4(-)$ via Schur functions*, Serdica Math. J. **33** (2007), 301–320.

[24] P. Pragacz, *Enumerative geometry of degeneracy loci*, Ann. Sc. Ec. Norm. Sup. **21** (1988), 413–454.

[25] P. Pragacz, *Cycles of isotropic subspaces and formulas for symmetric degeneracy loci*, in: "Topics in algebra" (S. Balcerzyk et al. eds.), Banach Center Publ. **26(2)**, 1990, 189–199.

[26] P. Pragacz, *Thom polynomials and Schur functions I*, math.AG/0509234.

[27] P. Pragacz, *Thom polynomials and Schur functions: the singularities $I_{2,2}(-)$*, Ann. Inst. Fourier **57** (2007), 1487–1508.

[28] P. Pragacz, *Thom polynomials and Schur functions: towards the singularities $A_i(-)$*, Preprint MPIM Bonn 2006 (139) – to appear in the Proceedings of the 9th Workshop on Real and Complex Singularities, São Carlos 2006, Contemporary Mathematics AMS.

[29] P. Pragacz, *Thom polynomials and Schur functions: the singularities $A_3(-)$*, in preparation.

[30] P. Pragacz, A. Thorup, *On a Jacobi-Trudi identity for supersymmetric polynomials*, Adv. in Math. **95** (1992), 8–17.

[31] P. Pragacz, A. Weber, *Positivity of Schur function expansions of Thom polynomials*, Fund. Math. **195** (2007), 85–95.

[32] R. Rimanyi, *Thom polynomials, symmetries and incidences of singularities*, Inv. Math. **143** (2001), 499–521.

[33] R. Thom, *Les singularités des applications différentiables*, Ann. Inst. Fourier **6** (1955–56), 43–87.

[34] B. Totaro, *The Chow ring of a classifying space* in: "Algebraic K-theory" (W. Raskind et al. eds.), Symp. Pure Math. **67** (1999), AMS, 249–281.

Piotr Pragacz
Institute of Mathematics of Polish Academy of Sciences
Śniadeckich 8
PL-00-956 Warszawa, Poland
e-mail: `P.Pragacz@impan.gov.pl`

Andrzej Weber
Department of Mathematics of Warsaw University
Banacha 2
PL-02-097 Warszawa, Poland
e-mail: `aweber@mimuw.edu.pl`

Algebraic Cycles, Sheaves, Shtukas, and Moduli
Trends in Mathematics, 131–183

Geometric Invariant Theory Relative to a Base Curve

Alexander H.W. Schmitt

Dedicated to Andrzej Białynicki-Birula on the occasion of his 70th birthday

Abstract. These are the lecture notes to the author's course "A relative version of Geometric Invariant Theory" taught during the mini-school "Moduli spaces" at the Banach Center in Warsaw which took place in April 2005.

We give an account of old and new results in Geometric Invariant Theory and present recent progress in the construction of moduli spaces of vector bundles and principal bundles with extra structure (called "augmented" or "decorated" vector/principal bundles).

Contents

Introduction

Suppose that X is a complex projective manifold, that G is a connected reductive linear algebraic group, and that $\varrho\colon G \longrightarrow \mathrm{GL}(V)$ is a representation of G. Using ϱ, we may associate to any principal G-bundle $\mathscr{P}$ on X a vector bundle $\mathscr{P}_\varrho$ with fiber V. We would like to study ϱ-pairs, i.e., triples $(\mathscr{P}, \mathscr{L}, \varphi)$ which are composed of a principal G-bundle on X, a line bundle $\mathscr{L}$ on X, and a homomorphism $\varphi\colon \mathscr{P}_\varrho \longrightarrow \mathscr{L}$. There is a natural equivalence relation on the set of all ϱ-pairs. The topological background data of a ϱ-pair are the element $\tau \in \pi_1(G)$ that classifies the topological principal G-bundle underlying $\mathscr{P}$ and the degree d of the line bundle $\mathscr{L}$. The precise program that we would like to carry out is the following:

- Define a (parameter dependent) notion of (semi)stability for ϱ-pairs.
- Show that, for $\tau \in \pi_1(G)$ and $d \in \mathbb{Z}$, the equivalence classes of stable ϱ-pairs $(\mathscr{P}, \mathscr{L}, \varphi)$, such that $\mathscr{P}$ has the topological type τ and the degree of $\mathscr{L}$ is d, are parameterized by a quasi-projective moduli scheme $\mathscr{M}$.
- Show that $\mathscr{M}$ may be compactified by a scheme $\overline{\mathscr{M}}$ whose points parameterize the semistable ϱ-pairs with the given topological background data with respect to some coarser equivalence relation, usually called S-equivalence.

The above problem has a considerable history in the mathematical literature (which we are not going to trace back). The motivation to study it ranges from such different fields as:

- Classification of complex algebraic varieties.
- Investigation of the (real analytic) spaces of representations of the fundamental group of X in a real form of G.
- Differential topology of (real) 4-manifolds.

The second and third topic belong to the larger field of gauge theory. If one uses that field as motivation, one finds ϱ-pairs which satisfy a canonical stability condition (see [26] for a state of the art account of these questions). The reader may also consult [8] and [37] for more examples and references.

So far, the program that we have formulated above has been worked out case by case for many different representations, mostly of $GL_r(\mathbb{C})$ or $GL_{r_1}(\mathbb{C}) \times \cdots \times GL_{r_t}(\mathbb{C})$. The construction of the moduli spaces in those examples follows a certain pattern which generalizes the strategy applied in the construction of the moduli spaces of semistable vector bundles and crucially uses Mumford's Geometric Invariant Theory (GIT). Furthermore, gauge theory has suggested a notion of semistability for the above ϱ-pairs. In view of these achievements, one would expect that one may perform the program that we have formulated in full generality, i.e., one should find a theory, such that, if the input data X, G, and ϱ are given, it puts out the semistability condition and grants the existence of the moduli spaces. For $GL_r(\mathbb{C})$, this is indeed possible, if one restricts to homogeneous representations (see [37] and [12]). It has also been completed for $GL_{r_1}(\mathbb{C}) \times \cdots \times GL_{r_t}(\mathbb{C})$ [40], although that case is already quite tricky. Before one may proceed to other structure groups, one needs an efficient formalism to deal with the principal bundles themselves. This has been conceived only recently in [36] and [38]. If one puts all these findings together, one may foresee the general solution.

We finally point out that the case of $X = \{pt\}$ as the base manifold is the investigation of the G-action on the projective space $\mathbb{P}(V)$ that results from the representation ϱ. Therefore, the above program is not only a sophisticated application of GIT but also a formal generalization of it. This is why we have dared to speak about a "relative version" (namely relative to X) of GIT.

The aim of the lectures and these notes was or is to introduce the reader to this kind of questions. Since GIT plays such a crucial role both as the main technical tool which will be applied in the proofs and as an important source of intuition, the first three lectures focus on GIT. We not only discuss the fundamental results from Hilbert to Mumford but also some more recent developments concerning the variation of GIT quotients (which has its counterpart in the setting of moduli spaces). In the last two lectures, we pass to the theory of ϱ-pairs, mainly for the structure group $GL_r(\mathbb{C})$. The base manifold will be a curve.

The notes are a slightly modified version of the slides that the author used during the oral presentation, i.e., those slides were supplemented by additional comments and references in order to fulfill the standards of publication in a proceedings volume. The author's original intent was to include new research results (namely for general structure groups) with complete proofs. Unfortunately, that project has grown out of size and will have to await another occasion for publication (see [41]).

1. Lecture I: Algebraic groups and their representations

As explained in the introduction, Geometric Invariant Theory deals with the actions of certain algebraic groups on algebraic varieties and the possibility of forming appropriate quotients. Thus, in a first step, we will have to introduce the necessary notions from the theory of (linear) algebraic groups. This will be done in the first section of this chapter. The most important class of actions one has to understand for studying Geometric Invariant Theory are linear actions of affine algebraic groups on vector spaces. These actions are representations of the corresponding algebraic groups and are presented in the second section. In the final section, we begin the investigation of the problem of forming quotients of vector spaces by linear actions of an algebraic group.

1.1. Basic definitions

The theory of algebraic groups is an important field of Algebraic Geometry in its own right. Standard references which include proofs of all the claims made below are the books [6], [20], and [46]. The more courageous reader may directly refer to SGA 3.

Definition 1.1.1. i) A **linear** or **affine algebraic group** is a tuple $(G, e, \mu, \mathrm{inv})$ with

- G an affine algebraic variety,
- $e \in G$, the **neutral element**,
- $\mu \colon G \times G \longrightarrow G$ a regular map, the **multiplication**, and
- $\mathrm{inv} \colon G \longrightarrow G$ a regular map, the **inversion**,

such that the axioms of a group are satisfied, i.e., the following diagrams are commutative:

$$
(\mathrm{Ass}) \quad
\begin{array}{ccc}
G \times G \times G & \xrightarrow{\mu \times \mathrm{id}_G} & G \times G \\
{\scriptstyle \mathrm{id}_G \times \mu}\downarrow & & \downarrow{\scriptstyle \mu} \\
G \times G & \xrightarrow{\mu} & G
\end{array}
\quad ; \quad
(\mathrm{Id}) \quad
\begin{array}{ccc}
G & \xrightarrow{(e_G, \mathrm{id}_G)} & G \times G \\
{\scriptstyle (\mathrm{id}_G, e_G)}\downarrow & \searrow{\scriptstyle \mathrm{id}_G} & \downarrow{\scriptstyle \mu} \\
G \times G & \xrightarrow{\mu} & G
\end{array}
\quad ;
$$

$$
(\mathrm{Inv}) \quad
\begin{array}{ccc}
G & \xrightarrow{(\mathrm{inv}, \mathrm{id}_G)} & G \times G \\
{\scriptstyle (\mathrm{id}_G, \mathrm{inv})}\downarrow & \searrow{\scriptstyle e_G} & \downarrow{\scriptstyle \mu} \\
G \times G & \xrightarrow{\mu} & G
\end{array}
\quad .
$$

In these diagrams, $e_G \colon G \longrightarrow G$ stands for the constant morphism $g \longmapsto e$, $g \in G$.

ii) Let G and H be linear algebraic groups. A **homomorphism from G to H** is a regular map $h\colon G \longrightarrow H$ which is at the same time a group homomorphism, i.e., the following diagram is commutative:

$$
\begin{array}{ccc}
G \times G & \xrightarrow{\ h \times h\ } & H \times H \\
\Big\downarrow{\scriptstyle \mu_G} & & \Big\downarrow{\scriptstyle \mu_H} \\
G & \xrightarrow{\quad h \quad} & H
\end{array}
$$

iii) A **(closed) subgroup** H of an algebraic group G is a closed subvariety of G which is also a subgroup.

Remark 1.1.2. A linear algebraic group G is non-singular as an algebraic variety.

Example 1.1.3. i) One checks that the kernel of a homomorphism $h\colon G \longrightarrow H$ between linear algebraic groups is an example for a subgroup of G.

ii) The **general linear group** $\mathrm{GL}_n(\mathbb{C})$ is a linear algebraic group:

- $\mathrm{GL}_n(\mathbb{C})$ is the *open* subvariety $\{\det \neq 0\} \subset M_n(\mathbb{C})$. We have the morphism

$$
\begin{array}{rccc}
\alpha\colon & \mathrm{GL}_n(\mathbb{C}) & \longrightarrow & M_n(\mathbb{C}) \times \mathbb{A}^1_{\mathbb{C}} \\
& g & \longmapsto & (g, \det(g)^{-1}).
\end{array}
$$

This yields the description

$$
\mathrm{GL}_n(\mathbb{C}) = \left\{\, (g,t) \in M_n(\mathbb{C}) \times \mathbb{A}^1_{\mathbb{C}} \,\middle|\, \det(g) \cdot t = 1 \,\right\},
$$

so that we have realized $\mathrm{GL}_n(\mathbb{C})$ as a closed subvariety of the affine variety $M_n(\mathbb{C}) \times \mathbb{A}^1_{\mathbb{C}}$;
- the neutral element is the identity matrix;
- multiplication is matrix multiplication, which is obviously regular;
- and the inversion is the formation of the inverse matrix. The regularity of that operation results from Cramer's rule.

The group $\mathbb{C}^\star = \mathrm{GL}_1(\mathbb{C})$ is just the multiplicative group of the field $\mathbb{C}$, viewed as an algebraic group.

The **special linear group** $\mathrm{SL}_n(\mathbb{C})$ is a closed subgroup of $\mathrm{GL}_n(\mathbb{C})$. It is described by the polynomial equation $\det = 1$; $\mathrm{SL}_n(\mathbb{C})$ is the kernel of the homomorphism $\det\colon \mathrm{GL}_n(\mathbb{C}) \longrightarrow \mathbb{C}^\star$ between linear algebraic groups.

ii) A homomorphism $\chi\colon G \longrightarrow \mathbb{C}^\star$ is called a **character of G**. The characters of G form an abelian group which is denoted by $X(G)$.

For $G = \mathrm{GL}_n(\mathbb{C})$ and any $r \in \mathbb{Z}$, the map $g \longmapsto \det(g)^r$ is a character of G. Conversely, one shows that any character of $\mathrm{GL}_n(\mathbb{C})$ is of that shape. This may be deduced from the fact that the coordinate algebra of $\mathrm{GL}_n(\mathbb{C})$ is isomorphic to the ring $\mathbb{C}[x_{ij},\ i,j = 1,\ldots,n;\ \det^{-1}]$.

iii) A linear algebraic group T which is isomorphic to $(\mathbb{C}^\star)^{\times n}$ is called **a(n) (algebraic) torus**. For its character group, we find $X(T) \cong X((\mathbb{C}^\star)^{\times n}) \cong \mathbb{Z}^n$. In the latter identification, a vector $\underline{\alpha} = (\alpha_1, \ldots, \alpha_n) \in \mathbb{Z}^n$ yields the character $(z_1, \ldots, z_n) \longmapsto z_1^{\alpha_1} \cdot \ldots \cdot z_n^{\alpha_n}$ of $(\mathbb{C}^\star)^{\times n}$.

iv) A **one parameter subgroup of** G is a homomorphism $\lambda\colon \mathbb{C}^\star \longrightarrow G$. The one parameter subgroups of a torus T also form a free abelian group $X_\star(T)$ of finite rank. Given a character χ and a one parameter subgroup λ of T, the composition $\chi \circ \lambda\colon \mathbb{C}^\star \longrightarrow \mathbb{C}^\star$ is given as $z \longmapsto z^\gamma$ for a uniquely determined integer γ. We set $\langle \lambda, \chi \rangle := \gamma$. In this way, we obtain the perfect pairing $\langle ., . \rangle\colon X_\star(T) \times X(T) \longrightarrow \mathbb{Z}$, i.e., the induced homomorphism $X_\star(T) \longrightarrow X(T)^\vee$ is an isomorphism.

v) It can be shown that any linear algebraic group is isomorphic to a closed subgroup of a general linear group.

1.2. Representations

The fundamental example of forming the quotient of a vector space by the linear action of a reductive affine algebraic group forms the technical heart of Geometric Invariant Theory. These linear actions are so-called representations of the affine algebraic group. Thus, we discuss this notion in the following paragraphs. The reductivity of an affine algebraic group can be characterized in terms of its representation theory. This motivates the notion of a linearly reductive affine algebraic group which will also be highlighted in this section.

A good introduction to the representation theory of general linear groups in characteristic zero are the lecture notes [24]. In positive characteristic, the representation theory of general linear groups becomes more involved [14]. More advanced topics in the representation theory of linear algebraic groups are contained in [22].

Definition 1.2.1. i) Suppose V is a finite-dimensional complex vector space and G is a linear algebraic group. We consider V as an affine algebraic variety. A **(left) action of** G **on** V is a regular map

$$\sigma\colon G \times V \longrightarrow V,$$

satisfying the axioms:

1. For any $g \in G$, the map $\sigma_g\colon V \longrightarrow V$, $v \longmapsto \sigma(g, v)$, is a linear isomorphism; $\sigma_e = \mathrm{id}_V$.
2. For any two elements $g_1, g_2 \in G$, one has $\sigma_{g_1 g_2} = \sigma_{g_1} \circ \sigma_{g_2}$.

Giving the action σ is the same as giving the homomorphism

$$\begin{array}{rccc} \varrho\colon & G & \longrightarrow & \mathrm{GL}(V) \\ & g & \longmapsto & \sigma_g. \end{array}$$

In this correspondence, one associates to a homomorphism ϱ the action

$$\begin{array}{rccc} \sigma\colon & G \times V & \longrightarrow & V \\ & (g, v) & \longmapsto & \varrho(g)(v). \end{array}$$

For an action $\sigma\colon G \times V \longrightarrow V$, we will abbreviate $\sigma(g, v)$ to $g \cdot v$, $g \in G$, $v \in V$. In the above situation, V is also said to be a **(left)** G**-module** and the homomorphism ϱ to be a **(rational) representation**.

ii) Let V and W be two G-modules. A linear map $l\colon V \longrightarrow W$ is said to be G**-equivariant** or a **homomorphism of** G**-modules**, if

$$l(g \cdot v) = g \cdot l(v), \quad \forall g \in G, \; v \in V.$$

Two representations $\varrho_i\colon G \longrightarrow \mathrm{GL}(V_i)$, $i = 1, 2$, are called **equivalent** or **isomorphic**, if there is an isomorphism of G-modules between V_1 and V_2.

Example 1.2.2. i) To a given family $\varrho_i\colon G \longrightarrow \mathrm{GL}(V_i)$, $i = 1, 2, \ldots, n$, of representations, we may associate new representations, using constructions from Linear Algebra, e.g.,

$$\varrho_1 \otimes \cdots \otimes \varrho_n\colon G \longrightarrow \mathrm{GL}(V_1 \otimes \cdots \otimes V_n),$$

by setting

$$(\varrho_1 \otimes \cdots \otimes \varrho_n)(g)(v_1 \otimes \cdots \otimes v_n) := \varrho_1(g)(v_1) \otimes \cdots \otimes \varrho_n(g)(v_n),$$

for $g \in G$ and $v_i \in V_i$, $i = 1, \ldots, n$.

Further representations are direct sums $\varrho_1 \oplus \cdots \oplus \varrho_n$, symmetric powers $\mathrm{Sym}^r(\varrho)$, or exterior powers $\bigwedge^r \varrho$.

For any representation $\varrho\colon G \longrightarrow \mathrm{GL}(V)$, its **dual** or **contragredient representation** $\varrho^\vee\colon G \longrightarrow \mathrm{GL}(V^\vee)$ on the dual space $V^\vee$ is defined by $\varrho^\vee(g)(l)\colon v \longmapsto l(\varrho(g)^{-1} \cdot (v))$, for $g \in G$, $l \in V^\vee$, and $v \in V$.

We derive the representations

$$\varrho_d^\vee := \mathrm{Sym}^d(\varrho^\vee)\colon G \longrightarrow \mathrm{GL}\big(\mathrm{Sym}^d(V^\vee)\big).$$

ii) For any representation $\varrho\colon \mathbb{C}^\star \longrightarrow \mathrm{GL}(V)$ of $\mathbb{C}^\star$, there are a basis $v_1, \ldots, v_n$ of V and integers $\gamma_1 \leq \cdots \leq \gamma_n$, with

$$\varrho(z)\left(\sum_{i=1}^n \alpha_i v_i\right) = \sum_{i=1}^n z^{\gamma_i} \alpha_i v_i,$$

for all $g \in G$.

If we use this basis to identify $\mathrm{GL}(V)$ with $\mathrm{GL}_n(\mathbb{C})$, then the image of ϱ lies in the group of diagonal matrices. Thus, we say that the representation is **diagonalizable**.

iii) Let $T = (\mathbb{C}^\star)^{\times n}$ be a torus and $\varrho\colon T \longrightarrow \mathrm{GL}(V)$ a representation of T on the vector space V. Then, ϱ is diagonalizable. More precisely, the T-module V is isomorphic to $\bigoplus_{\chi \in X(T)} V_\chi$ with

$$V_\chi := \Big\{\, v \in V \mid \varrho(t)(v) = \chi(t) \cdot v \quad \forall t \in T \,\Big\}, \quad \chi \in X(T).$$

iv) In any course on Linear Algebra, one considers the following actions of linear algebraic groups on vector spaces:

a) $G = \mathrm{GL}_m(\mathbb{C}) \times \mathrm{GL}_n(\mathbb{C})$, $V := M_{m,n}(\mathbb{C})$, and

$$\begin{aligned}
\sigma\colon \big(\mathrm{GL}_m(\mathbb{C}) \times \mathrm{GL}_n(\mathbb{C})\big) \times M_{m,n}(\mathbb{C}) &\longrightarrow M_{m,n}(\mathbb{C}) \\
(g, h; f) &\longmapsto g \cdot f \cdot h^{-1}.
\end{aligned}$$

b) $G = \mathrm{GL}_n(\mathbb{C})$, $V := M_n(\mathbb{C})$, and

$$\begin{aligned}
\sigma\colon \mathrm{GL}_n(\mathbb{C}) \times M_n(\mathbb{C}) &\longrightarrow M_n(\mathbb{C}) \\
(g, m) &\longmapsto g \cdot m \cdot g^{-1}.
\end{aligned}$$

c) $G = \mathrm{GL}_n(\mathbb{C})$, $V := \{\text{ symmetric } (n \times n)\text{-matrices}\}$, and

$$\sigma\colon \mathrm{GL}_n(\mathbb{C}) \times V \ \longrightarrow\ V$$
$$(g, m) \ \longmapsto\ g \cdot m \cdot g^t.$$

Definition 1.2.3. i) Let G be a linear algebraic group and V a G-module. A subspace $W \subseteq V$ is G-**invariant**, if $g \cdot w \in W$ for every $g \in G$ and every $w \in W$. We say that V is an **irreducible** or **simple G-module**, if $\{0\}$ and V are the only G-invariant subspaces. One calls V a **completely reducible** or **semisimple G-module**, if it is isomorphic to a direct sum of irreducible G-modules.

ii) A linear algebraic group is called **linearly reductive**, if every G-module is completely reducible.

Remark 1.2.4. A representation V is completely reducible, if and only if every G-invariant subspace W possesses a direct complement, i.e., there is a G-invariant subspace U, such that

$$V = U \oplus W$$

as G-module.

Example 1.2.5. i) The natural representation of $\mathrm{GL}_n(\mathbb{C})$ on $\mathbb{C}^n$ is obviously irreducible.

ii) The representation of $\mathrm{GL}_n(\mathbb{C})$ on $M_n(\mathbb{C})$ is not irreducible. The vector space $M_n^0(\mathbb{C})$ of matrices with trace zero is a submodule, and we may write $M_n(\mathbb{C}) \cong \langle \mathbb{E}_n \rangle \oplus M_n^0(\mathbb{C})$ as $\mathrm{GL}_n(\mathbb{C})$-module.

iii) As we have seen before, a torus T is linearly reductive. The irreducible modules are those of dimension zero and one.

iv) Every finite group is a linearly reductive linear algebraic group (known as THEOREM OF MASCHKE). To see this, let V be a G-module, W a G-invariant subspace, and $\widetilde{U}$ a vector space complement to W. Let $\widetilde{\pi}\colon V \longrightarrow W$ be the projection (which is not necessarily G-equivariant). Define

$$\pi\colon V \ \longrightarrow\ W$$
$$v \ \longmapsto\ \sum_{g \in G} g \cdot \left(\widetilde{\pi}(g^{-1} \cdot v) \right)$$

This is G-equivariant and surjective. Indeed, $\pi(w) = \#G \cdot w$ for $w \in W$. Now, $U := \ker(\pi)$ is the G-invariant complement we have been looking for.

v) Special and general linear groups are linearly reductive. The product of linearly reductive groups is linearly reductive, so that $\mathrm{GL}_m(\mathbb{C}) \times \mathrm{GL}_n(\mathbb{C})$ is also linearly reductive.

vi) In positive characteristic, the only linearly reductive algebraic groups are finite groups whose order is coprime to the characteristic and tori, or products of such groups. There is a notion of **reductivity** which is defined intrinsically (see [6], [20], and [46]). In characteristic zero, this notion is equivalent to "linear reductivity" (see [20], [23]). In positive characteristic, it is equivalent to "geometric reductivity" (see [43]) which is weaker than "linear reductivity", but suffices to develop Geometric Invariant Theory. In that weaker sense, special and general linear groups are reductive.

vii) The additive group $\mathbb{G}_a(\mathbb{C})$ of $\mathbb{C}$ is *not* linearly reductive. We have

$$\mathbb{G}_a(\mathbb{C}) \cong \left\{ \begin{pmatrix} 1 & \lambda \\ 0 & 1 \end{pmatrix} \,\Big|\, \lambda \in \mathbb{C} \right\}.$$

Then, $\left\langle \begin{pmatrix} 1 \\ 0 \end{pmatrix} \right\rangle$ is a $\mathbb{G}_a(\mathbb{C})$-invariant subspace of $\mathbb{C}^2$ without direct complement.

viii) For non-negative integers a and c, define the $\mathrm{GL}(V)$-module

$$V_{a,c} := V^{\otimes a} \otimes \left(\overset{\dim V}{\bigwedge} V \right)^{\otimes -c}.$$

Theorem. *Let W be an irreducible $\mathrm{GL}(V)$-module. Then, there exist non-negative integers a and c, such that W is a submodule (and thus a direct summand) of $V_{a,c}$*

ix) If the reductive group G is embedded into a general linear group $\mathrm{GL}(V)$, then the following result tells us that any representation of G may be extended to a representation of $\mathrm{GL}(V)$.

Proposition. *Let $\iota\colon G \subseteq \mathrm{GL}(V)$ be a closed subgroup and $\varrho\colon G \longrightarrow \mathrm{GL}(U)$ a representation of G. Then, there exists a representation $\widetilde{\varrho}\colon \mathrm{GL}(V) \longrightarrow \mathrm{GL}(W)$, such that ϱ is a direct summand of the representation $\widetilde{\varrho} \circ \iota$.*

x) We introduce an important class of particular representations which are building blocks for all representations.

Definition. i) A representation $\varrho\colon \mathrm{GL}(V) \longrightarrow \mathrm{GL}(W)$ is called **homogeneous of degree** $\alpha(\in \mathbb{Z})$, if

$$\varrho\big(z \cdot \mathrm{id}_V\big) = z^\alpha \cdot \mathrm{id}_W, \quad \forall z \in \mathbb{C}^\star.$$

ii) For non-negative integers a, b, and c, set

$$V_{a,b,c} := \big(V^{\otimes a}\big)^{\oplus b} \otimes \left(\overset{\dim V}{\bigwedge} V \right)^{\otimes -c}.$$

As an exercise, the reader may check – without using linear reductivity – that a representation may always be decomposed into a direct sum of homogeneous representations. This result, thus, holds in any characteristic.

Proposition. *Let W be a homogeneous $\mathrm{GL}(V)$-module. Then, there exist non-negative integers a, b, and c, such that W is a direct summand of $V_{a,b,c}$.*

Proof. We find non-negative integers a_i, and c_i, $i = 1,\ldots,b$, such that W is a direct summand of $\bigoplus_{i=1}^{b} V_{a_i,c_i}$. Note that $a_i - \dim(V) \cdot c_i = a_j - \dim(V) \cdot c_j$ for all $i,j \in \{1,\ldots,b\}$. Choose c so large that $-c_i + c > 0$, $i = 1,\ldots,b$. Then,

$$V_{a_i,b_i} = V^{\otimes a_i} \otimes \left(\overset{\dim V}{\bigwedge} V \right)^{\otimes -c_i+c} \otimes \left(\overset{\dim V}{\bigwedge} V \right)^{\otimes -c}$$

is a direct summand of

$$V^{\otimes a_i + \dim(V) \cdot (-c_i + c)} \otimes \left(\bigwedge^{\dim V} V \right)^{\otimes -c}, \quad i = 1, \ldots, b.$$

Since $a_1 + \dim(V) \cdot (-c_1 + c) = \cdots = a_b + \dim(V) \cdot (-c_b + c)$, we are done. $\square$

Proofs for the assertions in viii)-x) are contained in [24]. In positive characteristic, the latter result fails in general. It remains, however, true for homogeneous *polynomial* representations, provided the degree α is smaller than the characteristic of the base field. The reader will check that the proof of Proposition 5.3 in [24] works in that setting.

1.3. The problem of taking quotients

Let G be a linear algebraic group, $\varrho \colon G \longrightarrow \mathrm{GL}(V)$ a representation, and $\sigma \colon G \times V \longrightarrow V$ the resulting action of G on V. We have the equivalence relation

$$v_1 \sim v_2 \quad \Longleftrightarrow \quad \exists g \in G : \varrho(g)(v_1) = v_2.$$

Denote by $V/_\varrho G$ the set of equivalence classes. The fundamental question we would like to consider is:

Problem. Does $V/_\varrho G$ carry (in a natural way) the structure of an algebraic variety?

In particular, we expect a regular map

$$\pi \colon V \longrightarrow V/_\varrho G.$$

This map would be continuous. For $v \in V$, the fiber $\pi^{-1}(\pi(v))$ therefore would be a G-invariant closed subset which contains the orbit $G \cdot v$. This implies that π would be constant not only on orbits but also on their closures. In other words, the answer to the above problem is "no", if there are non-closed orbits. But non-closed orbits easily do occur:

Example 1.3.1. Look at the action

$$\begin{aligned} \sigma \colon \mathbb{C}^\star \times \mathbb{C}^n &\longrightarrow \mathbb{C}^n \\ (z, v) &\longmapsto z \cdot v \end{aligned}$$

which is associated to the representation $\varrho \colon \mathbb{C}^\star \longrightarrow \mathrm{GL}_n(\mathbb{C})$, $z \longmapsto z \cdot \mathbb{E}_n$. The orbits are $\{0\}$ and lines through the origin with the origin removed. Thus, $\{0\}$ is the only closed orbit, and 0 is contained in the closure of every orbit.

The notion of a quotient has, therefore, to be modified. The appropriate notion is introduced in the following definition.

Definition 1.3.2. A categorical quotient for the variety V with respect to the action σ is a pair $(V /\!\!/_\varrho G, \pi)$, consisting of an algebraic variety $V /\!\!/_\varrho G$ and a G-invariant morphism $\pi \colon V \longrightarrow V /\!\!/_\varrho G$, such that for every other variety Y and every G-invariant morphism $\varphi \colon V \longrightarrow Y$, there exists a unique morphism $\overline{\varphi} \colon V /\!\!/_\varrho G \longrightarrow Y$

with $\varphi = \overline{\varphi} \circ \pi$:

$$
\begin{array}{ccc}
V & \xrightarrow{\ \varphi\ } & Y \\
{\scriptstyle \pi}\downarrow & \nearrow & \\
V /\!\!/_{\varrho} G & {\scriptstyle \exists!\overline{\varphi}} &
\end{array}
$$

The next task is to characterize $V /\!\!/_{\varrho} G$ through its functions. We have

$$
\mathbb{C}[V] = \mathrm{Sym}^{\star}(V^{\vee}) = \bigoplus_{d \geq 0} \mathrm{Sym}^{d}(V^{\vee})
$$

and the action

$$
\begin{aligned}
\varrho^{\star} \colon G \times \mathrm{Sym}^{\star}(V^{\vee}) &\longrightarrow \mathrm{Sym}^{\star}(V^{\vee}) \\
(g, f) &\longmapsto \left(v \in V \longmapsto f(g^{-1} \cdot v) \right).
\end{aligned}
$$

Remark 1.3.3. The action of G on $\mathrm{Sym}^{\star}(V^{\vee})$ preserves the grading, i.e.,

$$
\varrho^{\star}\left(\mathrm{Sym}^{d}(V^{\vee})\right) \subseteq \mathrm{Sym}^{d}(V^{\vee}), \quad d \geq 0,
$$

and the restriction of $\varrho^{\star}$ to $\mathrm{Sym}^{d}(V^{\vee})$ is the representation $\varrho_{d}^{\vee}$.

Set

$$
\mathbb{C}[V]^{G} := \left\{ f \in \mathbb{C}[V]^{G} \,\middle|\, \varrho^{\star}(g)(f) = f, \ \forall g \in G \right\}.
$$

This ring is the potential coordinate algebra of the categorical quotient. In the next lecture, we shall investigate under which circumstances this construction really does work.

2. Lecture II: The basic results of Geometric Invariant Theory and examples

This section introduces the core results of Geometric Invariant Theory, namely the fundamental existence results on quotients and the Hilbert-Mumford criterion. The standard reference is, of course, Mumford's book [28]. Other, more user friendly treatises are the books [9], [23], and [30].

2.1. Finite generation of the ring of invariants: The theorem of Hilbert

Let G be a linear algebraic group, $\varrho \colon G \longrightarrow \mathrm{GL}(V)$ a representation, and $\sigma \colon G \times V \longrightarrow V$ the action of G on V. In this set-up, we have defined the **ring of invariants**

$$
\mathbb{C}[V]^{G} := \left\{ f \in \mathbb{C}[V] \,\middle|\, \varrho^{\star}(g)(f) = f, \ \forall g \in G \right\}.
$$

This is exactly the ring of regular functions on V which are constant on all G-orbits.

Theorem 2.1.1 (Hilbert). *Suppose that, in the above setting, G is reductive. Then, the ring $\mathbb{C}[V]^{G}$ is a finitely generated $\mathbb{C}$-algebra.*

The theorem implies that $\mathbb{C}[V]^G$ is the coordinate algebra of an affine algebraic variety, i.e., we may define

$$V /\!\!/_\varrho G := \mathrm{Specmax}\big(\mathbb{C}[V]^G\big).$$

Note that the inclusion $\mathbb{C}[V]^G \subset \mathbb{C}[V]$ gives rise to a G-invariant morphism

$$\pi \colon V \longrightarrow V /\!\!/_\varrho G.$$

From now on, we will assume that G is reductive.

Theorem 2.1.2. i) *The pair $(V /\!\!/_\varrho G, \pi)$ is the categorical quotient for the variety V with respect to the action σ.*

ii) *If W_1 and W_2 are two disjoint non-empty G-invariant closed subsets of V, then there is a G-invariant function $f \in \mathbb{C}[V]^G$, such that $f_{|W_1} \equiv 1$ and $f_{|W_2} \equiv 0$. In particular, the images of W_1 and W_2 under π are disjoint.*

Recall that the quotient morphism $V \longrightarrow V /\!\!/_\varrho G$ will, in general, not separate the G-orbits in V. The above theorem is therefore of great help in determining the fibers of the quotient map. We put the result into the following more transparent form.

Corollary 2.1.3. i) *Let $v \in V$. Then, the orbit closure $\overline{G \cdot v}$ contains a unique closed orbit.*

ii) *The map $\pi \colon V \longrightarrow V /\!\!/_\varrho G$ induces a bijection between the set of closed orbits in V and the points of $V /\!\!/_\varrho G$.*

In view of the provisos that we had formulated before, this is the best result we could have hoped for.

Example 2.1.4. For the action

$$\begin{aligned}
\sigma \colon \mathbb{C}^\star \times \mathbb{C}^n &\longrightarrow \mathbb{C}^n \\
(z, v) &\longmapsto z \cdot v,
\end{aligned}$$

we clearly have

$$\mathbb{C}[x_1, \ldots, x_n]^{\mathbb{C}^\star} = \mathbb{C},$$

so that $\mathbb{C}^n /\!\!/ \mathbb{C}^\star = \{\mathrm{pt}\}$.

Remark 2.1.5. If G is not reductive, then $\mathbb{C}[V]^G$ need not be finitely generated. The first counterexample was discovered by Nagata [29]. Nevertheless, $\mathbb{C}[V]^G$ is the algebra of regular functions of a quasi-affine variety Y, but there is only a G-invariant *rational map* $\pi \colon V \dashrightarrow Y$, so that Y need not be the categorical quotient (Winkelmann [50]).

2.2. Closed subvarieties

Here, we will demonstrate that forming the quotient commutes with closed embeddings. This property really requires linear reductivity, so that it is rarely available in positive characteristic.

Suppose $Y \subseteq V$ is a G-invariant closed subset. Then, its ideal $\mathscr{I}(Y) \subset \mathbb{C}[V]$ is G-invariant. We obtain the action

$$
\begin{aligned}
\overline{\varrho}^{\star} \colon G \times \mathbb{C}[Y] &\longrightarrow \mathbb{C}[Y](= \mathbb{C}[V]/\mathscr{I}(V)) \\
(g, f) &\longmapsto \left(y \longmapsto f(g^{-1} \cdot y) \right),
\end{aligned}
$$

and the surjection $\mathbb{C}[V] \longrightarrow \mathbb{C}[Y]$ is G-equivariant. Note that $\varrho^{\star}$ and $\overline{\varrho}^{\star}$ are actions of G on *infinite-dimensional* $\mathbb{C}$-vector spaces. However, they are **locally finite**, i.e., every element of, say, $\mathbb{C}[Y]$ is contained in a finite-dimensional G-invariant subspace W. Therefore, $\mathbb{C}[Y]$ may be decomposed into a direct sum of irreducible representations. In particular, $\mathbb{C}[Y]^{G}$ is a G-invariant direct summand of $\mathbb{C}[Y]$. Thus, we obtain a G-invariant, $\mathbb{C}$-linear projection

$$
R \colon \mathbb{C}[Y] \longrightarrow \mathbb{C}[Y]^{G}
$$

which is called the **Reynolds operator**.

Altogether, we find the commutative diagram

$$
\begin{array}{ccc}
\mathbb{C}[V] & \longrightarrow\!\!\!\!\to & \mathbb{C}[Y] \\
\downarrow & & \downarrow \\
\mathbb{C}[V]^{G} & \longrightarrow\!\!\!\!\to & \mathbb{C}[Y]^{G}.
\end{array}
$$

Then, $Y/\!\!/G := \mathrm{Specmax}(\mathbb{C}[Y]^{G})$ is the categorical quotient of Y with respect to the action of G, and we have the commutative diagram

$$
\begin{array}{ccc}
Y & \hookrightarrow & V \\
\downarrow & & \downarrow \\
Y/\!\!/G & \hookrightarrow & V/\!\!/_{\varrho}G.
\end{array}
$$

2.3. Semistable, polystable, and stable points

We will now distinguish the points in V according to the structure of their G-orbits. This will help to analyze the quotient morphism $V \longrightarrow V/\!\!/_{\varrho}G$ even better.

Definition 2.3.1. i) A point $v \in V$ is called **nullform**, if 0 lies in the closure of the G-orbit of v. Otherwise, $v \in V$ is called **semistable**.

ii) A point $v \in V$ is said to be **polystable**, if $v \neq 0$ and $G \cdot v$ is closed in V. (Note that v is semistable.)

iii) A point $v \in V$ is **stable**, if it is polystable, and $\dim(G \cdot v) = \dim(G)$ (i.e., the stabilizer of v is finite).

Remark 2.3.2. i) The quotient $V/\!\!/_\varrho G$ parameterizes the orbits of 0 $(= \{0\})$ and of polystable points.

ii) By the separation properties of the functions in $\mathbb{C}[V]^G$, we have:

$$v \in V \text{ is semistable} \iff \exists d > 0,\ \exists f \in \mathrm{Sym}^d(V^\vee)^G : f(v) \neq 0.$$
$$v \in V \text{ is a nullform} \iff \forall d > 0,\ \forall f \in \mathrm{Sym}^d(V^\vee)^G : f(v) = 0.$$

In particular, the set V^{ss}_ϱ of semistable points is open.

iii) The set V^{ps}_ϱ of polystable points need not be open.

Proposition 2.3.3. *The set V^{s}_ϱ of stable points is open, and so is its image V^{s}_ϱ/G in $V/\!\!/_\varrho G$. The pair $(V^{\mathrm{s}}_\varrho/G, \pi_{|V^{\mathrm{s}}_\varrho})$ is the categorical quotient for V^{s}_ϱ with respect to the induced G-action and an **orbit space**, i.e., the points in V^{s}_ϱ/G are in one to one correspondence to the G-orbits in V^{s}_ϱ.*

Therefore, we have discovered the open subset V^{s}_ϱ of stable points (which might be empty) for which we have the optimal results: In this case, the categorical quotient V^{s}_ϱ/G of V^{s}_ϱ exists and its set of closed points does equal the set of G-orbits in V^{s}_ϱ.

2.4. Quotients of projective varieties

Let G be a linear algebraic group and $\varrho\colon G \longrightarrow \mathrm{GL}(V)$ a representation. Define $\mathbb{P}(V) := (V^\vee \setminus \{0\}/\mathbb{C}^\star)$ (this is Grothendieck's convention for projectivization) and write $[l] \in \mathbb{P}(V)$ for the class of the element $l \in V^\vee \setminus \{0\}$. We get the action

$$\overline{\sigma}\colon G \times \mathbb{P}(V) \longrightarrow \mathbb{P}(V)$$
$$(g, [l]) \longmapsto \big[\varrho^\vee(g)(l)\big].$$

As an algebraic variety, $\mathbb{P}(V) = \mathrm{Projmax}(\mathrm{Sym}^\star(V))$. Recall

$$\mathrm{Sym}^\star(V)^G = \bigoplus_{d \geq 0} \mathrm{Sym}^d(V)^G,$$

i.e., $\mathrm{Sym}^\star(V)^G$ inherits a grading from $\mathrm{Sym}^\star(V)$. Since $\mathrm{Sym}^0(V)^G = \mathbb{C}$,

$$\mathbb{P}(V)/\!\!/_\varrho G := \mathrm{Projmax}\big(\mathrm{Sym}^\star(V)^G\big)$$

is a projective variety and the inclusion $\mathrm{Sym}^\star(V)^G \subset \mathrm{Sym}^\star(V)$ yields the G-invariant *rational map*

$$\overline{\pi}\colon \mathbb{P}(V) \dashrightarrow \mathbb{P}(V)/\!\!/_\varrho G.$$

Similarly, if $Z \subseteq \mathbb{P}(V)$ is a closed G-invariant subvariety, it is defined by a G-invariant homogeneous ideal $\mathscr{I}(Z) \subset \mathrm{Sym}^\star(V)$, and $\mathbb{C}[Z] := \mathrm{Sym}^\star(V)/\mathscr{I}(Z)$ is the homogeneous coordinate algebra of Z. We set

$$Z/\!\!/G := \mathrm{Projmax}\big(\mathbb{C}[Z]^G\big)$$

and obtain the commutative diagram

$$
\begin{array}{ccc}
Z & \lhook\joinrel\longrightarrow & \mathbb{P}(V) \\
\Big\downarrow{\scriptstyle \overline{\pi}_Z} & & \Big\downarrow{\scriptstyle \overline{\pi}} \\
Z /\!\!/ G & \lhook\joinrel\longrightarrow & \mathbb{P}(V) /\!\!/_{\varrho} G.
\end{array}
$$

Definition 2.4.1. Define

$$
Z^{\mathrm{ss}/\mathrm{ps}/\mathrm{s}} := \Big\{\, [l] \in Z \,\big|\, l \in V^{\vee} \setminus \{0\} \text{ is semistable/polystable/stable} \Big\}
$$

to be the **set of semistable, polystable, and stable points in** Z, respectively.

The central result on quotients of projective varieties is the following.

Proposition 2.4.2. i) *The sets Z^{ss} and Z^{s} are G-invariant open subsets of Z.*

ii) *The map $\overline{\pi}_Z$ is defined in Z^{ss}, and $(Z /\!\!/ G, \overline{\pi}_Z)$ is the categorical quotient for Z^{ss} with respect to the induced G-action.*

iii) *The map $\overline{\pi}_Z$ induces a bijection between the orbits of polystable points and the points of $Z /\!\!/ G$.*

iv) *The image Z^{s}/G of Z^{s} under $\overline{\pi}_Z$ is open, and $(Z^{\mathrm{s}}/G, \overline{\pi}_{Z|Z^{\mathrm{s}}})$ is the categorical quotient for Z^{s} with respect to the induced G-action and an orbit space.*

The results for projective varieties are of a different nature: In general, it will not be possible to form the categorical quotient of the whole variety Y or $\mathbb{P}(V)$ by the G-action. With the concept of a semistable point, we can only define a (possibly empty) G-invariant open subset Z^{ss}, such that the categorical quotient with respect to the induced G-action exists. If non-empty, this open subset is, however, large in the sense that it possesses a *projective* categorical quotient. As before, we also have the G-invariant open subset Z^{s} for which the best possible result can be achieved. One can view the results also in the following way: If we want to parameterize orbits by an algebraic variety, we have to restrict to the G-invariant open subset Z^{s} and obtain Z^{s}/G. This variety is only quasi-projective, and the space $Z /\!\!/ G$ provides a natural compactification.

The results of the last two sections, in particular the above proposition, motivate the following question.

Problem. How to find the semistable and stable points?

The answer will be discussed in the next section.

2.5. The Hilbert-Mumford criterion

The idea to find points in the closure of the orbit of, say, $v \in V$ is to find them via one parameter subgroups. Recall that a one parameter subgroup is a homomorphism

$$
\lambda \colon \mathbb{C}^{\star} \longrightarrow G.
$$

Together with the representation ϱ, we find the one parameter subgroup

$$
\mathbb{C}^{\star} \xrightarrow{\ \lambda\ } G \xrightarrow{\ \varrho\ } \mathrm{GL}(V).
$$

Since this representation of $\mathbb{C}^\star$ is diagonalizable, we find integers $\gamma_1 < \cdots < \gamma_t$ and a decomposition of V into non-trivial eigenspaces

$$V = V_1 \oplus \cdots \oplus V_t, \quad V_i := \Big\{ v \in V \,\Big|\, \varrho\big(\lambda(z)\big)(v) = z^{\gamma_i} \cdot v \ \forall z \in \mathbb{C}^\star \Big\}, \quad i = 1, \ldots, t.$$

Definition 2.5.1. For $v \in V \setminus \{0\}$, set

$$\mu(\lambda, v) := \max\Big\{ \gamma_i \,\Big|\, v \text{ has a non-trivial component in } V_i, \ i = 1, \ldots, t \Big\}.$$

We note the following evident property.

Lemma 2.5.2. *Suppose* $v = (v_1, \ldots, v_i \neq 0, 0, \ldots, 0)$. *Then,*
  i)
$$\mu(\lambda, v) \leq 0 \quad \Longleftrightarrow \quad \lim_{z \to \infty} \varrho\big(\lambda(z)\big)(v) \text{ exists.}$$
(*Note that this limit equals* v_i, *if "$=0$" holds, and* 0 *otherwise.*)
  ii)
$$\mu(\lambda, v) < 0 \quad \Longleftrightarrow \quad \lim_{z \to \infty} \varrho\big(\lambda(z)\big)(v) = 0.$$

We infer the following consequence of (semi)stability.

Corollary 2.5.3. *Let* $v \in V \setminus \{0\}$ *be a point.*
  i) *If* v *is semistable, then*
$$\mu(\lambda, v) \geq 0$$
for every one parameter subgroup λ *of* G.
  ii) *If* v *is stable, then*
$$\mu(\lambda, v) > 0$$
for every non-trivial one parameter subgroup λ *of* G.

Proof. i) is clear. For ii), we look at a one parameter subgroup λ with $\mu(\lambda, v) = 0$. Set $v' := \lim_{z \to \infty} \lambda(z) \cdot v$. Since the orbit of v is closed, there exists an element $g \in G$ with $v' = g \cdot v$, so that v' is also stable. Now, the image of λ lies in the G-stabilizer $G_{v'}$ of v'. Since $G_{v'}$ is finite, λ must be the trivial one parameter subgroup. $\square$

Hilbert discovered (in a specific setting) that any degeneration among orbits can be detected by one parameter subgroups. Mumford extended Hilbert's result to the general setting in which we are working. Their theorem is the converse to Corollary 2.5.3:

Theorem 2.5.4 (Hilbert-Mumford criterion). *A point* $v \in V \setminus \{0\}$ *is (semi)stable, if and only if*
$$\mu(\lambda, v)(\geq)0$$
holds for every non-trivial one parameter subgroup λ *of* G.

Richardson's idea of proof [5]. If $G = T$ is a torus, one may believe this (and it can be, in fact, proved by methods of Linear Algebra). For an arbitrary reductive group, one uses:

Theorem 2.5.5 (Cartan decomposition). *Let G be a reductive linear algebraic group and $T \subseteq G$ a* **maximal torus***, i.e., a subgroup which is isomorphic to a torus and maximal with respect to inclusion among all subgroups with this property. Then, there is a compact real subgroup H, such that*

$$G = H \cdot T \cdot H.$$

E.g.,

$$\mathrm{GL}_n(\mathbb{C}) = \mathrm{U}_n(\mathbb{C}) \cdot \big\{\, \text{Diagonal matrices} \,\big\} \cdot \mathrm{U}_n(\mathbb{C}).$$

Roughly speaking, the compact group H does not contribute anything to orbit degenerations, because the orbits of a compact group action are always closed. Therefore, the maximal tori are responsible for the orbit degenerations and they do contain all the one parameter subgroups of G, so that the result for tori may be applied. $\qquad\square$

Remark 2.5.6. i) The formalism we have discussed so far goes back to Hilbert (in the case of the $\mathrm{SL}_n(\mathbb{C})$-action on algebraic forms) and Mumford [28]. It is the rough version of **Geometric Invariant Theory (GIT)**.

ii) The Hilbert-Mumford criterion is crucial for applications.

iii) There are other theories and results which grant the existence of (Rosenlicht [34]) or define (e.g., Białynicki-Birula [3], Hausen [18]) G-invariant open subsets in V or $\mathbb{P}(V)$ or, more generally, in any quasi-projective G-variety, such that the categorical quotients of these open subsets do exist. However, there does not seem to be a numerical criterion such as the Hilbert-Mumford criterion.

2.6. Hypersurfaces in projective space (classical invariant theory)

We now come to one of the classical topics of invariant theory which is also the most basic example for the application of GIT to the classification of algebraic varieties.

Definition 2.6.1. An **algebraic form of degree d on $\mathbb{C}^n$** is a symmetric d-multi-linear form

$$\underbrace{\mathbb{C}^n \times \cdots \times \mathbb{C}^n}_{d \text{ times}} \longrightarrow \mathbb{C}.$$

Remark 2.6.2. i) An algebraic form of degree d is the same as a linear map

$$\varphi \colon \mathrm{Sym}^d(\mathbb{C}^n) \longrightarrow \mathbb{C}.$$

The algebraic forms of degree d on $\mathbb{C}^n$ are the elements of the vector space $\mathrm{Sym}^d(\mathbb{C}^n)^{\vee}$.

ii) Denote by $(e_1, \ldots, e_n)$ the standard basis of $\mathbb{C}^n$ and by $(x_1, \ldots, x_n)$ the dual basis of $\mathbb{C}^{n\vee}$. This yields the $\mathrm{GL}_n(\mathbb{C})$-module isomorphism

$$\mathrm{Sym}^d(\mathbb{C}^n)^{\vee} \longrightarrow \mathrm{Sym}^d(\mathbb{C}^{n\vee}) \tag{1}$$

$$\varphi \longmapsto \sum_{\underline{\nu}=(\nu_1,\ldots,\nu_d)\in\{1,\ldots,n\}^{\times d}} \varphi(e_{\nu_1} \otimes \cdots \otimes e_{\nu_d}) \cdot [x_{\nu_1} \otimes \cdots \otimes x_{\nu_d}].$$

We view $\mathrm{Sym}^d(\mathbb{C}^{n\vee})$ as the vector space of homogeneous polynomials of degree d in the variables $x_1, \ldots, x_n$, by identifying $[x_{\nu_1} \otimes \cdots \otimes x_{\nu_d}]$ with $x_{\nu_1} \cdot \cdots \cdot x_{\nu_d}$. Let $\mathbb{C}[x_1, \ldots, x_n]_d$ be the vector space of homogeneous polynomials of degree d. For an algebraic form φ of degree d on $\mathbb{C}^n$, its corresponding polynomial f, and $\underline{\alpha} := (\alpha_1, \ldots, \alpha_n)^t \in \mathbb{C}^n$, we find:

$$f(\alpha_1, \ldots, \alpha_n) = \varphi(\underline{\alpha} \otimes \cdots \otimes \underline{\alpha}).$$

We would like to study the action v_d of $\mathrm{SL}_n(\mathbb{C})$ on $\mathbb{C}[x_1, \ldots, x_n]_d$ by **substitution of variables**, i.e.,

$$v_d \colon \mathrm{SL}_n(\mathbb{C}) \times \mathbb{C}[x_1, \ldots, x_n]_d \longrightarrow \mathbb{C}[x_1, \ldots, x_n]_d$$
$$(g, f) \longmapsto (g \cdot f \colon \underline{\alpha} \longmapsto f(g^t \cdot \underline{\alpha})).$$

In the notation of Lecture I, Example 1.2.2, i), we have $v_d = \iota_d^\vee \circ .^{-1^t}$, $\iota \colon \mathrm{SL}_n(\mathbb{C}) \subset \mathrm{GL}_n(\mathbb{C})$ being the inclusion and $.^{-1^t} \colon \mathrm{SL}_n(\mathbb{C}) \longrightarrow \mathrm{SL}_n(\mathbb{C})$ being the automorphism that sends a matrix to the transpose of its inverse.

This is the topic of classical invariant theory. Famous representatives of that branch were Gordan and later Hilbert. See [9] and [47] for historical comments, including the symbolic method, and Hilbert's lecture notes [19] for an authentic reference. Here is a list of tasks related to the above group action.

Problems. 1. Describe the set $\mathbb{C}[x_1, \ldots, x_n]_d /_{v_d} \mathrm{SL}_n(\mathbb{C})$, i.e., find a "nice" representative in each orbit, a so-called **normal form**.
2. Describe $\mathbb{C}[x_1, \ldots, x_n]_d /\!/_{v_d} \mathrm{SL}_n(\mathbb{C})$. E.g., find generators and relations for the ring of invariants.
3. Find the (semi)stable forms and the nullforms.

Remark 2.6.3. The above problems concern the classification of certain algebraic varieties: Let $H \subset \mathbb{P}_{n-1}$ be a hypersurface of degree d. We may find a polynomial $f \in \mathbb{C}[x_1, \ldots, x_n]_d$ with $H = V(f) := \{ f = 0 \}$. Note that

$$V(f) = V(f'), \quad f, f' \in \mathbb{C}[x_1, \ldots, x_n]_d \quad \Longleftrightarrow \quad \exists \lambda \in \mathbb{C}^\star \colon f' = \lambda \cdot f.$$

Two hypersurfaces $H_1 = V(f_1)$ and $H_2 = V(f_2)$ are said to be **projectively equivalent**, if there is a matrix $g \in \mathrm{SL}_n(\mathbb{C})$, such that $[f_1] = [g \cdot f_2]$, i.e., there is an automorphism of $\mathbb{P}_{n-1}$, carrying H_2 into H_1. Thus, $\mathbb{P}(\mathbb{C}[x_1, \ldots, x_n]_d^\vee) /_{v_d} \mathrm{SL}_n(\mathbb{C})$ is the set of projective equivalence classes of hypersurfaces of degree d.

Projectively equivalent hypersurfaces are certainly isomorphic. On the other hand, two isomorphic smooth hypersurfaces of dimension at least three are projectively equivalent. In $\mathbb{P}_3$, the same holds for hypersurfaces of degree $d \neq 4$ (see [15], p. 178 (The condition $d \neq n+1$ in that reference can be removed for $n \geq 4$.)). Finally, the case of degree four surfaces in $\mathbb{P}_3$ belongs to the realm of K3-surfaces. Here, the notions of "isomorphy" and "projective equivalence" are still equivalent on the complement of countably many Zariski closed subsets [27].

One can show that non-singular hypersurfaces are stable ([28], Chapter 4, Proposition 4.2). Thus, the moduli space of smooth hypersurfaces exists as a quasi-projective variety and comes with a natural compactification via semistable hypersurfaces. Detailed examples for specific dimensions and degrees may be found below and in [23], [28], and [35].

2.7. Examples

We now present several mostly classical examples which illustrate the abstract formalism that we have introduced up to now.

Quadratic forms. To a homogeneous polynomial q of degree 2 corresponds the symmetric $(n \times n)$-matrix S_q with

$$q(\alpha_1, \ldots, \alpha_n) = \underline{\alpha}^t S_q \underline{\alpha}, \quad \forall \underline{\alpha} = (\alpha_1, \ldots, \alpha_n)^t \in \mathbb{C}^n.$$

One checks

$$S_{g \cdot q} = g \cdot S_q \cdot g^t \qquad \forall g \in \mathrm{GL}_n(\mathbb{C}).$$

Recall that, for a symmetric matrix $S \in M_n(\mathbb{C})$, there is a matrix $m \in \mathrm{GL}_n(\mathbb{C})$, such that gSg^t is a diagonal matrix with ones and zeroes on the diagonal. In other words, we have the following classification.

Lemma 2.7.1. *For $q \in \mathbb{C}[x_1, \ldots, x_n]_2$, there are a matrix $g \in \mathrm{GL}_n(\mathbb{C})$ and a natural number $m \in \{0, \ldots, n\}$ with*

$$g \cdot q = x_1^2 + \cdots + x_m^2.$$

Next, we look at the action of $\mathrm{SL}_n(\mathbb{C})$ on $\mathbb{C}[x_1, \ldots, x_n]_2$.

Definition 2.7.2. The **discriminant of the quadratic form** q is

$$\Delta(q) := \det(S_q).$$

Lemma 2.7.1 implies the following classification result for the $\mathrm{SL}_n(\mathbb{C})$-action.

Corollary 2.7.3. *Let $q \in \mathbb{C}[x_1, \ldots, x_n]_2$ and $\delta := \Delta(q)$. If $\delta \neq 0$, then q is equivalent to the form $q_\delta := \delta x_1^2 + x_2^2 + \cdots + x_n^2$. Otherwise, there is an $m \in \{0, \ldots, n-1\}$, such that q is equivalent to $x_1^2 + \cdots + x_m^2$.*

The corollary enables us to compute the ring of invariant functions and the categorical quotient.

Theorem 2.7.4. $W := \mathbb{C}[x_1, \ldots, x_n]_2 /\!\!/_{v_2} \mathrm{SL}_n(\mathbb{C}) = \mathrm{Specmax}(\mathbb{C}[\Delta]).$

Proof. Suppose $I \in \mathbb{C}[W]$. Write the "general" quadratic polynomial as

$$\sum_{1 \leq i \leq j \leq n} \kappa_{ij} x_i x_j.$$

The coordinate algebra of $\mathbb{C}[x_1, \ldots, x_n]_2$ is thus $\mathbb{C}[\kappa_{ij};\ 1 \leq i \leq j \leq n]$, and I is a polynomial in the κ_{ij}. Define $I_\Delta \in \mathbb{C}[\Delta]$, by replacing κ_{11} with Δ, κ_{ii}, $i = 2, \ldots, n$, with 1 and the remaining variables by 0. The polynomial $I - I_\Delta$ vanishes in q_δ for all δ. Since $I - I_\Delta \in \mathbb{C}[W]$, Corollary 2.7.3 implies $I - I_\Delta \equiv 0$. $\square$

Binary forms. We look at forms of degree ≥ 3. Write a binary form f of degree d as

$$f = a_0 x_1^d + a_1 x_1^{d-1} x_2 + \cdots + a_{d-1} x_1 x_2^{d-1} + a_d x_2^d.$$

Then, under the action of $\mathrm{SL}_2(\mathbb{C})$, it may be brought into one of the following shapes:

$$\lambda x_1^{d-i} x_2^i, \quad 2i \leq d, \ \lambda \in \mathbb{C}^\star, \quad \text{or}$$

$$\lambda x_1^{\mu_1} x_2^{\mu_2} (x_1 - x_2)^{\mu_3} \prod_{i=\mu_1+\mu_2+\mu_3+1}^{d} (x_1 - \beta_i x_2), \ \lambda \in \mathbb{C}^\star, \beta_i \in \mathbb{C} \setminus \{0, 1\}.$$

Let us determine the stable and semistable points and the nullforms. The property of being stable or semistable is invariant under the action of $\mathrm{SL}_2(\mathbb{C})$, and, by Lecture I, Example 1.2.2, ii), a one parameter subgroup may be diagonalized. By the Hilbert-Mumford criterion, we have to determine the forms $f = a_0 x_1^d + \cdots$ for which

$$\lim_{z \to \infty} \begin{pmatrix} z & \\ & z^{-1} \end{pmatrix} \cdot f$$
$$= \lim_{z \to \infty} \left(z^d a_0 x_1^d + z^{d-2} a_1 x_1^{d-1} x_2 + \cdots + z^{2-d} a_{d-1} x_1 x_2^{d-1} + z^{-d} a_d x_2^d \right)$$

exists or equals zero. We find out the following.

Lemma 2.7.5. i) *The limit exists, if and only if $a_0 = \cdots = a_{\lfloor \frac{d}{2} \rfloor} = 0$.*

ii) *The limit is zero, if and only if $a_0 = \cdots = a_{\lfloor \frac{d+1}{2} \rfloor} = 0$.*

This leads to the following intrinsic characterization of stable and semistable forms.

Corollary 2.7.6. i) *A binary form of degree d is stable, if and only if it doesn't have a zero of multiplicity $\geq \frac{d}{2}$.*

ii) *A binary form of degree d is semistable, if and only if it doesn't have a zero of multiplicity $> \frac{d}{2}$.*

In particular, if d is odd, then the notions "stable" and "semistable" agree.

Note that the last property is quite interesting, because in that case we have categorical quotients which are both projective and orbit spaces.

The invariant theory of matrices. We finally discuss some basic results related to the action of $\mathrm{GL}_n(\mathbb{C})$ on tuples of $(n \times n)$-matrices by simultaneous conjugation. We first interpret the results on the Jordan normal form in terms of Geometric Invariant Theory.

The group $\mathrm{GL}_n(\mathbb{C})$ acts on $M_n(\mathbb{C})$ by conjugation, i.e., $g \cdot m := g \cdot m \cdot g^{-1}$, $g \in \mathrm{GL}_n(\mathbb{C})$, $m \in M_n(\mathbb{C})$. Under this action, any matrix may be transformed into a matrix of the shape

$$
\begin{pmatrix}
\lambda_1 & 1 & & & & & & \\
 & \ddots & \ddots & & & & & \\
 & & \ddots & 1 & & & & \\
 & & & \lambda_1 & & & & \\
 & & & & \ddots & & & \\
 & & & & & \lambda_k & 1 & \\
 & & & & & & \ddots & \ddots \\
 & & & & & & & \ddots & 1 \\
 & & & & & & & & \lambda_k
\end{pmatrix}.
$$

The invariants of an $(n \times n)$-matrix m with eigenvalues $\lambda_1,\ldots,\lambda_n$ are

$$
\sigma_1(\lambda_1,\ldots,\lambda_n) = \lambda_1 + \cdots + \lambda_n, \ldots, \sigma_n(\lambda_1,\ldots,\lambda_n) = \lambda_1 \cdots \cdots \lambda_n.
$$

Instead of the elementary symmetric functions $\sigma_1,\ldots,\sigma_n$, one may also work with the **symmetric Newton functions**

$$
s_1,\ldots,s_n \quad \text{with} \quad s_i(\lambda_1,\ldots,\lambda_n) := \lambda_1^i + \cdots + \lambda_n^i, \quad i = 1,\ldots,n.
$$

As do the elementary symmetric functions, the Newton functions serve the purpose of generating the ring of symmetric functions:

Theorem 2.7.7.

$$
\mathbb{C}[s_1,\ldots,s_n] = \mathbb{C}[\sigma_1,\ldots,\sigma_n] = \mathbb{C}[\lambda_1,\ldots,\lambda_n]^{S_n}.
$$

Proof. [47], Proposition 1.1.2, p. 4.   □

In terms of matrices, this result reads as follows.

Corollary 2.7.8. *Let x_{ij}, $i,j = 1,\ldots,n$, be the coordinate functions on $M_n(\mathbb{C})$ and* $\mathbf{x} := (x_{ij})_{i,j}$. *Then,*

$$
\mathbb{C}\big[M_n(\mathbb{C})\big]^{\mathrm{GL}_n(\mathbb{C})} = \mathbb{C}[x_{ij}, \ i,j = 1,\ldots,n]^{\mathrm{GL}_n(\mathbb{C})} = \mathbb{C}\big[\mathrm{Trace}(\mathbf{x}),\ldots,\mathrm{Trace}(\mathbf{x}^n)\big].
$$

Next, we consider the action of $\mathrm{GL}_n(\mathbb{C})$ on $M_n(\mathbb{C})^{\oplus s}$ which is given as

$$
g \cdot (m_1,\ldots,m_s) := \big(g \cdot m_1 \cdot g^{-1}, \ldots, g \cdot m_s \cdot g^{-1}\big),
$$
$$
g \in \mathrm{GL}_n(\mathbb{C}), \ (m_1,\ldots,m_s) \in M_n(\mathbb{C})^{\oplus s}.
$$

Theorem 2.7.9. *Let x^i_{jk}, $i = 1,\ldots,s$, $j,k = 1,\ldots,n$, be the coordinate functions on $M_n(\mathbb{C})^{\oplus s}$, and set $\mathbf{x}_i := (x^i_{jk})_{j,k}$, $i = 1,\ldots,s$. Then, the invariant ring $\mathbb{C}[M_n(\mathbb{C})^{\oplus s}]^{\mathrm{GL}_n(\mathbb{C})}$ is generated by the invariants*

$$
\mathrm{Trace}\big(\mathbf{x}_{i_1} \cdots \cdots \mathbf{x}_{i_l}\big).
$$

It suffices to take the invariants with $l \le n^2 + 1$.

Proof. See [17], [32], and [44].   □

Example 2.7.10. i) For $n = s = 2$, one finds

$$\mathbb{C}[M_2(\mathbb{C})^{\oplus 2}]^{\mathrm{GL}_2(\mathbb{C})} = \mathbb{C}[\widetilde{T}_1, \ldots, \widetilde{T}_5].$$

Here, we use the following (algebraically independent) invariants:

$$
\begin{aligned}
\widetilde{T}_1(m_1, m_2) \quad &:= \quad \mathrm{Trace}(m_1), \qquad \widetilde{T}_2(m_1, m_2) \quad := \quad \mathrm{Det}(m_1), \\
\widetilde{T}_3(m_1, m_2) \quad &:= \quad \mathrm{Trace}(m_2), \qquad \widetilde{T}_4(m_1, m_2) \quad := \quad \mathrm{Det}(m_2), \\
\text{and} \quad \widetilde{T}_5(m_1, m_2) \quad &:= \quad \mathrm{Trace}(m_1 m_2), \quad (m_1, m_2) \in M_2(\mathbb{C}) \oplus M_2(\mathbb{C}).
\end{aligned}
$$

ii) In general, one uses the Hilbert-Mumford criterion to prove that $(m_1, \ldots, m_s) \in M_n(\mathbb{C})^{\oplus s}$ is a nullform, if and only if $m_1, \ldots, m_s$ may be simultaneously brought into upper triangular form, such that the diagonal entries are zero.

3. Lecture III: Some advanced results of Geometric Invariant Theory

We now take a more general viewpoint: We look at an action of a reductive linear algebraic group on a projective algebraic variety. By means of linearizations, we can use our former results to find open subsets of the projective variety of which we may take the quotients as projective varieties. The concept of a linearization was introduced by Mumford in [28]. The choice of a linearization is a parameter in the theory, and its significance has been investigated only recently by Dolgachev/Hu [10], Ressayre [33], and Thaddeus [49]. We will present these new findings in a quite elementary fashion.

3.1. Linearizations

In many applications to moduli problems (see Lecture IV), one faces the problem of taking the quotient of a projective variety by the action of a reductive linear algebraic group. The concept of a linearization reduces this problem to the results of Lecture II.

Definition 3.1.1. Let X be a projective variety and G a linear algebraic group.

i) An **action of G on X** is a regular map

$$\alpha \colon G \times X \longrightarrow X,$$

such that

1. $\alpha_g \colon X \longrightarrow X$, $x \longmapsto g \cdot x := \alpha(g, x)$ is a regular map; $\alpha_e = \mathrm{id}_X$;
2. For g_1 and $g_2 \in G$, one has $\alpha_{g_1 g_2} = \alpha_{g_1} \circ \alpha_{g_2}$.

ii) A **linearization of the action** α is a pair $l = (\varrho, \iota)$ which consists of a representation $\varrho \colon G \longrightarrow \mathrm{GL}(V)$ and a G-equivariant closed embedding $\iota \colon X \hookrightarrow \mathbb{P}(V)$.

Remark 3.1.2. i) Recall that we assume G to be reductive. Thus, a linearization $l = (\varrho, \iota)$ of α provides the open subset $X_l^s \subseteq X$ of l-stable points, such that the orbit space X_l^s/G carries in a natural way the structure of a quasi-projective variety, and the open subset $X_l^{ss} \subseteq X$ of l-semistable points, such that the categorical quotient $X /\!\!/_l G := X_l^{ss} /\!\!/ G$ exists as a projective variety and compactifies X_l^s/G.

ii) In the above situation, we may associate to a linearization $l = (\varrho, \iota)$ its **kth symmetric power** $l_k := \mathrm{Sym}^k(l) = (\mathrm{Sym}^k(\varrho), \iota_k)$ where

$$\iota_k \colon X \overset{\iota}{\lhook\joinrel\longrightarrow} \mathbb{P}(V) \overset{v_k}{\lhook\joinrel\longrightarrow} \mathbb{P}\big(\mathrm{Sym}^k(V)\big), \quad v_k \text{ being the } k\text{-fold Veronese embedding.}$$

One verifies

$$X_{l_k}^{(s)s} = X_l^{(s)s}, \quad \forall k > 0.$$

iii) Let $l = (\varrho, \iota)$ be a linearization of the G-action α, and $\chi \colon G \longrightarrow \mathbb{C}^\star$ a character of G. Then, $l^\chi := (\varrho \otimes \chi, \iota)$ is another linearization of the G-action α. Finally, we let $l_k^\chi = (\mathrm{Sym}^k(\varrho) \otimes \chi, \iota_k)$ be the symmetric power of the linearization l modified by the character χ.

iv) Suppose $\mathrm{Sym}^\star(V)^G$ is generated by homogeneous elements $f_1, \ldots, f_t$ of degree $d_1, \ldots, d_t$ and let k be a common multiple of $d_1, \ldots, d_t$. Then, we find the closed embedding

$$\bar{\iota}(l, k) \colon X /\!\!/_l G \lhook\joinrel\longrightarrow \mathbb{P}(V) /\!\!/_\varrho G \overset{\bar{v}_k}{\lhook\joinrel\longrightarrow} \mathbb{P}\big(\mathrm{Sym}^k(V)^G\big)$$

and the line bundle $\mathscr{L}_k := \bar{\iota}(l, k)^\star(\mathscr{O}_{\mathbb{P}(\mathrm{Sym}^k(V))}(1))$. For any other set of generators and any other common multiple m of the degrees of these generators, one verifies

$$\mathscr{L}_k^{\otimes m} \cong \mathscr{L}_m^{\otimes k}.$$

Thus, we obtain the *polarized* quotient $(X /\!\!/_l G, [\mathscr{L}_l])$ [1]. We see that, in ii), l and l_k supply the "same" polarized quotient.

Let G and H be two reductive linear algebraic groups. Suppose we are given an action

$$\alpha \colon (G \times H) \times X \longrightarrow X$$

and a linearization $l = (\varrho, \iota)$ of this action. Let $m := (\varrho_{|G \times \{e\}}, \iota)$ be the induced linearization of the G-action. For any $k > 0$, we get an induced representation

$$\bar{\varrho}_k \colon H \longrightarrow \mathrm{GL}\big(\mathrm{Sym}^k(V)^G\big),$$

an induced action $\bar{\alpha}$ on $X /\!\!/_l G$, and $\bar{\iota}(l, k)$ is an equivariant embedding. We view $n_k := (\bar{\varrho}_k, \bar{\iota}(l, k))$ as a linearization of $\bar{\alpha}$.

Proposition 3.1.3. *In the above setting, let*

$$\pi_m \colon X \dashrightarrow X /\!\!/_m G$$

be the quotient map. Then, for any $k > 0$,

$$X_l^{ss} = \pi_m^{-1}\big(X_{n_k}^{ss}\big) \big(\subseteq X_m^{ss}\big).$$

---

[1] Here, the convention is that $[\mathscr{L}] = [\mathscr{M}]$, if there are positive integers p, q with $\mathscr{L}^{\otimes p} \cong \mathscr{M}^{\otimes q}$; we set $\mathscr{L}_l := \mathscr{L}_k$ for some $k > 0$.

In particular,

$$X/\!\!/_l(G \times H) = \left(X/\!\!/_m G\right)/\!\!/_{n_k} H.$$

Proof. This is fairly easy to verify. The reader may consult [31], Proposition 1.3.1.
$\square$

Remark 3.1.4. It is a good trick to use this procedure also the "other way round", i.e., first take the H-quotient and then the G-quotient.

Apparently, we have, in general, infinitely many possibilities of linearizing a given action on a projective variety. Thus, we formulate the following question.

Problem. Do there exist infinitely many different GIT quotients?

In the rest of this lecture, we will demonstrate that the answer to this question is "no" and analyze the relationships between different quotients.

3.2. Polarized $\mathbb{C}^\star$-quotients

This is the easiest framework in which one may study the above problem. Yet, it is also of importance for the general case as we shall see. We first note the following result.

Proposition 3.2.1. *Let $\varrho\colon G \longrightarrow \mathrm{GL}(V)$ be a representation of the reductive group G, $\alpha\colon G \times \mathbb{P}(V) \longrightarrow \mathbb{P}(V)$ the induced G-action, and $l = (\varrho, \mathrm{id}_{\mathbb{P}(V)})$ the canonical linearization of this G-action. Then, up to isomorphy, all polarized GIT quotients are given by*

$$\left(\mathbb{P}(V)/\!\!/_{l_k^\chi} G, [\mathscr{L}_{l_k^\chi}]\right), \quad k > 0, \ \chi \in X(G).$$

Now, we may investigate the case of an action $\overline{\lambda}\colon \mathbb{C}^\star \times \mathbb{P}(V) \longrightarrow \mathbb{P}(V)$ more closely. Suppose $\overline{\lambda}$ comes from an action $\lambda\colon \mathbb{C}^\star \times V^\vee \longrightarrow V^\vee$, and let l be the canonical linearization as above. Using the above proposition, we will determine all GIT quotients of $\mathbb{P}(V)$ by the $\mathbb{C}^\star$-action $\overline{\lambda}$.

Recall from Lecture I, Example 1.2.2, ii), that $V^\vee$ decomposes as

$$V^\vee = \bigoplus_{i=1}^{m} V_i^\vee.$$

Here, $V_i^\vee$ denotes the non-trivial eigenspace of the character $\chi_{d_i}\colon \mathbb{C}^\star \longrightarrow \mathbb{C}^\star$, $z \longmapsto z^{d_i}$, and we assume $d_1 < \cdots < d_m$. Suppose $x = [l] \in \mathbb{P}(V)$ for $l \in V^\vee \setminus \{0\}$. Set

$$d_{\min}^l(x) \quad := \quad \min\{\, d_i \,|\, l \text{ has a non-trivial component in } V_i^\vee \,\}$$

$$d_{\max}^l(x) \quad := \quad \max\{\, d_i \,|\, l \text{ has a non-trivial component in } V_i^\vee \,\}.$$

With these quantities, we may characterize the semistable and polystable points as follows:

Proposition 3.2.2. i) *The point $x \in \mathbb{P}(V)$ is l-semistable, if and only if*

$$d^l_{\min}(x) \leq 0 \leq d^l_{\max}(x).$$

ii) *The point $x \in \mathbb{P}(V)$ is l-polystable, if and only if*

$$\text{either } d^l_{\min}(x) = 0 = d^l_{\max}(x) \text{ or } d^l_{\min}(x) < 0 < d^l_{\max}(x).$$

Proof. Suppose l has the coordinates $(l_1, \ldots, l_n)$ with respect to a basis of eigenvectors for $V^\vee$, such that the corresponding weights are non-decreasing. For $z \in \mathbb{C}^\star$, we find

$$z \cdot (l_1, \ldots, l_n) = (0, \ldots, 0, z^{d^l_{\min}(x)} \cdot l_{i_0}, \ldots, z^{d^l_{\max}(x)} \cdot l_{i_r}, 0, \ldots, 0).$$

This formula clearly implies our claim. $\qquad\square$

Set $l^d_k := l^{\chi - d}_k$, $d \in \mathbb{Z}$, $k > 0$. For a point $x \in \mathbb{P}(V)$, it is obvious that

$$d^{l^d_k}_{\min}\big(v_k(x)\big) = k \cdot d^l_{\min}(x) - d, \qquad d^{l^d_k}_{\max}\big(v_k(x)\big) = k \cdot d^l_{\max}(x) - d.$$

Together with Proposition 3.2.2, this implies the following.

Proposition 3.2.3. i) *A point $x \in \mathbb{P}(V)$ is l^d_k-semistable, if and only if*

$$d^l_{\min}(x) \leq \frac{d}{k} \leq d^l_{\max}(x).$$

ii) *A point $x \in \mathbb{P}(V)$ is l^d_k-polystable, if and only if*

$$\text{either } d^l_{\min}(x) = \frac{d}{k} = d^l_{\max}(x) \text{ or } d^l_{\min}(x) < \frac{d}{k} < d^l_{\max}(x).$$

Note that for any $x \in \mathbb{P}(V)$, there are $d \in \mathbb{Z}$ and $k \in \mathbb{Z}_{>0}$, such that x is l^d_k-polystable.

To an integer i with $1 \leq i \leq 2m$, we assign the following subset of $\mathbb{Q}$:

$$I_i := \begin{cases} \mathbb{Q} \setminus [d_1, d_m] & \text{if } i = 2m \\ \{d_{\frac{i+1}{2}}\} & \text{if } i \text{ is odd} \\ (d_{\frac{i}{2}}, d_{\frac{i}{2}+1}) & \text{if } i \neq 2m \text{ is even.} \end{cases}$$

These subsets parameterize the different notions of semistability:

Corollary 3.2.4. i) *We have*

$$\mathbb{P}(V)^{\mathrm{ss}}_{l^d_k} = \mathbb{P}(V)^{\mathrm{ss}}_{l^{d'}_{k'}},$$

if and only if there is an index $i \in \{1, \ldots, 2m\}$, such that I_i contains both d/k and d'/k'.

ii) *For i even, d, k with $d/k \in I_i$ and $d^\pm_i, k^\pm_i$ with $d^-_i/k^-_i = d_{i/2}$ and $d^+_i/k^+_i = d_{i/2+1}$*

$$\mathbb{P}(V)^{\mathrm{ss}}_{l^d_k} \subset \mathbb{P}(V)^{\mathrm{ss}}_{l^{d-}_{k-}} \qquad \text{and} \qquad \mathbb{P}(V)^{\mathrm{ss}}_{l^d_k} \subset \mathbb{P}(V)^{\mathrm{ss}}_{l^{d+}_{k+}}.$$

There are, thus, $2m$ notions of semistability for the given action $\overline{\lambda}$. (Note that, for the notion corresponding to I_{2m}, there a no semistable points at all, whereas the other notions do yield semistable points.) These yield the unpolarized quotients, and, by the second statement in the corollary, there is the "flip" diagram

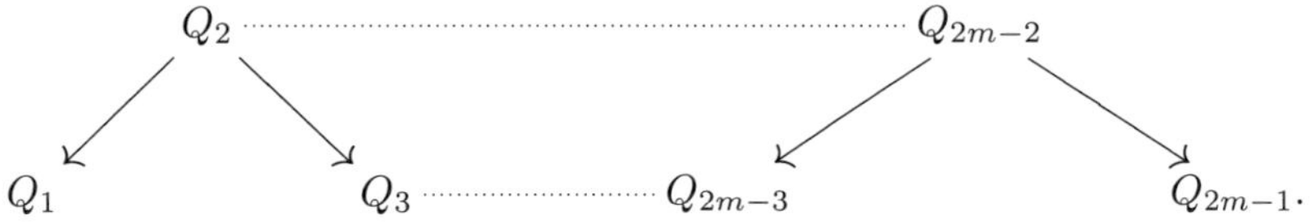

Remark 3.2.5. i) For $i = 1, \ldots, 2m$, there exists d, such that $Q_i = Q_2^d$, $Q_k^d :=$ $\mathbb{P}(V) /\!\!/_{l_k^d} \mathbb{C}^\star$.

ii) Let $i \in \{3, \ldots, 2m-3\}$ be an odd index, and set $Q_i^{\pm} := Q_{i\pm1}$. Note that the maps $\pi_i^{\pm} : Q_i^{\pm} \longrightarrow Q_i$ are isomorphisms outside the closed subset $\mathbb{P}(V_i) \subset Q_i$. The exceptional loci of π_i^+ and π_i^- are

$$
\begin{aligned}
\mathbb{P}_i^+ \quad &:= \quad \left\{\, v \in \mathbb{P}(V) \,\big|\, \lim_{z \to 0} z \cdot v \in \mathbb{P}(V_i) \right\} /\!\!/ \mathbb{C}^\star \\
&= \quad \left\{\, v = [l_i, \ldots, l_m] \in \mathbb{P}(V_i \oplus \cdots \oplus V_m) \,\big|\, l_i \neq 0 \right\} /\!\!/ \mathbb{C}^\star
\end{aligned}
$$

and

$$
\begin{aligned}
\mathbb{P}_i^- \quad &:= \quad \left\{\, v \in \mathbb{P}(V) \,\big|\, \lim_{z \to \infty} z \cdot v \in \mathbb{P}(V_i) \right\} /\!\!/ \mathbb{C}^\star \\
&= \quad \left\{\, v = [l_1, \ldots, l_i] \in \mathbb{P}(V_1 \oplus \cdots \oplus V_i) \,\big|\, l_i \neq 0 \right\} /\!\!/ \mathbb{C}^\star.
\end{aligned}
$$

Observe that $\mathbb{P}_i^+$ and $\mathbb{P}_i^-$ are weighted projective bundles over $\mathbb{P}(V_i)$. The quotients Q_i^- and Q_i^+ are birationally equivalent, and the birational transformation $Q_i^- \dashrightarrow Q_i^+$ is a weighted blow down followed by a weighted blow up.

iii) One may give an intrinsic description of the sets of semistable points in terms of the fixed point locus of the $\mathbb{C}^\star$-action (Białynicki-Birula/Sommese [4], Gross [16]): The connected components of the fixed point locus are $F_i := \mathbb{P}(V_i)$, $i = 1, \ldots, m$. Define, for $i = 1, \ldots, m$,

$$
\begin{aligned}
X_i^+ \quad &:= \quad \left\{\, v \in \mathbb{P}(V) \,\big|\, \lim_{z \to 0} z \cdot v \in F_i \right\} \\
&= \quad \mathbb{P}(V_i \oplus \cdots \oplus V_m) \setminus \mathbb{P}(V_{i+1} \oplus \cdots \oplus V_m) \\
X_i^- \quad &:= \quad \left\{\, v \in \mathbb{P}(V) \,\big|\, \lim_{z \to \infty} z \cdot v \in F_i \right\} \\
&= \quad \mathbb{P}(V_1 \oplus \cdots \oplus V_i) \setminus \mathbb{P}(V_1 \oplus \cdots \oplus V_{i-1})
\end{aligned}
$$

and, for $i < j$,

$$
\begin{aligned}
C_{ij} \quad &:= \quad (X_i^+ \setminus F_i) \cap (X_j^- \setminus F_j) \\
&= \quad \mathbb{P}(V_i \oplus \cdots \oplus V_j) \setminus \left(\mathbb{P}(V_{i+1} \oplus \cdots \oplus V_j) \cup \mathbb{P}(V_i \oplus \cdots \oplus V_{j-1}) \right).
\end{aligned}
$$

Corollary. i) *If* $k \cdot d_i - d \neq 0$ *for* $i = 1, \ldots, m$, *and* $i_0 := \max\{\, k \cdot d_i - d < 0 \,\}$, *then*

$$
\mathbb{P}(V)_{l_k^d}^{\mathrm{ss}} = \bigcup_{\substack{1 \leq i \leq i_0 \\ i_0 + 1 \leq j \leq m}} C_{ij}.
$$

ii) *If $k \cdot d_{i_0} - d = 0$, then*

$$\mathbb{P}(V)^{\mathrm{ss}}_{l^d_k} = X^-_{i_0} \cup X^+_{i_0} \cup \bigcup_{\substack{1 \le i \le i_0 - 1 \\ i_0 + 1 \le j \le m}} C_{ij}.$$

Example 3.2.6 (Induced polarizations). Let V be a finite-dimensional $\mathbb{C}$-vector space, and let λ be an action of $\mathbb{C}^\star$ on $V^\vee$, such that

$$V^\vee = V_1^\vee \oplus V_2^\vee,$$

$V_i^\vee$ being the eigenspace to the weight $z \longmapsto z^{d_i}$, $i = 1, 2$, and $d_1 < d_2$. For $d/k = d_i$, $i = 1$ or $i = 2$, the polarized quotient is $(\mathbb{P}(V_i), [\mathscr{O}_{\mathbb{P}(V_i)}(1)])$. For $d_1 < d/k < d_2$, the unpolarized quotient is $\mathbb{P}(V_1) \times \mathbb{P}(V_2)$, and the projection map

$$\pi^k_d \colon \mathbb{P}(V) \setminus \big(\mathbb{P}(V_1) \cup \mathbb{P}(V_2)\big) \longrightarrow \mathbb{P}(V_1) \times \mathbb{P}(V_2)$$

is the obvious one.

Claim. *The induced polarization $[\mathscr{L}^k_d]$ on $\mathbb{P}(V_1) \times \mathbb{P}(V_2)$ is given as*

$$\big[\mathscr{L}^k_d\big] = \big[\mathscr{O}_{\mathbb{P}(V_1) \times \mathbb{P}(V_2)}(kd_2 - d, -kd_1 + d)\big].$$

For given positive integers m, n, there are integers $d \in \mathbb{Z}$ and $k \in \mathbb{Z}_{>0}$, such that

$$\big[\mathscr{L}^k_d\big] = \big[\mathscr{O}_{\mathbb{P}(V_1) \times \mathbb{P}(V_2)}(m, n)\big].$$

Proof. For a representative $\mathscr{L} = \mathscr{O}_{\mathbb{P}(V_1) \times \mathbb{P}(V_2)}(m, n)$ of the polarization, we find

$$\pi^{k\,\star}_d \big(H^0(\mathscr{L})\big) = \mathrm{Sym}^m(V_1) \otimes \mathrm{Sym}^n(V_2).$$

This is an eigenspace to the character $\chi_{-(md_1 + nd_2) + ((m+n)/k)d}$. This character must be trivial, so that $-(md_1 + nd_2) + ((m+n)/k)d = 0$, that is,

$$m(kd_1 - d) + n(kd_2 - d) = 0.$$

For the second assertion, we have to find positive integers k and r and an integer d with

$$\begin{aligned} kd_2 - d &= rm \\ -kd_1 + d &= rn, \end{aligned}$$

but this is easy. $\qquad\qquad\square$

3.3. There are only finitely many GIT quotients

Let X be a projective algebraic variety and $\alpha \colon G \times X \longrightarrow X$ an action of the reductive group G on X. We may now answer the question raised in Section 3.1 in general:

Theorem 3.3.1 (Białynicki-Birula, Dolgachev/Hu). *For fixed α, there are only finitely many open subsets $U \subseteq X$ of the form X^{ss}_l, for l a linearization of α as above.*

Proof. This theorem was proved by Białynicki-Birula [2] in a setting which is far more general than the GIT which we are considering here and, independently and in the same framework as ours, by Dolgachev and Hu. The proof of the latter authors also builds on techniques developed by Białynicki-Birula but is quite involved. Adapting the strategy in [2] to GIT yields the following elementary proof.

Step 1. For $G = \mathbb{C}^\star$, we know the result: Let $F_1,\ldots,F_m$ be the connected components of the fixed point locus of the $\mathbb{C}^\star$-action. Then, any set of semistable points may be described in terms of a decomposition

$$\{\, 1,\ldots,m \,\} = P_1 \sqcup \cdots \sqcup P_s.$$

(Indeed, we have just seen this for $X = \mathbb{P}_n$ and, by Definition 2.4.1, it is also clear in general.) Since there are only finitely many possibilities for such a decomposition, we are done.

Step 2. For a torus $T \cong \mathbb{C}^{\star \times n}$, the assertion follows by induction: Write $T = \mathbb{C}^\star \times T'$ and use the fact that the quotient may be taken in two steps (Proposition 3.1.3).

Step 3. Fix a maximal torus $T \subset G$, and let l_T be the induced linearization of the resulting T-action. By the Hilbert-Mumford criterion, $x \in X$ is l-semistable, if, for any one parameter subgroup $\lambda \colon \mathbb{C}^\star \longrightarrow G$, one finds $\mu(\lambda, x) \geq 0$. The image of λ lies in a maximal torus T' of G. From the theory of algebraic groups, one knows that there is a $g \in G$, such that $g \cdot T' \cdot g^{-1} = T$, i.e., $g \cdot \lambda \cdot g^{-1}$ is a one parameter subgroup of T. We see

$$x \in X_l^{\mathrm{ss}} \iff \mu(g \cdot \lambda \cdot g^{-1}, x) \geq 0 \quad \text{for all } \lambda \colon \mathbb{C}^\star \longrightarrow T \text{ and all } g \in G.$$

One easily checks $\mu(g \cdot \lambda \cdot g^{-1}, x) = \mu(\lambda, g \cdot x)$, so that

$$x \in X_l^{\mathrm{ss}} \iff \mu(\lambda, g \cdot x) \geq 0 \quad \text{for all } \lambda \colon \mathbb{C}^\star \longrightarrow T \text{ and all } g \in G,$$

i.e.,

$$X_l^{\mathrm{ss}} = \bigcap_{g \in G} \left(g \cdot X_{l_T}^{\mathrm{ss}} \right).$$

Since there are only finitely many options for $X_{l_T}^{\mathrm{ss}}$, by Step 2, we are done. $\square$

3.4. The master space construction

Here, we will discuss how GIT quotients to different linearizations are related. These results have not been included into text books, so far.

Let $\varrho_1 \colon G \longrightarrow \mathrm{GL}(V_1)$ and $\varrho_2 \colon G \longrightarrow \mathrm{GL}(V_2)$ be two representations of the reductive group G, providing an action of G on $\mathbb{P}(V_1) \times \mathbb{P}(V_2)$. For every pair (m, n) of positive integers, we find the linearization $l_{m,n} = (\varrho_{m,n}, \iota_{m,n})$ of this action. Here, $\varrho_{m,n} := \mathrm{Sym}^m(\varrho_1) \otimes \mathrm{Sym}^n(\varrho_2)$, and $\iota_{m,n} \colon \mathbb{P}(V_1) \times \mathbb{P}(V_2) \hookrightarrow \mathbb{P}(\mathrm{Sym}^m(V_1) \otimes \mathrm{Sym}^n(V_2))$ is the product of the mth Veronese emdedding of $\mathbb{P}(V_1)$ with the nth Veronese embedding of $\mathbb{P}(V_2)$ followed by the Segre embedding.

On the other hand, we may form the representation $\tau := \varrho_1 \oplus \varrho_2 \colon G \longrightarrow \mathrm{GL}(V_1 \oplus V_2)$. This representation gives an action $\overline{\tau}$ of G on $\mathbb{P}(V_1 \oplus V_2)$ and a linearization l_τ of this action.

Furthermore, we introduce the auxiliary representation $\lambda\colon \mathbb{C}^\star \longrightarrow \mathrm{GL}(V_1 \oplus V_2)$, $z \mapsto z^{-1}\,\mathrm{id}_{V_1} \oplus z\,\mathrm{id}_{V_2}$. This representation yields a $\mathbb{C}^\star$-action $\overline{\lambda}$ on $\mathbb{P}(V_1 \oplus V_2)$ which commutes with $\overline{\tau}$, so that we find the action

$$
\begin{aligned}
\overline{\lambda} \times \overline{\tau}\colon (\mathbb{C}^\star \times G) \times \mathbb{P}(V_1 \oplus V_2) &\longrightarrow \mathbb{P}(V_1 \oplus V_2) \\
(z, g, x) &\longmapsto z \cdot (g \cdot x).
\end{aligned}
$$

For $k > 0$ and $d \in \mathbb{Z}$, there is the linearization $L_k^d = l_k^d \cdot \mathrm{Sym}^k(l_\tau)$. By Example 3.2.6 and Proposition 3.1.3,

$$
\begin{aligned}
\big(\mathbb{P}(V_1) \times \mathbb{P}(V_2)\big) /\!\!/_{l_{m,n}} G \; &= \; \big(\mathbb{P}(V_1 \oplus V_2) /\!\!/_{l_k^d}\mathbb{C}^\star\big) /\!\!/_{l_{m,n}} G \\
&= \; \mathbb{P}(V_1 \oplus V_2) /\!\!/_{L_k^d}(\mathbb{C}^\star \times G)
\end{aligned}
$$

with $m = k - d$ and $n = k + d$. Another application of Proposition 3.1.3 (see Remark 3.1.4) shows that all the quotients $\big(\mathbb{P}(V_1) \times \mathbb{P}(V_2)\big)/\!\!/_{l_{m,n}} G$, $m, n \in \mathbb{Z}_{>0}$, are $\mathbb{C}^\star$-quotients of $\mathbb{P}(V_1 \oplus V_2)/\!\!/_{l_\tau} G$, the $\mathbb{C}^\star$-action coming from $\overline{\lambda}$. All the possible linearizations of this $\mathbb{C}^\star$-action are of the form $l'^{\,d}_k$, $k > 0$, $d \in \mathbb{Z}$, with l' the linearization induced by l. Thus, we may derive structural results for the quotients $\big(\mathbb{P}(V_1) \times \mathbb{P}(V_2)\big)/\!\!/_{l_{m,n}} G$ from those for $\mathbb{C}^\star$-quotients

More generally, let $\overline{\sigma}\colon G \times X \longrightarrow X$ be an action of G on the projective variety X, and suppose we are given two linearizations $l_i = (\varrho_i, \iota_i)$, $i = 1, 2$, of $\overline{\sigma}$ with the representations $\varrho_1\colon G \longrightarrow \mathrm{GL}(V_1)$ and $\varrho_2\colon G \longrightarrow \mathrm{GL}(V_2)$ and the equivariant embeddings $\iota_1\colon X \hookrightarrow \mathbb{P}(V_1)$ and $\iota_2\colon X \hookrightarrow \mathbb{P}(V_2)$. Define $\mathscr{L}_i := \iota_i^\star(\mathscr{O}_{\mathbb{P}(V_i)}(1))$, $i = 1, 2$. Finally, we obtain the equivariant embedding $\iota\colon X \hookrightarrow \mathbb{P}(V_1) \times \mathbb{P}(V_2)$, $x \mapsto (\iota_1(x), \iota_2(x))$. We find the linearizations $l_{m,n} := l_1^{\otimes m} \otimes l_2^{\otimes n}$. For $m > 0$ and $n > 0$, we set $\eta := m/n$. Then,

$$
X^{(s)s}_{l_{m,n}} = X^{(s)s}_{l_{m',n'}}, \quad \text{if } m/n = m'/n',
$$

so that we may write $X^{(s)s}_\eta$ for $X^{(s)s}_{l_{m,n}}$. We also define $X^{(s)s}_0 := X^{(s)s}_{l_1}$ and $X^{(s)s}_\infty := X^{(s)s}_{l_2}$. Define $M := \mathbb{P}(\mathscr{L}_1 \oplus \mathscr{L}_2)$. The given linearizations l_1 and l_2 yield a G-action on M and an equivariant embedding $\kappa\colon M \hookrightarrow \mathbb{P}(V_1 \oplus V_2)$. As above, there is a linearized $\mathbb{C}^\star$-action on M, and we conclude that all the quotients $X/\!\!/_{l_{m,n}} G$, $\eta := m/n \in [0, \infty]$, are $\mathbb{C}^\star$-quotients of $M/\!\!/ G$.

Remark 3.4.1. The construction of $M/\!\!/ G$ goes back to Thaddeus [49]. One refers to $M/\!\!/ G$ as the **master space**.

Together with the results on $\mathbb{C}^\star$-action, we derive the following statement:

Theorem 3.4.2. *There are finitely many critical values* $\eta_1, \ldots, \eta_m \in (0, \infty) \cap \mathbb{Q}$, *such that, with* $\eta_0 := 0$ *and* $\eta_{m+1} := \infty$, *the following properties hold true:*
   i) *For* $i = 0, \ldots, m$ *and* $\eta, \eta' \in (\eta_i, \eta_{i+1})$:

$$
X^{(s)s}_\eta = X^{(s)s}_{\eta'}.
$$

ii) *For $i = 0, \ldots, m$ and $\eta \in (\eta_i, \eta_{i+1})$:*

$$X_\eta^{\mathrm{s}} \;\supset\; X_{\eta_i}^{\mathrm{s}} \cup X_{\eta_{i+1}}^{\mathrm{s}}$$
$$X_\eta^{\mathrm{ss}} \;\subset\; X_{\eta_i}^{\mathrm{ss}} \cap X_{\eta_{i+1}}^{\mathrm{ss}}.$$

Set $Q_i := X_{\eta_i}^{\mathrm{ss}}/\!/G$, $i = 0, \ldots, m+1$, and $\widetilde{Q}_i := X_\eta^{\mathrm{ss}}/\!/G$ for $\eta \in (\eta_i, \eta_{i+1})$, $i = 0, \ldots, m$. These quotients fit into the diagram

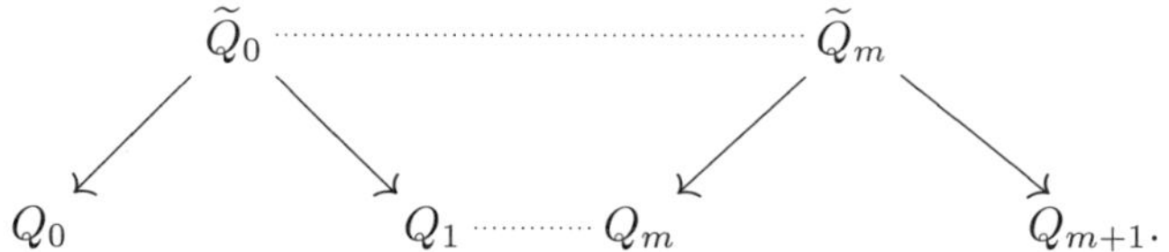

Here, the maps $\widetilde{Q}_i \dashrightarrow \widetilde{Q}_{i+1}$ are birational blow downs of a weighted projective bundle followed by a birational weighted projective blow up, $i = 0, \ldots, m-1$.

4. Lecture IV: Decorated principal bundles, semistable vector bundles

In this section, we work relative to a fixed smooth projective curve X over $\mathbb{C}$. If you prefer the language of complex analytic geometry, you may think of X as a compact Riemann surface.

4.1. The moduli problem of decorated principal bundles

In this section, we introduce the objects that we wish to classify. The resulting classification problem formally takes the place of the problem of forming the quotient of a variety by a group action in GIT.

Let $\mathscr{P}$ be a principal G-bundle[2] over X. Then, we may trivialize $\mathscr{P}$ in both the strong and the étale topology (the former may be more appealing to the intuition). Let F be an algebraic variety together with a G-action $\alpha \colon G \times F \longrightarrow F$. Then, we find the G-action

$$\begin{aligned}
(\mathscr{P} \times F) \times G &\longrightarrow \mathscr{P} \times F \\
((p, f), g) &\longmapsto (p \cdot g, g^{-1} \cdot f)
\end{aligned}$$

from the right, and the quotient

$$\mathscr{P}(F) := (\mathscr{P} \times F)/G \longrightarrow X$$

exists and is a fiber bundle over X with fiber F which is locally trivial in the strong and the étale topology.

The objects we would like to consider are pairs $(\mathscr{P}, \sigma)$ which consist of a principal G-bundle $\mathscr{P}$ and a section $\sigma \colon X \longrightarrow \mathscr{P}(F)$. Two such pairs $(\mathscr{P}_1, \sigma_1)$

---

[2]The reader who is not familiar with principal bundles may consult the references [1], [11], [42], and [45].

and $(\mathscr{P}_2, \sigma_2)$ are said to be **equivalent**, if there is an isomorphism $\psi\colon \mathscr{P}_1 \longrightarrow \mathscr{P}_2$ with

$$\sigma_2 = \psi(F) \circ \sigma_1, \quad \psi(F)\colon \mathscr{P}_1(F) \longrightarrow \mathscr{P}_2(F) \text{ being the induced isomorphism.}$$

In this and the next lecture, we will start with a representation $\varrho\colon G \longrightarrow \mathrm{GL}(V)$ and look at the induced action $\alpha\colon G \times \mathbb{P}(V) \longrightarrow \mathbb{P}(V)$. Abbreviate $\mathscr{P}_\varrho := \mathscr{P}(V)$.

Remark 4.1.1. A pair $(\mathscr{P}, \sigma\colon X \longrightarrow \mathbb{P}(\mathscr{P}_\varrho))$ is a relative version of a point x in the G-variety $\mathbb{P}(V)$: Note that G acts on itself by conjugation, and $\mathscr{G} := \mathscr{P}(G) \longrightarrow X$ is a group scheme over X (indeed, it can easily be seen to be the bundle of local G-bundle automorphisms of $\mathscr{P}$). The projective bundle $\mathbb{P}(\mathscr{P}_\varrho)$ comes with the action

$$\alpha_X\colon \mathscr{G} \times_X \mathbb{P}(\mathscr{P}_\varrho) \longrightarrow \mathbb{P}(\mathscr{P}_\varrho).$$

The section $\sigma\colon X \longrightarrow \mathbb{P}(\mathscr{P}_\varrho)$ is then a family of points in the $\mathscr{G}_{|\{x\}}$-varieties $\mathbb{P}(\mathscr{P}_{\varrho|\{x\}})$, $x \in X$. More generally, we could allow any projective algebraic manifold X to be the base variety. The case $X = \{\mathrm{pt}\}$ corresponds to GIT on a projective space which we have outlined in Lecture II.

Note that, in the above definition, we cannot replace a principal G-bundle over X by a group scheme $\mathscr{G} \longrightarrow X$ with fiber G: $\mathscr{G}$ is an $\mathrm{Aut}(G)$-bundle, and $G \longrightarrow \mathrm{Aut}(G)$ is, in general, neither injective nor surjective.

To give a section

$$\sigma\colon X \longrightarrow \mathscr{P}\big(\mathbb{P}(V)\big) = \mathbb{P}\big(\mathscr{P}_\varrho\big),$$

one has to give a line bundle $\mathscr{L}$ on X and a surjection

$$\varphi\colon \mathscr{P}_\varrho \longrightarrow \mathscr{L},$$

and two pairs $(\mathscr{L}, \varphi)$ and $(\mathscr{L}', \varphi')$ give the same section, if and only if there is an isomorphism $\chi\colon \mathscr{L} \longrightarrow \mathscr{L}'$, such that

$$\varphi' = \chi \circ \varphi.$$

Surjectivity is an open condition on parameter spaces. Hence, in order to find compact (projective) moduli spaces, we introduce more general objects:

Definition 4.1.2. i) A ϱ**-pair** is a triple $(\mathscr{P}, \mathscr{L}, \varphi)$, consisting of a principal G-bundle $\mathscr{P}$, a line bundle $\mathscr{L}$, and a non-zero homomorphism $\varphi\colon \mathscr{P}_\varrho \longrightarrow \mathscr{L}$.

ii) The **type of the ϱ-pair** $(\mathscr{P}, \mathscr{L}, \varphi)$ is the pair $(\tau, \deg(\mathscr{L}))$, where $\tau \in \pi_1(G)$ classifies $\mathscr{P}$ as a topological G-bundle.

iii) Two ϱ-pairs $(\mathscr{P}_1, \mathscr{L}_1, \varphi_1)$ and $(\mathscr{P}_2, \mathscr{L}_2, \varphi_2)$ are said to be **equivalent**, if there are isomorphisms $\psi\colon \mathscr{P}_1 \longrightarrow \mathscr{P}_2$ and $\chi\colon \mathscr{L}_1 \longrightarrow \mathscr{L}_2$, such that

$$\varphi_2 = \chi \circ \varphi_1 \circ \psi_\varrho^{-1}, \quad \psi_\varrho\colon \mathscr{P}_{1,\varrho} \longrightarrow \mathscr{P}_{2,\varrho} \text{ being the induced isomorphism.}$$

The Classification Problem. Fix the type (τ, d), $\tau \in \pi_1(G)$, $d \in \mathbb{Z}$, and classify ϱ-pairs of type (τ, d) up to equivalence.

The basic difficulty in attacking this problem is that, even if we fix the type of the ϱ-pairs under consideration, they cannot be parameterized in a reasonable way by an algebraic variety (see Section 4.3). Thus, we will have to define a priori a concept of semistability which meets the following requirements:

- There exist a projective variety $\mathfrak{P}$ and an open subset $U \subseteq \mathfrak{P}$ which (over-)parameterizes the semistable ϱ-pairs of given type.
- There are a vector space Y and a $\mathrm{GL}(Y)$-action on $\mathfrak{P}$ which leaves U invariant, such that two points p_1, $p_2 \in U$ lie in the same orbit, if and only if they correspond to equivalent ϱ-pairs.
- There is a linearization l of the $\mathrm{GL}(Y)$-action on $\mathfrak{P}$, such that the set of l-semistable points is U.

If we can achieve this, the projective variety

$$\mathscr{M}(\varrho)_{\tau/d} := \mathfrak{P} /\!\!/_l \mathrm{GL}(Y)$$

will be the moduli space for semistable ϱ-pairs of type (τ, d).

4.2. Examples

The following two examples of specific groups and specific representations illustrate how the above abstract classification problem plays a role in the classification of certain projective algebraic varieties.

Families of hypersurfaces. We take $G = \mathrm{GL}_n(\mathbb{C})$ and $\varrho\colon G \longrightarrow \mathrm{GL}(\mathrm{Sym}^d(\mathbb{C}^n))$. Here, we will work with vector bundles of rank n rather than with principal G-bundles.

Let $(\mathscr{E}, \mathscr{L}, \varphi)$ be a ϱ-pair. This defines a geometric object: For this, let

$$P(\mathscr{E}) := \mathbb{P}(\mathscr{E}^\vee) = \mathscr{P}\!\mathit{roj}\big(\mathscr{S}\mathit{ym}^\star(\mathscr{E}^\vee)\big) \xrightarrow{\ \pi\ } X$$

be the projectivization of $\mathscr{E}^\vee$ in Grothendieck's sense. Let $D \subseteq P(\mathscr{E})$ be an effective divisor. Its associated line bundle is of the form

$$\mathscr{O}_{P(\mathscr{E})}(D) = \mathscr{O}_{P(\mathscr{E})}(d) \otimes \pi^\star(\mathscr{L})$$

for a unique positive integer d and a unique line bundle $\mathscr{L}$ on X. Thus, D is the zero divisor of a section

$$s\colon \mathscr{O}_{P(\mathscr{E})} \longrightarrow \mathscr{O}_{P(\mathscr{E})}(d) \otimes \pi^\star(\mathscr{L}).$$

We project this homomorphism to X in order to obtain

$$\pi_\star(s)\colon \mathscr{O}_X \longrightarrow \mathscr{S}\mathit{ym}^d(\mathscr{E}^\vee) \otimes \mathscr{L}.$$

Now, as representations,

$$\mathrm{Sym}^d(\mathbb{C}^{n\vee}) \cong \mathrm{Sym}^d(\mathbb{C}^n)^\vee \quad \text{(see (1) in Remark 2.6.2).}$$

Thus, s corresponds to a non-trivial homomorphism

$$\varphi\colon \mathscr{S}\mathit{ym}^d(\mathscr{E}) \longrightarrow \mathscr{L}.$$

We also have a map $\pi_D \colon D \longrightarrow X$, and its fibers are

$$\pi_D^{-1}(\{x\}) = \begin{cases} \text{hypersurface of degree } d, & \text{if } \varphi \text{ is surjective in } x \\ \mathbb{P}(\mathscr{E}\langle x\rangle^\vee), & \text{else} \end{cases}.$$

Hence, a ϱ-pair $(\mathscr{E}, \mathscr{L}, \varphi)$ basically describes a family of hypersurfaces of degree d (inside $P(\mathscr{E})$), and equivalence is a relative version of projective equivalence.

Figure 1 is an illustration of the real part of the affine part of a surface which is fibered over $\mathbb{P}_1$ in plane cubic curves. It was generated with Polyray[3].

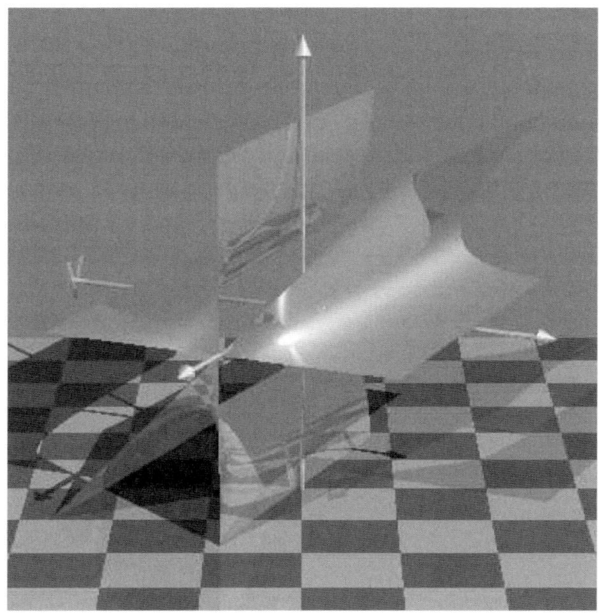

FIGURE 1. The surface $\{\,(t-1)(x^2 y - x y^2) + t = 0\,\}$

Dimensional reduction. Here, we choose $G = \mathrm{GL}_{n_1}(\mathbb{C}) \times \mathrm{GL}_{n_2}(\mathbb{C})$ as the group and $\varrho \colon G \longrightarrow \mathrm{GL}(\mathrm{Hom}(\mathbb{C}^{n_2}, \mathbb{C}^{n_1}))$ as the representation.

Let $\mathscr{E}$ and $\mathscr{F}$ be two vector bundles on X of rank n_1 and n_2, respectively. Suppose that, on $X \times \mathbb{P}_1$, we are given an extension

$$\varepsilon \colon \{0\} \longrightarrow \pi_X^\star(\mathscr{F}) \longrightarrow \mathscr{A} \longrightarrow \pi_X^\star(\mathscr{E}) \otimes \pi_{\mathbb{P}_1}^\star\big(\mathscr{O}_{\mathbb{P}_1}(2)\big) \longrightarrow \{0\}.$$

Then,

$$\varepsilon \in \mathrm{Ext}^1\big(\pi_X^\star(\mathscr{E}) \otimes \pi_{\mathbb{P}_1}^\star(\mathscr{O}_{\mathbb{P}_1}(2)), \pi_X^\star(\mathscr{F})\big) = H^0\big(\mathscr{E}^\vee \otimes \mathscr{F}\big) = \mathrm{Hom}(\mathscr{E}, \mathscr{F}).$$

---

[3] http://wims.unice.fr/wims/en_tool~geometry~polyray.en.html

Thus, a ϱ-pair $((\mathscr{E},\mathscr{F}),\mathscr{O}_X,\varphi)$ describes extensions and vector bundles on the smooth projective surface $X \times \mathbb{P}_1$. Since the ϱ-pair lives in one dimension lower, namely on the curve X, one speaks of **dimensional reduction**.

The corresponding projective algebraic manifolds are the projective bundles $\mathbb{P}(\mathscr{A})$ over $X \times \mathbb{P}_1$.

4.3. Bounded families of vector bundles

The main difficulty in parameterizing decorated vector bundles with certain properties consists in parameterizing the occurring vector bundles themselves. In this section, we will work out the corresponding conditions which permit to do so.

Definition 4.3.1. Fix integers $r > 0$ and d (the topological invariants). Let $\mathfrak{S}$ be a set of isomorphism classes of vector bundles $\mathscr{E}$ on X of rank r and degree d. We say that $\mathfrak{S}$ is **bounded**, if there exist an algebraic variety S and a vector bundle $\mathscr{E}_S$ on $S \times X$, such that for every class $[\mathscr{E}] \in \mathfrak{S}$, there is a point $s \in S$ with

$$\mathscr{E} \cong \mathscr{E}_{S|\{s\}\times X}.$$

The (relative) Serre vanishing theorem and the base change theorem for cohomology imply:

Proposition 4.3.2. *Let x_0 be a point in X and $\mathscr{O}_X(1) := \mathscr{O}_X(x_0)$. If $\mathfrak{S}$ is bounded, then there is a natural number n_0, such that, for every vector bundle $\mathscr{E}$ with $[\mathscr{E}] \in \mathfrak{S}$ and for every $n \geq n_0$:*

- *$\mathscr{E}(n) := \mathscr{E} \otimes \mathscr{O}_X(1)^{\otimes n}$ is globally generated.*
- *$H^1(\mathscr{E}(n)) = \{0\}$.*

Remark 4.3.3. For $r \geq 2$, the set of isomorphy classes of vector bundles of rank r and degree d is not(!) bounded. Indeed, in the set of vector bundles

$$\mathscr{E}_s := \mathscr{O}_X(-s) \oplus \mathscr{O}_X(d+s) \oplus \mathscr{O}_X^{\oplus(r-2)}, \;\; s \in \mathbb{N},$$

the bundle $\mathscr{E}_{n+1}(n)$ is not globally generated, $n \geq 0$.

By the Riemann-Roch theorem,

$$h^0(\mathscr{E}(n)) - h^1(\mathscr{E}(n)) = rn + d + r(1-g).$$

Fix $n \geq n_0$ and a complex vector space Y of dimension $rn + d + r(1-g)$. Our observations yield the following necessary condition for boundedness:

Corollary 4.3.4. *If $\mathfrak{S}$ is bounded, then every vector bundle $\mathscr{E}$ with $[\mathscr{E}] \in \mathfrak{S}$ may be written as a quotient*

$$q \colon Y \otimes \mathscr{O}_X(-n) \longrightarrow \mathscr{E},$$

such that $H^0(q(n)) \colon Y \longrightarrow H^0(\mathscr{E}(n))$ is an isomorphism.

The following theorem introduces the most fundamental of all parameter spaces for vector bundles.

Theorem 4.3.5 (Grothendieck's quot scheme). *Fix $r > 0$ and d, and let $\mathscr{A}$ be a coherent $\mathscr{O}_X$-module. Then, there are a projective scheme $\mathfrak{Q}$ and a flat family*

$$q_{\mathfrak{Q}} \colon \pi_X^\star(\mathscr{A}) \longrightarrow \mathscr{F}_{\mathfrak{Q}}$$

on $\mathfrak{Q} \times X$, such that for every sheaf $\mathscr{F}$ of rank r and degree d and every quotient

$$q \colon \mathscr{A} \longrightarrow \mathscr{F},$$

there is a point $t \in \mathfrak{Q}$ with[4]

$$q \sim q_{\mathfrak{Q}|\{t\}\times X}.$$

In particular, the condition obtained in Proposition 4.3.2 is equivalent to the boundedness of $\mathfrak{S}$. After these preparations, we may formulate the necessary and sufficient criterion for boundedness which we are going to use in our applications:

Proposition 4.3.6. *The family $\mathfrak{S}$ is bounded, if and only if there exists a constant C, such that*

$$\mu_{\max}(\mathscr{E}) := \max\left\{ \mu(\mathscr{F}) := \frac{\deg(\mathscr{F})}{\mathrm{rk}(\mathscr{F})} \mid \{0\} \subsetneq \mathscr{F} \subseteq \mathscr{E} \text{ a subbundle} \right\} \leq \mu(\mathscr{E}) + C.$$

Proof. We start with the direction "$\Longrightarrow$". Fix an n_0, such that $h^1(\mathscr{E}(n_0)) = 0$ for every $\mathscr{E}$ with $[\mathscr{E}] \in \mathfrak{S}$. If there were no bound on $\mu_{\max}(\mathscr{E})$, then we would find an extension

$$\{0\} \longrightarrow \mathscr{F} \longrightarrow \mathscr{E} \longrightarrow \mathscr{Q} \longrightarrow \{0\}$$

with $[\mathscr{E}] \in \mathfrak{S}$ and $\mu(\mathscr{Q}) < -n_0 + g - 1$. We compute

$$\frac{h^1(\mathscr{E}(n_0))}{\mathrm{rk}(\mathscr{Q})} = \frac{h^0(\mathscr{E}^\vee(-n_0) \otimes \omega_X)}{\mathrm{rk}(\mathscr{Q})} \geq \frac{h^0(\mathscr{Q}^\vee(-n_0) \otimes \omega_X)}{\mathrm{rk}(\mathscr{Q})} \geq -\mu(\mathscr{Q}) - n_0 + g - 1 > 0,$$

a contradiction.

Next, we prove the direction "$\Longleftarrow$". Let n be such that

$$H^1(\mathscr{E}(n)) = H^0(\mathscr{E}^\vee(-n) \otimes \omega_X)^\vee = \mathrm{Hom}(\mathscr{E}(n), \omega_X)^\vee \neq \{0\},$$

and let $\varphi \colon \mathscr{E}(n) \longrightarrow \omega_X$ be a non-trivial homomorphism. Then, we get the extension

$$\{0\} \longrightarrow \mathscr{F} := \ker(\varphi) \longrightarrow \mathscr{E}(n) \longrightarrow \mathscr{L} := \varphi(\mathscr{E}(n)) \longrightarrow \{0\}$$

and

$$\begin{aligned}
rn + d \;&=\; \deg(\mathscr{E}(n)) = \deg(\mathscr{F}) + \deg(\mathscr{L}) = (r-1)\mu(\mathscr{F}) + \deg(\mathscr{L}) \;\leq\; \\
&\leq\; (r-1)\frac{d}{r} + (r-1)n + (r-1)C + 2g - 2.
\end{aligned}$$

So, if $n > n_0 := -(d/r) + (r-1)C + 2g - 2$, this inequality will be violated and we must have $H^1(E(n)) = \{0\}$.

---

[4]The equivalence relation used in the following line is: $q \colon \mathscr{A} \longrightarrow \mathscr{F} \sim q' \colon \mathscr{A} \longrightarrow \mathscr{F}'$, if there is an isomorphism $\psi \colon \mathscr{F} \longrightarrow \mathscr{F}'$ with $q' = \psi \circ q$, i.e., if $\ker(q) = \ker(q')$.

Similarly, one finds an n_0, such that, for every $[\mathscr{E}] \in \mathfrak{S}$, every $n \geq n_0$, and every $x \in X$, one has

$$H^1(\mathscr{E}(n)(-x)) = \{0\}.$$

Thus, we arrive at the exact sequence

$$H^0(\mathscr{E}(n)(-x)) \longrightarrow H^0(\mathscr{E}(n)) \longrightarrow \mathscr{E}(n)\langle x \rangle \longrightarrow \{0\}\ .$$

This shows that $\mathscr{E}(n)$ is globally generated at $x \in X$. Since x can be any point on the curve, the proof is now complete. $\qquad\square$

4.4. The moduli space of semistable vector bundles

In this section, we sketch one of the best known constructions of a moduli space with GIT, namely, of the moduli space of semistable vector bundles on X, originally due to Seshadri. On the one hand, this serves as a nice illustration of the techniques which we have presented up to now. On the other hand, it will help to build our intuition how the semistability concept for decorated vector bundles may look like. More details are contained in the books [21] and [25].

Definition 4.4.1. A vector bundle $\mathscr{E}$ on X is called **(semi)stable**, if for every non-trivial proper subbundle $\{0\} \subsetneq \mathscr{F} \subsetneq \mathscr{E}$, the inequality

$$\mu(\mathscr{F})(\leq)\mu(\mathscr{E})$$

is satisfied.

By Proposition 4.3.6, the family of isomorphism classes of semistable vector bundles of fixed rank and degree is bounded. Given $r > 0$ and $d \in \mathbb{Z}$, we may therefore choose an n, such that every semistable vector bundle $\mathscr{E}$ of rank r and degree d has the following properties:

- $\mathscr{E}(n)$ is globally generated.
- $H^1(\mathscr{E}(n)) = \{0\}$.

Fix a complex vector space Y of dimension $p = rn + d + r(1 - g)$, and let $\mathfrak{Q}$ be the quot scheme parameterizing the quotients of $Y \otimes \mathscr{O}_X(-n)$ of rank r and degree d. One can then show the following.

Proposition 4.4.2. *There are open subsets $U^{(\mathrm{s})\mathrm{s}} \subset \mathfrak{Q}$, such that $[q\colon Y \otimes \mathscr{O}_X(-n) \longrightarrow \mathscr{F}] \in U^{(\mathrm{s})\mathrm{s}}$, if and only if*

- *$\mathscr{F}$ is a (semi)stable vector bundle.*
- *$H^0(q(n))\colon Y \longrightarrow H^0(\mathscr{F}(n))$ is an isomorphism.*

Next, we have the action

$$\alpha\colon \mathrm{GL}(Y) \times \mathfrak{Q} \longrightarrow \mathfrak{Q}$$

$$g \cdot [q\colon Y \otimes \mathscr{O}_X(-n) \longrightarrow \mathscr{E}] := \left[\, Y \otimes \mathscr{O}_X(-n) \xrightarrow{g^{-1}\otimes\mathrm{id}} Y \otimes \mathscr{O}_X(-n) \xrightarrow{q} \mathscr{E} \,\right].$$

Clearly, U^{ss} and U^{s} are $\mathrm{GL}(Y)$-invariant. One easily checks the following:

Lemma 4.4.3. *Two points $[q_{1,2}\colon Y \otimes \mathscr{O}_X(-n) \longrightarrow \mathscr{E}_{1,2}]$ from U^{ss} lie in the same $\mathrm{GL}(Y)$-orbit, if and only if $\mathscr{E}_1$ and $\mathscr{E}_2$ are isomorphic.*

Remark 4.4.4. Since the action of the center $\mathbb{C}^{\star} \cdot \mathrm{id}_Y \subset \mathrm{GL}(Y)$ is trivial, we may restrict to the induced action of $\mathrm{SL}(Y)$.

Theorem 4.4.5 (Simpson). *There is a linearization l of the $\mathrm{SL}(Y)$-action on $\mathfrak{Q}$, such that*

$$\mathfrak{Q}_l^{(\mathrm{s})\mathrm{s}} \cap \overline{U^{\mathrm{ss}}} = U^{(\mathrm{s})\mathrm{s}}.$$

Thus, we may define

$$\mathscr{M}^{(\mathrm{s})\mathrm{s}}(r,d) := U^{(\mathrm{s})\mathrm{s}} /\!\!/ \mathrm{SL}(Y).$$

The moduli space $\mathscr{M}^{\mathrm{s}}(r,d)$ is a (smooth) subvariety of the projective variety $\mathscr{M}^{\mathrm{ss}}(r,d)$ which parameterizes isomorphy classes of stable vector bundles of rank r and degree d on X. The moduli space $\mathscr{M}^{\mathrm{ss}}(r,d)$ parameterizes S-equivalence classes of semistable vector bundles of rank r and degree d on X. We have to explain the concept of S-equivalence.

Proposition 4.4.6 (Jordan-Hölder filtration). *Every semistable vector bundle possesses a filtration*

$$\{0\} =: \mathscr{E}_0 \subsetneq \mathscr{E}_1 \subsetneq \cdots \subsetneq \mathscr{E}_s \subsetneq \mathscr{E}_{s+1} := \mathscr{E}$$

by subbundles with $\mu(\mathscr{E}_i) = \mu(\mathscr{E})$, $i = 1,\ldots,s$, such that $\mathscr{E}_{i+1}/\mathscr{E}_i$ is stable, $i = 0,\ldots,s$.
   The **associated graded object**

$$\mathrm{gr}(\mathscr{E}) := \bigoplus_{i=0}^{s} \mathscr{E}_{i+1}/\mathscr{E}_i$$

is – up to isomorphy – independent of the filtration.

This result motivates the next definition.

Definition 4.4.7. Two semistable vector bundles $\mathscr{E}_1$ and $\mathscr{E}_2$ are said to be **S-equivalent**, if

$$\mathrm{gr}(\mathscr{E}_1) \cong \mathrm{gr}(\mathscr{E}_2).$$

So, as in the abstract GIT setting, we have the class of stable vector bundles whose set of isomorphism classes can be parameterized nicely by an algebraic variety, the moduli space. But, in general (precisely when r and d are not coprime), it is only quasi-projective. We compactify it with semistable vector bundles. In order to do so, we have to alter the equivalence relation on semistable but not stable bundles. The resulting relation of S-equivalence reflects the possible degenerations among the $\mathrm{SL}(Y)$-orbits in U^{ss}.

Illustration. In order to illustrate the relationship between the notion of a semi-stable vector bundle and the Hilbert-Mumford criterion, we give a sample computation, using an older approach by Gieseker.

To simplify matters even further, we choose a line bundle $\mathcal{N}$ on X and look at vector bundles $\mathcal{E}$ with $\det(\mathcal{E}) \cong \mathcal{N}$. There is a closed subscheme $U_{\mathcal{N}}^{\mathrm{ss}} \subset U^{\mathrm{ss}}$ which parameterizes those semistable vector bundles with determinant $\mathcal{N}$. To a point

$$q \colon Y \otimes \mathcal{O}_X(-n) \longrightarrow \mathcal{E}$$

in $U_{\mathcal{N}}^{\mathrm{ss}}$, we associate

$$\bigwedge^r (q(n)) \colon \bigwedge^r Y \otimes \mathcal{O}_X \longrightarrow \det(\mathcal{E}(n)) \cong \mathcal{N}(rn)$$

and

$$f := H^0\left(\bigwedge^r (q(n))\right) \in \mathbb{H} := \mathrm{Hom}\left(\bigwedge^r Y, H^0(\mathcal{N}(rn))\right).$$

The assignment $q \mapsto [f]$ induces an injective and $\mathrm{SL}(Y)$-equivariant morphism

$$F \colon U_{\mathcal{N}}^{\mathrm{ss}} \longrightarrow \mathbb{P}(\mathbb{H}^{\vee}).$$

On the right-hand space, we have a natural notion of semistability which we might test with the Hilbert-Mumford criterion.

One parameter subgroups of a special linear group. Before we can seriously evaluate the Hilbert-Mumford criterion, we have to pause a moment in order to discuss the structure of one parameter subgroups of $\mathrm{SL}(Y)$.

Let $\lambda \colon \mathbb{C}^\star \longrightarrow \mathrm{SL}(Y)$ be a one parameter subgroup of $\mathrm{SL}(Y)$. Then, we know from Lecture I, Example 1.2.2, ii), that there are a basis $\underline{y} = (y_1, \ldots, y_p)$ for Y and an integral weight vector $\underline{\gamma} = (\gamma_1, \ldots, \gamma_p)$ with

- $\gamma_1 \leq \cdots \leq \gamma_p$, $\sum_{i=1}^{p} \gamma_i = 0$, and
- $\lambda(z) \cdot \sum_{i=1}^{p} c_i y_i = \sum_{i=1}^{p} z^{\gamma_i} c_i y_i$, $\forall z \in \mathbb{C}^\star$.

Conversely, the datum of a basis $\underline{y}$ for Y and of a weight vector as above determine a one parameter subgroup $\lambda(\underline{y}, \underline{\gamma})$ of $\mathrm{SL}(Y)$.

Finally, we define the **basic weight vectors**

$$\gamma^{(i)} := \big(\underbrace{i - p, \ldots, i - p}_{i\times}, \underbrace{i, \ldots, i}_{(p-i)\times} \big), \quad i = 1, \ldots, p - 1.$$

Lemma 4.4.8. *For a weight vector* $\underline{\gamma} = (\gamma_1, \ldots, \gamma_p)$ *with* $\gamma_1 \leq \cdots \leq \gamma_p$ *and* $\sum_{i=1}^{p} \gamma_i = 0$, *one finds*

$$\underline{\gamma} = \sum_{i=1}^{p-1} \frac{\gamma_{i+1} - \gamma_i}{p} \cdot \gamma^{(i)}.$$

Back to our problem. Let $\lambda = \lambda(\underline{y}, \underline{\gamma})$ be a one parameter subgroup of $\mathrm{SL}(Y)$. Set

$$I := \left\{ \underline{i} = (i_1, \ldots, i_r) \in \{1, \ldots, p\}^{\times r} \,\middle|\, i_1 < \cdots < i_r \right\}.$$

Then, the elements

$$y_{\underline{i}} = y_{i_1} \wedge \cdots \wedge y_{i_r}, \quad \underline{i} \in I,$$

form a basis for $\bigwedge^r Y$ which consists of eigenvectors for the one parameter subgroup $\lambda(\underline{y}, \underline{\gamma})$. For a point $[f] \in \mathbb{P}(\mathbb{H}^\vee)$, one checks

$$\mu\big(\lambda(\underline{y}, \underline{\gamma}), [f]\big) = -\min\left\{ \gamma_{i_1} + \cdots + \gamma_{i_r} \,\middle|\, f(y_{\underline{i}}) \neq 0, \ i \in I \right\}.$$

A closer inspection of this formula gives the following result.

Lemma 4.4.9. *For a point $q \in U^{\mathrm{ss}}$ and a basis $\underline{y}$ of Y, define $\underline{i}^0 = (i_1^0, \ldots, i_r^0)$ with*

$$i_j^0 := \min\left\{ k = 1, \ldots, p \,\middle|\, \mathrm{rk}\big(q(\langle y_1, \ldots, y_k \rangle \otimes \mathcal{O}_X(-n))\big) = j \right\},$$

$j = 1, \ldots, r$. Then, for any weight vector $\underline{\gamma}$ as above,

$$\mu\big(\lambda(\underline{y}, \underline{\gamma}), F(q)\big) = -\gamma_{i_1^0} - \cdots - \gamma_{i_r^0}.$$

In particular, for $\underline{\gamma} = \sum_{i=1}^{p-1} \alpha_i \gamma^{(i)}$,

$$\mu\big(\lambda(\underline{y}, \underline{\gamma}), F(q)\big) = \sum_{i=1}^{p-1} \alpha_i \cdot \mu\big(\lambda(\underline{y}, \gamma^{(i)}), F(q)\big).$$

As a consequence of this lemma, we have to work only with the basic weight vectors. For $q \colon Y \otimes \mathcal{O}_X(-n) \longrightarrow \mathcal{E}$ and $i \in \{1, \ldots, p-1\}$, let $\mathcal{F}_i$ be the subbundle generated by

$$q\big(\langle y_1, \ldots, y_i \rangle \otimes \mathcal{O}_X(-n)\big).$$

Then,

$$\mu\big(\lambda(\underline{y}, \gamma^{(i)}), F(q)\big) = \mathrm{rk}(\mathcal{F}_i) \cdot p - r \cdot i.$$

Since $i \leq h^0(\mathcal{F}_i)$, we see:

Proposition 4.4.10. *The point $F(q) \in \mathbb{P}(\mathbb{H}^\vee)$ is (semi)stable, if and only if*

$$\frac{h^0(\mathcal{F}(n))}{\mathrm{rk}(\mathcal{F})} (\leq) \frac{h^0(\mathcal{E}(n))}{\mathrm{rk}(\mathcal{E})}$$

for every non-trivial subbundle $\{0\} \subsetneq \mathcal{F} \subsetneq \mathcal{E}$.

A difficult argument shows that one may restrict to subbundles with $h^1(\mathcal{F}(n)) = 0$. Then, the condition from the proposition becomes

$$\frac{h^0(\mathcal{F}(n))}{\mathrm{rk}(\mathcal{F})} = \mu(\mathcal{F}) + n + 1 - g (\leq) \mu(\mathcal{E}) + n + 1 - g = \frac{h^0(\mathcal{E}(n))}{\mathrm{rk}(\mathcal{F})}.$$

Corollary 4.4.11. *For $q \in U_{\mathcal{N}}^{(\mathrm{s})\mathrm{s}}$, the point $F(q)$ is (semi)stable.*

In the final step, one has to show the following:

Proposition 4.4.12. *The morphism $F \colon U_{\mathcal{N}}^{\mathrm{ss}} \longrightarrow \mathbb{P}(\mathbb{H}^\vee)^{\mathrm{ss}}$ is proper.*

5. Lecture V: Semistable decorated principal bundles

In this lecture, we will first introduce the concept of semistability for ϱ-pairs with the structure group $\mathrm{GL}_r(\mathbb{C})$ and then discuss elements of the construction of their moduli spaces from our paper [37] (which also contains additional information on the subject). At the end of the lecture, we define the notion of semistability for ϱ-pairs with semisimple structure group.

5.1. Decorated vector bundles

Before we come to the definition of semistability for decorated vector bundles, we will rewrite the Hilbert-Mumford criterion for $\mathrm{SL}_r(\mathbb{C})$.

We fix a representation $\varrho\colon \mathrm{GL}_r(\mathbb{C}) \longrightarrow \mathrm{GL}(V)$ which we assume to be homogeneous, i.e., we assume that there is an integer α, such that

$$\varrho\bigl(z \cdot \mathbb{E}_n\bigr) = z^\alpha \cdot \mathrm{id}_V, \quad \forall z \in \mathbb{C}^\star.$$

A one parameter subgroup $\lambda\colon \mathbb{C}^\star \longrightarrow \mathrm{SL}_r(\mathbb{C})$ leads to a decomposition

$$\mathbb{C}^r =: W = W_{\gamma_1} \oplus \cdots \oplus W_{\gamma_{s+1}}$$

where W_{γ_i} is the eigenspace to the character $z \longmapsto z^{\gamma_i}$, $i = 1, \ldots, s+1$, and $\gamma_1 < \cdots < \gamma_{s+1}$. Set $W_i := W_{\gamma_1} \oplus \cdots \oplus W_{\gamma_i}$, $i = 1, \ldots, s$, in order to find the (partial) flag

$$W^\bullet(\lambda) : \{0\} \subsetneq W_1 \subsetneq \cdots \subsetneq W_s \subsetneq W.$$

Furthermore, we set

$$\alpha_i := \frac{\gamma_{i+1} - \gamma_i}{r} \in \mathbb{Q}_{>0}, \quad i = 1, \ldots, s,$$

and $\underline{\alpha}(\lambda) := (\alpha_1, \ldots, \alpha_s)$.

Definition 5.1.1. The pair $(W^\bullet(\lambda), \underline{\alpha}(\lambda))$ is called the **weighted flag of** λ.

Weighted flags are the true test objects for the Hilbert-Mumford criterion for actions of the group $\mathrm{SL}_r(\mathbb{C})$:

Proposition 5.1.2. *Suppose λ and λ' are two one parameter subgroups of $\mathrm{SL}_r(\mathbb{C})$ which define the same weighted flag in $\mathbb{C}^r$. Then, for any point $x \in \mathbb{P}(V)$,*

$$\mu(\lambda, x) = \mu(\lambda', x).$$

The concept of a weighted flag may be easily generalized to the setting of vector bundles:

Definition 5.1.3. i) Let $\mathscr{E}$ be a vector bundle on X. Then, a **weighted filtration of** $\mathscr{E}$ is a pair $(\mathscr{E}^\bullet, \underline{\alpha})$ which consists of a filtration

$$\mathscr{E}^\bullet : \{0\} \subsetneq \mathscr{E}_1 \subsetneq \cdots \subsetneq \mathscr{E}_s \subsetneq \mathscr{E}$$

of $\mathscr{E}$ by subbundles and a vector $\underline{\alpha} = (\alpha_1, \ldots, \alpha_s)$ with $\alpha_i \in \mathbb{Q}_{>0}$, $i = 1, \ldots, s$.

ii) For a weighted filtration $(\mathscr{E}^\bullet, \underline{\alpha})$ of $\mathscr{E}$, we set

$$M(\mathscr{E}^\bullet, \underline{\alpha}) := \sum_{i=1}^{s} \alpha_i \bigl(\deg(\mathscr{E}) \cdot \mathrm{rk}(\mathscr{E}_i) - \deg(\mathscr{E}_i) \cdot \mathrm{rk}(\mathscr{E})\bigr).$$

The weighted filtrations of vector bundles will be the test objects for the semistability concept for decorated vector bundles. Next, we will define the quantity $\mu(\mathscr{E}^\bullet, \underline{\alpha}; \varphi)$. We proceed as follows: Choose a basis $\underline{w} = (w_1, \ldots, w_r)$ for $W := \mathbb{C}^r$, define $W_i := \langle\, w_1, \ldots, w_{\mathrm{rk}(\mathscr{E}_i)}\,\rangle$, $i = 1, \ldots, s$, and choose an open subset $\varnothing \subsetneq U \subset X$, such that

- $\varphi_{|U} \colon \mathscr{E}_{\varrho|U} \longrightarrow \mathscr{L}_{|U}$ is surjective,
- there is a trivialization $\psi \colon \mathscr{E}_{|U} \longrightarrow W \otimes \mathscr{O}_U$ with $\psi(\mathscr{E}_{i|U}) = W_i \otimes \mathscr{O}_U$, $i = 1, \ldots, s$.

Then, we get the morphism

$$\sigma \colon U \xrightarrow{\ \varphi_{|U}\ } \mathbb{P}\big(\mathscr{E}_{\varrho|U}\big) \overset{\text{``}\psi\text{''}}{\cong} \mathbb{P}(V) \times U \longrightarrow \mathbb{P}(V).$$

Finally, define

$$\underline{\gamma} := \sum_{i=1}^{s} \alpha_i \gamma^{(\mathrm{rk}(\mathscr{E}_i))}.$$

Definition 5.1.4.

$$\mu(\mathscr{E}^\bullet, \underline{\alpha}; \varphi) := \max\Big\{\, \mu\big(\lambda(\underline{w}, \underline{\gamma}), \sigma(x)\big) \,\big|\, x \in U \,\Big\}.$$

One verifies:

Proposition 5.1.5. *The quantity $\mu(\mathscr{E}^\bullet, \underline{\alpha}; \varphi)$ is well defined.*

Example 5.1.6. Assume that $V = W_{a,b,c}$ for $W = \mathbb{C}^r$ and appropriate integers a, b, and c (see Example 1.2.5, x). We may give other expressions of the number $\mu(\mathscr{E}^\bullet, \underline{\alpha}; \varphi)$. We first define the *associated weight vector* (*of the weighted filtration* $(\mathscr{E}^\bullet, \underline{\alpha})$ *of* $\mathscr{E}$) as

$$\big(\underbrace{\gamma_1, \ldots, \gamma_1,}_{(\mathrm{rk}\,\mathscr{E}_1)\times}\ \underbrace{\gamma_2, \ldots, \gamma_2}_{(\mathrm{rk}\,\mathscr{E}_2 - \mathrm{rk}\,\mathscr{E}_1)\times},\ \ldots,\ \underbrace{\gamma_{s+1}, \ldots, \gamma_{s+1}}_{(\mathrm{rk}\,\mathscr{E} - \mathrm{rk}\,\mathscr{E}_s)\times}\big) := \sum_{j=1}^{s} \alpha_j \cdot \gamma^{(\mathrm{rk}\,\mathscr{E}_j)}.$$

(Note that we recover $\alpha_j = (\gamma_{j+1} - \gamma_j)/r$, $j = 1, \ldots, s$.)

Then,

$$\mu\big(\mathscr{E}^\bullet, \underline{\alpha}; \varphi\big)$$

$$= -\min\Big\{\, \gamma_{\iota_1} + \cdots + \gamma_{\iota_a} \,\big|\, (\iota_1, \ldots, \iota_a) \in \{\, 1, \ldots, s+1 \,\}^{\times a} : \varphi_{|(\mathscr{E}_{\iota_1} \otimes \cdots \otimes \mathscr{E}_{\iota_a})^{\oplus b}} \not\equiv 0 \,\Big\}.$$

For $\underline{\iota} = (\iota_1, \ldots, \iota_a) \in \{\, 1, \ldots, s+1 \,\}^{\times a}$, we may also write

$$-\big(\gamma_{\iota_1} + \cdots + \gamma_{\iota_a}\big) = \sum_{j=1}^{s} \alpha_j \big(\nu_j(\underline{\iota}) \cdot r - a \cdot \mathrm{rk}(\mathscr{E}_j)\big), \quad \nu_j(\underline{\iota}) := \#\{\, \iota_k \leq j \mid k = 1, \ldots, a \,\}.$$

$$(2)$$

Definition 5.1.7. Let δ be a positive rational number. A ϱ-pair $(\mathscr{E}, \mathscr{L}, \varphi)$ is said to be δ-(semi)stable, if

$$M\big(\mathscr{E}^\bullet, \underline{\alpha}\big) + \delta \cdot \mu\big(\mathscr{E}^\bullet, \underline{\alpha}, \varphi\big)(\geq)0$$

holds for every weighted filtration $(\mathscr{E}^\bullet, \underline{\alpha})$ of $\mathscr{E}$.

This semistability concept is so to speak the Hilbert-Mumford criterion for decorated vector bundles. It mixes the semistability concept for vector bundles with GIT along the fibers of $\mathbb{P}(\mathscr{E}_\varrho) \longrightarrow X$.

Main Theorem 5.1.8. *Let $\varrho\colon \mathrm{GL}_r(\mathbb{C}) \longrightarrow \mathrm{GL}(V)$ be a homogeneous representation. Fix integers d, l, and a positive rational number δ.*

i) There exists a projective moduli space $\mathscr{M}(\varrho)_{d/l}^{\delta\text{-ss}}$ which parameterizes (S-equivalence classes) of δ-semistable ϱ-pairs $(\mathscr{E}, \mathscr{L}, \varphi)$ where $\deg(\mathscr{E}) = d$, and $\deg(\mathscr{L}) = l$.

ii) There is an open subspace $\mathscr{M}(\varrho)_{d/l}^{\delta\text{-s}} \subseteq \mathscr{M}(\varrho)_{d/l}^{\delta\text{-ss}}$ which parameterizes the equivalence classes of stable ϱ-pairs.

This result should be viewed as the analog of the GIT theorem that the set of semistable points in a projective space admits a projective categorical quotient and that the set of stable points a (quasi-projective) categorical quotient which is also an orbit space. It is a general existence theorem and the beginning of investigations in concrete examples.

5.2. Examples

There are several devices to simplify the concept of δ-semistability in terms of the representation ϱ (see Section 3.1 of [37]). Here, we give two examples of semistability concepts for decorated vector bundles in the simplified form.

Bradlow pairs. Set $\varrho = \mathrm{id}_{\mathrm{GL}_r(\mathbb{C})}$. Thus, a ϱ-pair is a triple $(\mathscr{E}, \mathscr{L}, \varphi)$ where $\mathscr{E}$ is a vector bundle of rank r, $\mathscr{L}$ is a line bundle, and $\varphi\colon \mathscr{E} \longrightarrow \mathscr{L}$ is a non-trivial homomorphism.

The simplified semistability concept takes the following form:

A ϱ-pair $(\mathscr{E}, \mathscr{L}, \varphi)$ is δ-(semi)stable, if and only

$$\mu(\mathscr{F}) \quad (\leq) \quad \mu(\mathscr{E}) - \frac{\delta}{\mathrm{rk}(\mathscr{E})}, \quad \text{if } \mathscr{F} \subseteq \ker(\varphi)$$

$$\mu(\mathscr{F}) - \frac{\delta}{\mathrm{rk}(\mathscr{F})} \quad (\leq) \quad \mu(\mathscr{E}) - \frac{\delta}{\mathrm{rk}(\mathscr{E})}, \quad \text{if } \mathscr{F} \nsubseteq \ker(\varphi).$$

This stability concept was formulated by Bradlow [7]. It is the first example of a notion of semistability which depends on a parameter.

Conic bundles. This time, we work with $\varrho\colon \mathrm{GL}_3(\mathbb{C}) \longrightarrow \mathrm{GL}(\mathrm{Sym}^2(\mathbb{C}^3))$, i.e., a ϱ-pair consists of a vector bundle $\mathscr{E}$ of rank 3, a line bundle $\mathscr{L}$, and a non-zero homomorphism $\varphi\colon \mathscr{S}ym^2(\mathscr{E}) \longrightarrow \mathscr{L}$. If φ is everywhere surjective, then such a ϱ-pair describes a conic bundle $\pi\colon \mathscr{C} \longrightarrow X$, i.e., a surface which is fibered over X in plane conics (see also Example 2.7).

In order to explain semistability for a ϱ-pair $(\mathscr{E}, \mathscr{L}, \varphi\colon \mathscr{S}ym^2(\mathscr{E}) \longrightarrow \mathscr{L})$, we need the following:

Definition 5.2.1. i) For a subbundle $\{0\} \subsetneq \mathscr{F} \subsetneq \mathscr{E}$, we set

$$
c_\varphi(\mathscr{F}) := \left\{
\begin{array}{ll}
2 & \text{if } \varphi_{|\mathscr{F} \otimes \mathscr{F}} \not\equiv 0 \\
1 & \text{if } \varphi_{|\mathscr{F} \otimes \mathscr{F}} \equiv 0 \text{ and } \varphi_{\mathscr{F} \otimes \mathscr{E}} \not\equiv 0 \\
0 & \text{if } \varphi_{|\mathscr{F} \otimes \mathscr{E}} \equiv 0
\end{array}
\right. .
$$

ii) A filtration $\mathscr{E}^\bullet : \{0\} \subsetneq \mathscr{E}_1 \subsetneq \mathscr{E}_2 \subsetneq \mathscr{E}$ is called **critical**, if

$$
\varphi_{|\mathscr{E}_1 \otimes \mathscr{E}_2} \equiv 0, \quad \varphi_{|\mathscr{E}_1 \otimes \mathscr{E}} \not\equiv 0, \quad \text{and } \varphi_{|\mathscr{E}_2 \otimes \mathscr{E}_2} \not\equiv 0.
$$

Then, a ϱ-pair is δ-(semi)stable, if and only if

$$
\mu(\mathscr{F}) - \delta \cdot \frac{c_\varphi(\mathscr{F})}{\operatorname{rk}(\mathscr{F})} (\leq) \mu(\mathscr{E}) - \delta \cdot \frac{2}{3},
$$

for every subbundle $\{0\} \subsetneq \mathscr{F} \subsetneq \mathscr{E}$, and

$$
\deg(\mathscr{E}_1) + \deg(\mathscr{E}_2)(\leq)\deg(\mathscr{E}),
$$

for every critical filtration $\mathscr{E}^\bullet : \{0\} \subsetneq \mathscr{E}_1 \subsetneq \mathscr{E}_2 \subsetneq \mathscr{E}$.

The stability of conic bundles was investigated by Gómez and Sols [13]. It is the first and probably easiest example where filtrations of higher length are necessary in order to define semistability.

5.3. Boundedness

By Lecture I, Example 1.2.5, x), Proposition, the homogeneous representation ϱ is a direct summand of the representation $\varrho_{a,b,c}$ of $\mathrm{GL}_r(\mathbb{C})$ on the vector space

$$
W_{a,b,c} = \left((\mathbb{C}^r)^{\otimes a} \right)^{\oplus b} \otimes \left(\overset{r}{\bigwedge} \mathbb{C}^r \right)^{\otimes -c},
$$

for suitable non-negative integers a, b, and c.

According to Lecture IV, Proposition 4.3.6, the following result shows that the set of isomorphism classes of vector bundles which occur in δ-semistable ϱ-pairs with fixed numerical data form a bounded family in the sense of Definition 4.3.1.

Proposition 5.3.1. *There is a non-negative constant C_1, depending only on r, a, and δ, such that for every δ-semistable $\varrho_{a,b,c}$-pair $(\mathscr{E}, \mathscr{L}, \varphi)$ with $\deg(\mathscr{E}) = d$ and every non-trivial proper subbundle $\mathscr{F}$ of $\mathscr{E}$*

$$
\mu(\mathscr{F}) \leq \frac{d}{r} + C_1.
$$

Proof. Let $\mathscr{E}^\bullet : \{0\} \subsetneq \mathscr{F} \subsetneq \mathscr{E}$ be any subbundle. Using Example 5.1.6, one easily estimates

$$
\mu_{\varrho_{a,b,c}}(\mathscr{E}^\bullet, (1); \varphi) \leq a(r-1).
$$

Hence, δ-semistability gives

$$
d\operatorname{rk}(\mathscr{F}) - \deg(\mathscr{F})r + \delta a(r-1) \geq d\operatorname{rk}(\mathscr{F}) - \deg(\mathscr{F})r + \delta\mu_{\varrho_{a,b,c}}(\mathscr{E}^\bullet, (1); \varphi) \geq 0,
$$

i.e.,

$$\mu(\mathscr{F}) \leq \frac{d}{r} + \frac{\delta \cdot a \cdot (r-1)}{r \cdot \mathrm{rk}(\mathscr{F})} \leq \frac{d}{r} + \frac{\delta \cdot a \cdot (r-1)}{r},$$

so that the theorem holds for $C_1 := \delta \cdot a \cdot (r-1)/r$. $\qquad\square$

In fact, a much stronger boundedness result is true:

Theorem 5.3.2 (Langer/Schmitt). *There is a non-negative constant C_2, depending only on r, a, b, c, d, and l, such that for every $\delta \in \mathbb{Q}_{>0}$ and every δ-semistable $\varrho_{a,b,c}$-pair $(\mathscr{E}, \mathscr{L}, \varphi)$ with $\deg(\mathscr{E}) = d$ and $\deg(\mathscr{L}) = l$ and every non-trivial proper subbundle $\mathscr{F}$ of $\mathscr{E}$*

$$\mu(\mathscr{F}) \leq \frac{d}{r} + C_2.$$

Proof. This result was first published in [39]. Later, the author learned a much easier argument from Adrian Langer. It is given in [12]. $\qquad\square$

5.4. The parameter space

To simplify matters, we fix a line bundle $\mathscr{L}_0$ on X and look only at ϱ-pairs of the form $(\mathscr{E}, \mathscr{L}_0, \varphi)$.

By Proposition 5.3.1, the occurring vector bundles can be parameterized by some quot scheme $\mathfrak{Q}$. Recall that we have the universal quotient

$$q_{\mathfrak{Q}} \colon Y \otimes \pi_X^\star\big(\mathscr{O}_X(-n)\big) \longrightarrow \mathfrak{E}_{\mathfrak{Q}}$$

on $\mathfrak{Q} \times X$.

For $m \gg 0$,

$$\mathfrak{F}_m := \pi_{\mathfrak{Q}\star}\Big((Y^{\otimes a})^{\oplus b} \otimes \pi_X^\star\big(\mathscr{O}_X(a(m-n))\big)\Big)$$

will be a vector bundle on $\mathfrak{Q}$, and so will be

$$\mathfrak{G}_m := \pi_{\mathfrak{Q}\star}\Big(\det(\mathfrak{E}_{\mathfrak{Q}})^{\otimes c} \otimes \pi_X^\star\big(\mathscr{L}_0(am)\big)\Big).$$

We form the projective bundle

$$\pi \colon \mathfrak{P} := \mathbb{P}\big(\mathscr{H}\!om(\mathfrak{F}_m, \mathfrak{G}_m)^{\vee}\big) \longrightarrow \mathfrak{Q}.$$

On $\mathfrak{P} \times X$, we have the universal quotient

$$q_{\mathfrak{P}} := (\pi \times \mathrm{id}_X)^\star(q_{\mathfrak{Q}}) \colon Y \otimes \pi_X^\star\big(\mathscr{O}_X(-n)\big) \longrightarrow \mathfrak{E}_{\mathfrak{P}}$$

and the tautological homomorphism

$$\widetilde{f}_{\mathfrak{P}} \colon (Y^{\otimes a})^{\oplus b} \otimes \pi_X^\star\big(\mathscr{O}_X(a(m-n))\big) \longrightarrow \det(\mathfrak{E}_{\mathfrak{P}})^{\otimes c} \otimes \pi_X^\star\big(\mathscr{L}_0(am)\big) \otimes \pi_{\mathfrak{P}}^\star\big(\mathscr{O}_{\mathfrak{P}}(1)\big).$$

Set $f_{\mathfrak{P}} := \widetilde{f}_{\mathfrak{P}} \otimes \mathrm{id}_{\mathscr{O}_X(-am)}$.

There is a closed subscheme $\mathfrak{T} \subseteq \mathfrak{P}$ where $f_{\mathfrak{P}}$ factorizes over

$$\big((\mathfrak{E}_{\mathfrak{T}})^{\otimes a}\big)^{\oplus b}, \quad \mathfrak{E}_{\mathfrak{T}} := \mathfrak{E}_{\mathfrak{P}|\mathfrak{T} \times X}.$$

Thus, on $\mathfrak{T} \times X$, we have the universal quotient

$$q_{\mathfrak{T}} \colon Y \otimes \pi_X^\star\big(\mathscr{O}_X(-n)\big) \longrightarrow \mathfrak{E}_{\mathfrak{T}}$$

and the universal homomorphism

$$\varphi_{\mathfrak{P}} \colon \left((\mathfrak{E}_{\mathfrak{T}})^{\otimes a} \right)^{\oplus b} \longrightarrow \det(\mathfrak{E}_{\mathfrak{T}})^{\otimes c} \otimes \pi_X^\star(\mathscr{L}_0) \otimes \mathfrak{N}_{\mathfrak{T}}, \quad \mathfrak{N}_{\mathfrak{T}} := \pi_{\mathfrak{P}}^\star\big(\mathscr{O}_{\mathfrak{P}}(1)\big)_{|\mathfrak{T} \times X}.$$

We call $(\mathfrak{E}_{\mathfrak{T}}, \pi_X^\star(\mathscr{L}_0) \otimes \mathfrak{N}_{\mathfrak{T}}, \varphi_{\mathfrak{T}})$ the **universal family**.

There is an open subset $U \subset \mathfrak{T}$ consisting of those points $([q \colon Y \otimes \mathscr{O}_X(-n) \longrightarrow \mathscr{E}], \mathscr{L}_0, \varphi)$, such that $\mathscr{E}$ is a vector bundle and $H^0(q(n))$ is an isomorphism. There is also a natural $\mathrm{GL}(Y)$-action on $\mathfrak{T}$ which leaves U invariant and induces equivalence of ϱ-pairs on U. Moreover, $\mathbb{C}^\star \cdot \mathrm{id}_Y$ acts trivially, so that we have to investigate the $\mathrm{SL}(Y)$-action.

To further simplify matters, we fix a line bundle $\mathscr{N}$ on X and look only at those ϱ-pairs $(\mathscr{E}, \mathscr{L}_0, \varphi)$ with $\det(\mathscr{E}) \cong \mathscr{N}$. Again, the ϱ-pairs with this condition on the determinant belong to a closed subscheme $U_{\mathscr{N}} \subset U$. We set

$$\mathbb{H} := \mathrm{Hom}\left(\bigwedge^r Y, H^0(\mathscr{N}(rn)) \right)^\vee$$

and

$$\mathbb{K} := \mathrm{Hom}\left((Y^{\otimes a})^{\oplus b}, H^0(\mathscr{N}^{\otimes c} \otimes \mathscr{L}_0(an)) \right)^\vee.$$

This time, we obtain a Gieseker morphism

$$F \colon U_{\mathscr{N}} \longrightarrow \mathbb{P}(\mathbb{H}) \times \mathbb{P}(\mathbb{K}).$$

Let $\varrho_1 \colon \mathrm{SL}(Y) \longrightarrow \mathrm{GL}(\mathbb{H})$ and $\varrho_2 \colon \mathrm{SL}(Y) \longrightarrow \mathrm{GL}(\mathbb{K})$ be the obvious representations which give the $\mathrm{SL}(Y)$-action on the right-hand space. For positive integers m and n, we have the linearization $l_{m,n} = (\varrho_{m,n}, \iota_{m,n})$ of that action with $\varrho_{m,n} := \mathrm{Sym}^m(\varrho_1) \otimes \mathrm{Sym}^n(\varrho_2)$ and the Veronese embeddings combined with the Segre embedding

$$\iota_{m,n} \colon \mathbb{P}(\mathbb{H}) \times \mathbb{P}(\mathbb{K}) \hookrightarrow \mathbb{P}\big(\mathrm{Sym}^m(\mathbb{H}) \otimes \mathrm{Sym}^n(\mathbb{K})\big).$$

We choose the linearization parameters m and n in such a way that

$$\varepsilon := \frac{m}{n} := \frac{p - a\delta}{r\delta}.$$

Below, we will illustrate the relationship between the notion of δ-semistability of a ϱ-pair and the Hilbert-Mumford criterion for the linearization $l_{m,n}$.

5.5. Evaluation of the Hilbert-Mumford criterion

The linearization of the group action determines a notion of (semi)stability which we test with the Hilbert-Mumford criterion. In this section, we want to discuss some elements of the proof that this notion of semistability on the Gieseker space equals the notion of δ-(semi)stability for decorated vector bundles. More precisely, we want to illustrate the implication:

The Gieseker point $F(t)$ of $t = ([q \colon Y \otimes \mathscr{O}_X(-n) \longrightarrow \mathscr{E}], \mathscr{L}_0, \varphi) \in U$ is (semi)stable $\implies (\mathscr{E}, \mathscr{L}_0, \varphi)$ is δ-(semi)stable.

We will check the condition of δ-(semi)stability for a weighted filtration $(\mathscr{E}^\bullet, \underline{\alpha})$, such that $\mathscr{E}_j(n)$ is globally generated and $H^1(\mathscr{E}_j(n)) = \{0\}$, $j = 1, \ldots, s$. As in the case of vector bundles without extra structure, one may show that this suffices to establish δ-(semi)stability of $(\mathscr{E}, \mathscr{L}_0, \varphi)$.

First, we have to cook up the correct one parameter subgroup to put into the Hilbert-Mumford criterion. To this end, let $\underline{y} = (y_1, \ldots, y_p)$ be a basis of Y, such that there are indices $l_1, \ldots, l_s$ with $Y_{\underline{y}}^{(l_j)} = H^0(\mathscr{E}_j(n))$ (under the isomorphism $H^0(q(n)))$, $j = 1, \ldots, s$, and define

$$\widetilde{\gamma} := \sum_{j=1}^{s} \alpha_j \gamma^{(l_j)}.$$

We also set

$$\mathrm{gr}_j(Y, \underline{y}) := Y_{\underline{y}}^{(l_j)} / Y_{\underline{y}}^{(l_{j-1})} = H^0(\mathscr{E}_j/\mathscr{E}_{j-1}(n)), \quad j = 1, \ldots, s+1, \ l_0 := 0, \ l_{s+1} := p.$$

The fixed basis $\underline{y}$ for Y provides us with the isomorphism $Y \cong \bigoplus_{j=1}^{s+1} \mathrm{gr}_j(Y, \underline{y})$. For every index $\underline{\iota} \in J := \{1, \ldots, s\}^{\times a}$, we set

$$Y_{\underline{\iota}, \underline{y}} := \mathrm{gr}_{\iota_1}(Y, \underline{y}) \otimes \cdots \otimes \mathrm{gr}_{\iota_a}(Y, \underline{y}).$$

Moreover, for $k \in \{1, \ldots, b\}$ and $\underline{\iota} \in J$, we let $Y_{\underline{\iota}, \underline{y}}^k$ be the subspace of $Y_{a,b} := (Y^{\otimes a})^{\oplus b}$ which is $Y_{\underline{\iota}, \underline{y}}$ living in the kth copy of $Y^{\otimes a}$ in $Y_{a,b}$. The spaces $Y_{\underline{\iota}, \underline{y}}^k$, $k \in \{1, \ldots, b\}$, $\underline{\iota} \in J$, are eigenspaces for the action of the one parameter subgroups $\lambda(\underline{y}, \gamma_p^{(l_j)})$, $j = 1, \ldots, s$. Define

$$\nu_j(\underline{\iota}) := \#\{\iota_i \le j \mid \underline{\iota} = (\iota_1, \ldots, \iota_a), \ i = 1, \ldots, a\}. \tag{3}$$

Then, $\lambda(\underline{y}, \gamma_p^{(l_j)})$ acts on $Y_{\underline{\iota}, \underline{y}}^k$ with weight $\nu_j(\underline{\iota}) \cdot p - a \cdot l_j$.

Suppose the second component of $F(t)$ is represented by the homomorphism

$$L \colon (Y^{\otimes a})^{\oplus b} \longrightarrow H^0(\mathscr{N}^{\otimes c} \otimes \mathscr{L}_0(an)).$$

With (3), one readily verifies

$$\mu\big(\lambda(\underline{y}, \widetilde{\gamma}), [L]\big)$$
$$= -\min\Big\{ \sum_{j=1}^{s} \alpha_j\big(\nu_j(\underline{\iota})p - al_j\big) \,\Big|\, k \in \{1, \ldots, b\}, \underline{\iota} \in J : Y_{\underline{\iota}, \underline{y}}^k \not\subseteq \ker(L) \Big\}. \tag{4}$$

We observe that for every $k \in \{1, \ldots, b\}$ and every $\underline{\iota} \in J$:

$$Y_{\underline{\iota}, \underline{y}}^k \not\subseteq \ker(L) \quad \Longrightarrow \quad \varphi_{|(\mathscr{E}_{\iota_1} \otimes \cdots \otimes \mathscr{E}_{\iota_a})^{\oplus b}} \not\equiv 0. \tag{5}$$

This is because $(Y_{\underline{y}}^{(l_{\iota_1})} \otimes \cdots \otimes Y_{\underline{y}}^{(l_{\iota_a})})^{\oplus b}$ generates the bundle $(\mathscr{E}_{\iota_1}(n) \otimes \cdots \otimes \mathscr{E}_{\iota_a}(n))^{\oplus b}$.

Now, let $k_0 \in \{1, \ldots, b\}$ and $\underline{\iota}_0 \in J$ be such that the minimum in (4) is achieved by $\sum_{j=1}^{s} \alpha_j (\nu_j(\underline{\iota}_0) \cdot p - a \cdot l_j)$ and $Y_{\underline{\iota}_0, \underline{y}}^{k_0} \not\subseteq \ker(L)$. The (semi)stability of t gives:

$$
\begin{aligned}
0 \quad (\le) \quad & \frac{1}{n}\mu_{l_{m,n}}\big(\lambda(\underline{y}, \widetilde{\gamma}), F(t)\big) \\[2mm]
= \quad & \varepsilon\mu\big(\lambda(\underline{y}, \widetilde{\gamma}), F_1(t)\big) + \mu\big(\lambda(\underline{y}, \widetilde{\gamma}), F_2(t)\big) \\[2mm]
= \quad & \varepsilon \sum_{j=1}^{s} \alpha_j \big(p\,\mathrm{rk}\,\mathscr{E}_j - h^0(\mathscr{E}_j(n))r\big) + \sum_{j=1}^{s} \alpha_j\big(\nu_j(\underline{\iota}_0)p - al_j\big) \\[2mm]
= \quad & \frac{p - a\delta}{r\delta} \sum_{j=1}^{s} \alpha_j \big(p\,\mathrm{rk}\,\mathscr{E}_j - h^0(\mathscr{E}_j(n))r\big) + \sum_{j=1}^{s} \alpha_j\big(\nu_j(\underline{\iota}_0)p - ah^0(\mathscr{E}_j(n))\big) \\[2mm]
= \quad & \sum_{j=1}^{s} \alpha_j \left(\frac{p^2\,\mathrm{rk}\,\mathscr{E}_j}{r\delta} - \frac{pa\,\mathrm{rk}\,\mathscr{E}_j}{r} - \frac{ph^0(\mathscr{E}_j(n))}{\delta}\right) + \sum_{j=1}^{s} \alpha_j \nu_j(\underline{\iota}_0)p.
\end{aligned}
$$

We multiply this inequality by $r\delta/p$ and find

$$
0(\le) \sum_{j=1}^{s} \alpha_j \big(p\,\mathrm{rk}\,\mathscr{E}_j - h^0(\mathscr{E}_j(n))r\big) + \delta \sum_{j=1}^{s} \alpha_j\big(\nu_j(\underline{\iota}_0)r - a\,\mathrm{rk}\,\mathscr{E}_j\big).
$$

Since $h^1(\mathscr{E}_j(n)) = 0$, $j = 1, \ldots, s$, we have $p\,\mathrm{rk}\,\mathscr{E}_j - h^0(\mathscr{E}_j(n))r = d\,\mathrm{rk}\,\mathscr{E}_j - \deg(\mathscr{E}_j)r$, $j = 1, \ldots, s$. Moreover, $\mathrm{rk}\,\mathscr{E}_j = i_j$, by definition. Therefore, (5) and Example 5.1.6 – observing Equation (2) – imply that

$$
\mu_{\varrho_{a,b,c}}(\mathscr{E}^\bullet, \underline{\alpha}; \varphi) \ge \sum_{j=1}^{s} \alpha_j\big(\nu_j(\underline{\iota}_0)r - ai_j\big).
$$

Hence, we finally see

$$
M\big(\mathscr{E}^\bullet, \underline{\alpha}\big) + \delta \cdot \mu_{\varrho_{a,b,c}}\big(\mathscr{E}^\bullet, \underline{\alpha}; \varphi\big)(\ge)0.
$$

To complete the GIT construction, one also has to prove that a δ-(semi)stable decorated vector bundle gives rise to a (semi)stable Gieseker point and that the Gieseker map between the corresponding semistable loci in U and $\mathbb{P}(\mathbb{H}) \times \mathbb{P}(\mathbb{K})$ is proper. The arguments ascertaining these facts are similar in nature but technically slightly more involved (see [37]). $\qquad\square$

5.6. The chain of moduli spaces

Theorem 5.3.2 and our discussions in Lecture III yield the following result:

Theorem 5.6.1. *Fix the input data a, b, c, d, and l. Then, there is a finite set $\{\,\widehat{\delta}_1, \ldots, \widehat{\delta}_m\,\}$ of rational numbers*

$$
0 =: \widehat{\delta}_0 < \widehat{\delta}_1 < \cdots < \widehat{\delta}_m < \widehat{\delta}_{m+1} := \infty,
$$

such that, for a ϱ-pair $(\mathscr{E}, \mathscr{L}, \varphi)$ with $\deg(\mathscr{E}) = d$ and $\deg(\mathscr{L}) = l$, the following properties hold true:

178   A.H.W. Schmitt

i) *Suppose there is an index $i \in \{0, \ldots, m\}$ with $\widehat{\delta}_i < \delta_1 < \delta_2 < \widehat{\delta}_{i+1}$. Then, $(\mathscr{E}, \mathscr{L}, \varphi)$ is δ_1-(semi)stable, if and only if it is δ_2-(semi)stable. In particular, there is a canonical isomorphism*

$$\mathscr{M}(\varrho)_{d/l}^{\delta_1\text{-ss}} \cong \mathscr{M}(\varrho)_{d/l}^{\delta_2\text{-ss}}.$$

ii) *Assume $\widehat{\delta}_i < \delta < \widehat{\delta}_{i+1}$ for some index $i \in \{1, \ldots, m-1\}$. If $(\mathscr{E}, \mathscr{L}, \varphi)$ is δ-semistable, then $(\mathscr{E}, \mathscr{L}, \varphi)$ is also $\widehat{\delta}_i$- and $\widehat{\delta}_{i+1}$-semistable, so that there are canonical morphisms*

$$\mathscr{M}(\varrho)_{d/l}^{\delta\text{-ss}} \longrightarrow \mathscr{M}(\varrho)_{d/l}^{\widehat{\delta}_i\text{-ss}} \quad \text{and} \quad \mathscr{M}(\varrho)_{d/l}^{\delta\text{-ss}} \longrightarrow \mathscr{M}(\varrho)_{d/l}^{\widehat{\delta}_{i+1}\text{-ss}}.$$

Conversely, if $(\mathscr{E}, \mathscr{L}, \varphi)$ is $\widehat{\delta}_i$- or $\widehat{\delta}_{i+1}$-stable, then $(\mathscr{E}, \mathscr{L}, \varphi)$ is also δ-stable.

iii) *Suppose $\delta > \widehat{\delta}_m$. If $(\mathscr{E}, \mathscr{L}, \varphi)$ is δ-semistable, it is also $\widehat{\delta}_m$-semistable, so that there is a natural morphism*

$$\mathscr{M}(\varrho)_{d/l}^{\delta\text{-ss}} \longrightarrow \mathscr{M}(\varrho)_{d/l}^{\widehat{\delta}_m\text{-ss}}.$$

Conversely, if $(\mathscr{E}, \mathscr{L}, \varphi)$ is $\widehat{\delta}_m$-stable, then $(\mathscr{E}, \mathscr{L}, \varphi)$ is also δ-stable.

iv) *Suppose $0 < \delta < \widehat{\delta}_1$. If $(\mathscr{E}, \mathscr{L}, \varphi)$ is δ-semistable, then $\mathscr{E}$ is a semistable vector bundle. Letting $\widehat{\mathscr{M}}_0$ be the moduli space of semistable vector bundles of rank r and degree d, we find a canonical morphism*

$$\mathscr{M}(\varrho)_{d/l}^{\delta\text{-ss}} \longrightarrow \widehat{\mathscr{M}}_0.$$

If $\mathscr{E}$ is a stable vector bundle, then $(\mathscr{E}, \mathscr{L}, \varphi)$ is δ-stable.

We set $\widehat{\mathscr{M}}_i := \mathscr{M}(\varrho)_{d/l}^{\widehat{\delta}_i\text{-ss}}$, $i = 1, \ldots, m$, $\mathscr{M}_i := \mathscr{M}(\varrho)_{d/l}^{\delta\text{-ss}}$ for some δ with $\widehat{\delta}_{i-1} < \delta < \widehat{\delta}_i$, $i = 1, \ldots, m$, and $\mathscr{M}_\infty := \mathscr{M}(\varrho)_{d/l}^{\delta\text{-ss}}$ for some δ with $\delta > \widehat{\delta}_m$. Our theorem is then summarized by the following picture

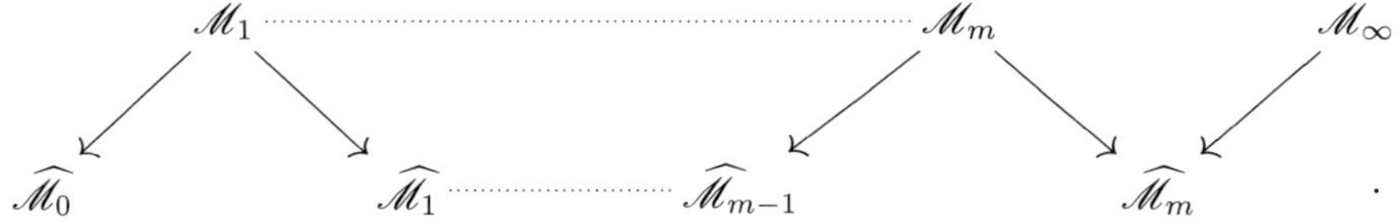

Remark 5.6.2. This "chain of flips" has first figured in the setting of Bradlow pairs (see Example 4.2) of the kind $(\mathscr{E}, \mathscr{O}_X, \varphi)$ with $\mathrm{rk}(\mathscr{E}) = 2$ and $\det(\mathscr{E})$ a fixed line bundle $\mathscr{N}$ of odd degree in the work of Thaddeus [48]. In that important application of decorated vector bundles, the moduli space $\mathscr{M}_\infty$ is empty, and $\widehat{\mathscr{M}}_m$ is a projective space. The moduli space $\mathscr{M}_1$ is a projective bundle over the moduli space $\mathscr{M}(2, \mathscr{N})$ of stable vector bundles of rank 2 with determinant $\mathscr{N}$ (note that "stable" = "semistable", because $\mathrm{rk}(\mathscr{E}) = 2$ and $\deg(\mathscr{E})$ is odd). Furthermore, it is possible to explicitly analyze all the maps in the above flip diagram. This enables Thaddeus to transfer the simple information on a projective space to important information on the moduli space of stable vector bundles. For example, one easily concludes that the Picard group of $\mathscr{M}(2, \mathscr{N})$ is isomorphic to the group of integers.

5.7. Decorated principal bundles

In this final section, we begin the discussion of decorated principal G-bundles. We assume that the structure group G is a semisimple linear algebraic group, i.e., a connected reductive linear algebraic group with finite center (such as $\mathrm{SL}_n(\mathbb{C})$, $\mathrm{SO}_n(\mathbb{C})$, or $\mathrm{Sp}_{2n}(\mathbb{C})$). To begin with, we sketch how principal G-bundles may be treated as decorated vector bundles.

To do so, we fix a faithful representation $\iota\colon G \hookrightarrow \mathrm{GL}(V)$. By means of the representation ι, any principal G-bundle $\mathscr{P}$ on X gives rise to a principal $\mathrm{GL}(V)$-bundle which we denote by $\iota_\star(\mathscr{P})$. To a principal G-bundle $\mathscr{P}$ on X, we now associate:

- the vector bundle $\mathscr{E} := \mathscr{P}(V)$ with fiber V (compare with the introduction). Then, we may view $\iota_\star(\mathscr{P})$ as the frame bundle of $\mathscr{E}$, i.e., $\iota_\star(\mathscr{P}) = \mathscr{I}som(V \otimes \mathscr{O}_X, \mathscr{E})$;
- a section $\sigma\colon X \longrightarrow \mathscr{I}som(V \otimes \mathscr{O}_X, \mathscr{E})/G$. In fact, we have the commutative diagram

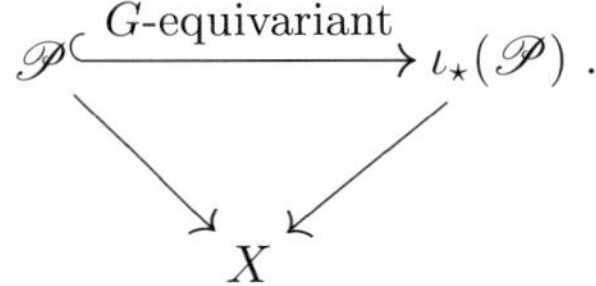

If we take the G-quotient on both sides in the top line, we find σ.

Conversely, to a pair $(\mathscr{E}, \sigma)$ as above, we associate the fiber product

$$
\begin{array}{ccc}
\mathscr{P} & \longrightarrow & \mathscr{I}som(V \otimes \mathscr{O}_X, \mathscr{E}) \\
\downarrow & & \quad\;\; \downarrow\; {}_{G\text{-}}\;\text{bundle} \\
X & \xrightarrow{\;\sigma\;} & \mathscr{I}som(V \otimes \mathscr{O}_X, \mathscr{E})/G.
\end{array}
$$

Note that $\mathscr{P}$ is a principal G-bundle on X. There is an obvious equivalence relation on the set of pairs $(\mathscr{E}, \sigma)$. The upshot is the following.

Proposition 5.7.1. *The above constructions give rise to a bijection*

$$
\left\{ \begin{array}{c} \textit{Isomorphism classes of} \\ \textit{principal } G\textit{-bundles} \end{array} \right\} \longleftrightarrow \left\{ \begin{array}{c} \textit{Equivalence classes of} \\ \textit{pairs } (\mathscr{E}, \sigma) \end{array} \right\}.
$$

Remark 5.7.2. The reader may consult reference [36] for a precise account on how pairs $(\mathscr{E}, \sigma)$ may be interpreted as decorated vector bundles in the sense which we have used, so far.

Now, let $\varrho\colon G \longrightarrow \mathrm{GL}(W)$ be a representation of G. By Lecture I, Example 1.2.5, ix), Proposition, we may – after possibly adding a direct summand – assume that ϱ is the restriction of a representation $\widetilde{\varrho}\colon \mathrm{GL}(V) \longrightarrow \mathrm{GL}(W)$. A similar reasoning as above shows that we can identify ϱ-pairs $(\mathscr{P}, \mathscr{L}, \varphi)$ with tuples $(\mathscr{E}, \sigma, \mathscr{L}, \widetilde{\varphi})$ where $(\mathscr{E}, \mathscr{L}, \widetilde{\varphi})$ is a $\widetilde{\varrho}$-pair.

Definition 5.7.3. i) Let $\lambda\colon \mathbb{C}^{\star} \longrightarrow G$ be a one parameter subgroup of G. Define the subgroup

$$Q_G(\lambda) := \Big\{ g \in G \,\Big|\, \lim_{z\to\infty} \lambda(z) \cdot g \cdot \lambda(z)^{-1} \text{ exists in } G \Big\}.$$

ii) A **reduction of the principal G-bundle** $\mathscr{P}$ **to** λ consists of a section $\beta\colon X \longrightarrow \mathscr{P}/Q_G(\lambda)$. The composed section

$$\beta'\colon X \longrightarrow \mathscr{P}/Q_G(\lambda) \hookrightarrow \mathscr{I}som(V \otimes \mathscr{O}_X, \mathscr{E})/Q_{\mathrm{GL}(V)}(\lambda)$$

yields a weighted filtration

$$(\mathscr{E}_\beta^\bullet, \underline{\alpha}_\beta) = \big(\{0\} \subsetneq \mathscr{E}_1 \subsetneq \cdots \subsetneq \mathscr{E}_s \subsetneq \mathscr{E}, (\alpha_1, \ldots, \alpha_s)\big)$$

of $\mathscr{E}$.

iii) Fix a positive rational number δ. A ϱ-pair $(\mathscr{P}, \mathscr{L}, \varphi)$ is said to be δ-**(semi)stable**, if

$$M(\mathscr{E}_\beta^\bullet, \underline{\alpha}_\beta) + \delta \cdot \mu\big(\mathscr{E}_\beta^\bullet, \underline{\alpha}_\beta; \varphi\big)(\geq)0$$

holds for every non-trivial one parameter subgroup $\lambda\colon \mathbb{C}^{\star} \longrightarrow G$ and every reduction β of $\mathscr{P}$ to λ.

Remark 5.7.4. i) The subgroup $Q_G(\lambda)$ is a parabolic subgroup of G, and any parabolic subgroup of G is of the form $Q_G(\lambda)$ for an appropriate one parameter subgroup $\lambda\colon \mathbb{C}^{\star} \longrightarrow G$ (see [46]).

ii) Let $(V^\bullet(\lambda) : \{0\} \subsetneq V_1 \subsetneq \cdots \subsetneq V_s \subsetneq V, \underline{\alpha}(\lambda))$ be the weighted flag of λ (see Definition 5.1.1). The bundle $\mathscr{I}som(V \otimes \mathscr{O}_X, \mathscr{E})/Q_{\mathrm{GL}(V)}(\lambda)$ is the bundle of flags in the fibers of $\mathscr{E}$ of the same type as $V^\bullet(\lambda)$, i.e., the bundle whose sections over a subset $Y \subsetneq V$ correspond to filtrations $\mathscr{E}^\bullet : \{0\} \subsetneq \mathscr{E}_1 \subsetneq \cdots \subsetneq \mathscr{E}_s \subsetneq \mathscr{E}_{|Y}$ where $\mathrm{rk}(\mathscr{E}_i) = \dim(V_i)$, $i = 1, \ldots, s$. This explains that β' gives a filtration $\mathscr{E}_\beta^\bullet$. The tuple $\underline{\alpha}_\beta$ is simply $\underline{\alpha}(\lambda)$.

We haven't written down the existence of moduli spaces for δ-(semi)stable decorated principal G-bundles as a theorem, because the complete proofs haven't been finished up to now. The author intends to supply them and extensions of the theory in [41]. The idea is, of course, to use our description of decorated principal G-bundles as decorated vector bundles in order to reduce everything to the theory of decorated vector bundles which we have already developed.

Acknowledgments

The author would like to thank the organizers of the mini school, Adrian Langer, Piotr Pragacz, and Halszka Tutaj-Gasińska, for the invitation to give the talks and the hospitality during the visit to Warsaw. He is also very grateful to Adrian Langer for his careful reading of the manuscript.

He also acknowledges support by the DFG via a Heisenberg fellowship and via the Schwerpunkt program "Globale Methoden in der komplexen Geometrie – Global Methods in Complex Geometry".

References

[1] V. Balaji, *Lectures on principal bundles*, available at `http://www.cimat.mx/Eventos/c_vectorbundles/`, 17 pp.

[2] A. Białynicki-Birula, *Finiteness of the number of maximal open subsets with good quotients*, Transform. Groups **3** (1998), 301–19.

[3] A. Białynicki-Birula, J.B. Carrell, W.M. McGovern, *Algebraic quotients. Torus actions and cohomology. The adjoint representation and the adjoint action*, Encyclopaedia of Mathematical Sciences, vol. 131, *Invariant Theory and Algebraic Transformation Groups*, II, Springer-Verlag, Berlin, 2002, iv+242 pp.

[4] A. Białynicki-Birula, A.J. Sommese, *Quotients by $\mathbb{C}^* \times \mathbb{C}^*$ actions*, Trans. Amer. Math. Soc. **289** (1985), 519–43.

[5] D. Birkes, *Orbits of linear algebraic groups*, Annals of Math. **93** (1971), 459–75.

[6] A. Borel, *Linear algebraic groups*, Second edition, Graduate Texts in Mathematics, 126, Springer-Verlag, New York, 1991, xii+288 pp.

[7] S.B. Bradlow, *Special metrics and stability for holomorphic bundles with global sections*, J. Differential Geom. **33** (1991), 169–213.

[8] S.B. Bradlow, G.D. Daskalopoulos, O. García-Prada, R. Wentworth, *Stable augmented bundles over Riemann surfaces* in *Vector Bundles in Algebraic Geometry* (Durham, 1993), 15–67, London Math. Soc. Lecture Note Ser., 208, Cambridge Univ. Press, Cambridge, 1995.

[9] J.A. Dieudonné, J.B. Carrell, *Invariant theory – Old and new*, Academic Press, New York-London, 1971, viii+85 pp.

[10] I. Dolgachev, Y. Hu, *Variation of geometric invariant theory quotients*, with an appendix by Nicolas Ressayre, Inst. Hautes Études Sci. Publ. Math. **87** (1998), 5–56.

[11] T.L. Gómez, *Lectures on principal bundles over projective varieties*, this volume.

[12] T.L. Gómez, A. Langer, A. Schmitt, I. Sols, *Moduli spaces for principal bundles in large characteristic*, to appear in the proceedings of the Allahabad International Workshop on Teichmüller Theory and Moduli Problems 2006, 77 pp.

[13] T.L. Gómez, I. Sols, *Stability of conic bundles*, with an appendix by I. Mundet i Riera, Internat. J. Math. **11** (2000), 1027–55.

[14] J.A. Green, *Polynomial representations of* GL_n, Lecture Notes in Mathematics, 830, Springer-Verlag, Berlin-New York, 1980, vi+118 pp.

[15] Ph. Griffiths, J. Harris, *Principles of Algebraic Geometry*, reprint of the 1978 original, Wiley Classics Library, John Wiley & Sons, Inc., New York, 1994, xiv+813 pp.

[16] D. Gross, *Compact quotients by $\mathbb{C}^*$-actions*, Pacific J. Math. **114** (1984), 149–64.

[17] G.B. Gurevich, *Foundations of the theory of algebraic invariants*, Translated by J.R.M. Radok and A.J.M. Spencer, P. Noordhoff Ltd., Groningen, 1964, viii+429 pp.

[18] J. Hausen, *Geometric invariant theory based on Weil divisors*, Compos. Math. **140** (2004), 1518–36.

[19] D. Hilbert, *Theory of algebraic invariants*, translated from the German and with a preface by Reinhard C. Laubenbacher, edited and with an introduction by Bernd Sturmfels, Cambridge University Press, Cambridge, 1993, xiv+191 pp.

[20] J.E. Humphreys, *Linear algebraic groups*, Graduate Texts in Mathematics, 21, Springer-Verlag, New York-Heidelberg, 1975, xiv+247 pp.

[21] D. Huybrechts, M. Lehn, *The geometry of moduli spaces of sheaves*, Aspects of Mathematics, E31, Friedr. Vieweg & Sohn, Braunschweig, 1997, xiv+269 pp.

[22] J.C. Jantzen, *Representations of algebraic groups*, Second edition, Mathematical Surveys and Monographs, 107, American Mathematical Society, Providence, RI, 2003, xiv+576 pp.

[23] H. Kraft, *Geometrische Methoden in der Invariantentheorie*, Aspects of Mathematics, D1, Friedr. Vieweg & Sohn, Braunschweig, 1984, x+308 pp.

[24] H. Kraft, C. Procesi, *Classical Invariant Theory – a Primer*, Lecture Notes, Version 2000, 128 pages. Available at http://www.math.unibas.ch.

[25] J. Le Potier, *Lectures on vector bundles*, Translated by A. Maciocia, Cambridge Studies in Advanced Mathematics, 54, Cambridge University Press, Cambridge, 1997, viii+251 pp.

[26] M. Lübke, A. Teleman, *The universal Kobayashi-Hitchin correspondence on Hermitian manifolds*, Mem. Amer. Math. Soc. **183** (2006), 97 pp.

[27] A.L. Mayer, *Families of K-3 surfaces*, Nagoya Math. J. **48** (1972), 1–17.

[28] D. Mumford et al., *Geometric Invariant Theory*, Third edition, Ergebnisse der Mathematik und ihrer Grenzgebiete (2), 34, Springer-Verlag, Berlin, 1994, xiv+292 pp.

[29] M. Nagata, *Lectures on Hilbert's fourteenth problem*, Tata Institute of Fundamental Research, Bombay 1965, ii+78+iii pp.

[30] P.E. Newstead, *Introduction to moduli problems and orbit spaces*, Tata Institute of Fundamental Research Lectures on Mathematics and Physics, 51, Tata Institute of Fundamental Research, Bombay, by the Narosa Publishing House, New Delhi, 1978, vi+183 pp.

[31] Ch. Okonek, A. Teleman, A. Schmitt, *Master spaces for stable pairs*, Topology **38** (1999), 117–39.

[32] C. Procesi, *The invariant theory of $n \times n$ matrices*, Advances in Math. **19** (1976), 306–81.

[33] N. Ressayre, *The GIT-equivalence for G-line bundles*, Geom. Dedicata **81** (2000), 295–324.

[34] M. Rosenlicht, *A remark on quotient spaces*, An. Acad. Brasil. Ci. **35** (1963), 487–9.

[35] A.H.W. Schmitt, *Quaternary cubic forms and projective algebraic threefolds*, Enseign. Math. (2) **43** (1997), 253–70.

[36] A.H.W. Schmitt, *Singular principal bundles over higher-dimensional manifolds and their moduli spaces*, Int. Math. Res. Not. **2002:23** (2002), 1183–209.

[37] A.H.W. Schmitt, *A universal construction for moduli spaces of decorated vector bundles over curves*, Transform. Groups **9** (2004), 167–209.

[38] A.H.W. Schmitt, *A closer look at semistability for singular principal bundles*, Int. Math. Res. Not. **2004:62** (2004), 3327–66.

[39] A.H.W. Schmitt, *Global boundedness for decorated sheaves*, Internat. Math. Res. Not. **2004:68** (2004), 3637–71.

[40] A.H.W. Schmitt, *Moduli for decorated tuples of sheaves and representation spaces for quivers*, Proc. Indian Acad. Sci. Math. Sci. **115** (2005), 15–49.

[41] A.H.W. Schmitt, *A relative version of Geometric Invariant Theory*, manuscript in preparation, approx. 300 pp.

[42] J.-P. Serre, *Espaces fibrés algébriques*, Séminaire Chevalley, 1958, also Documents Mathématiques **1** (2001), 107–40.

[43] C.S. Seshadri, *Geometric reductivity (Mumford's conjecture) – revisited, Commutative algebra and algebraic geometry*, 137-45, Contemp. Math., 390, Amer. Math. Soc., Providence, RI, 2005.

[44] K.S. Sibirskiĭ, *Unitary and orthogonal invariants of matrices*, Russian, Dokl. Akad. Nauk SSSR **172** (1967), 40–43; English translation in Soviet Math. Dokl. **8** (1967), 36–40.

[45] Ch. Sorger, *Lectures on moduli of principal G-bundles over algebraic curves*, School on Algebraic Geometry (Trieste, 1999), 1–57, ICTP Lect. Notes, 1, Abdus Salam Int. Cent. Theoret. Phys., Trieste, 2000.

[46] T.A. Springer, *Linear algebraic groups*, Second edition, Progress in Mathematics, 9, Birkhäuser Boston, Inc., Boston, MA, 1998, xiv+334 pp.

[47] B. Sturmfels, *Algorithms in invariant theory*, Texts and Monographs in Symbolic Computation, Springer-Verlag, Vienna, 1993, vi+197 pp.

[48] M. Thaddeus, *Stable pairs, linear systems and the Verlinde formula*, Invent. Math. **117** (1994), 317–53.

[49] M. Thaddeus, *Geometric Invariant Theory and flips*, J. Amer. Math. Soc. **9** (1996), 691–723.

[50] J. Winkelmann, *Invariant rings and quasiaffine quotients*, Math. Z. **244** (2003), 163–74.

Alexander H.W. Schmitt
Freie Universität Berlin
Institut für Mathematik
Arnimallee 3
D-14195 Berlin, Deutschland

Algebraic Cycles, Sheaves, Shtukas, and Moduli
Trends in Mathematics, 185–215
© 2007 Birkhäuser Verlag Basel/Switzerland

Some Applications of Algebraic Cycles to Affine Algebraic Geometry

Vasudevan Srinivas

Abstract. In this series of talks, I will discuss some applications of the theory of algebraic cycles to affine algebraic geometry (i.e., to commutative algebra).

1. The Chow ring and Chern classes

First, we recall the definition of the graded *Chow ring* $CH^*(X) = \bigoplus_{p \geq 0} CH^p(X)$ of a non-singular variety X over a field k (see [9] for more details; see also [3]). We will usually (but not always) take k to be algebraically closed; X need not be irreducible. The graded components $CH^p(X)$ generalize the more familiar notion of the *divisor class group*, which is just the group $CH^1(X)$.

If $Z \subset X$ is irreducible, let $\mathcal{O}_{Z,X}$ be the local ring of Z on X (*i.e.*, the local ring of the generic point of Z, in the terminology of Hartshorne's book [12]). The codimension of Z in X, denoted $\mathrm{codim}_X Z$, is the dimension of the local ring $\mathcal{O}_{Z,X}$. Now let

$$Z^p(X) = \text{Free abelian group on irreducible subvarieties of } X \text{ of codimension } p$$
$$= \text{Group of algebraic cycles on } X \text{ of codimension } p.$$

For an irreducible subvariety $Z \subset X$, let $[Z]$ denote its class in $Z^p(X)$ (where $p = \mathrm{codim}_X Z$).

Let $Y \subset X$ be irreducible of codimension $p - 1$, and let $k(Y)^*$ denote the multiplicative group of non-zero rational functions on Y ($k(Y)$, which is the field of rational functions on Y, is the residue field of $\mathcal{O}_{Y,X}$). For each irreducible divisor $Z \subset Y$, we have a homomorphism $\mathrm{ord}_Z : k(Y)^* \to \mathbb{Z}$, given by

$$\mathrm{ord}_Z(f) = \ell(\mathcal{O}_{Z,Y}/a\mathcal{O}_{Z,Y}) - \ell(\mathcal{O}_{Z,Y}/b\mathcal{O}_{Z,Y}),$$

for any expression of f as a ratio $f = a/b$ with $a, b \in \mathcal{O}_{Z,Y} \setminus \{0\}$. Here $\ell(M)$ denotes the length of an Artinian module M.

Lectures at *IMPANGA 100*, Warsaw, 28th–29th October, 2005.

For $f \in k(Y)^*$, let $(f)_Y$ denote the divisor of f on Y, defined by

$$(f)_Y = \sum_{Z \subset Y} \operatorname{ord}_Z(f) \cdot [Z],$$

where Z runs over all irreducible divisors in Y; the sum has only finitely many non-zero terms, and is hence well defined. Clearly we may also view $(f)_Y$ as an element of $Z^p(X)$.

Let $R^p(X) \subset Z^p(X)$ be the subgroup generated by cycles $(f)_Y$ as (Y, f) ranges over all irreducible subvarieties Y of X of codimension $p - 1$, and all $f \in k(Y)^*$. We refer to elements of $R^p(X)$ as cycles *rationally equivalent to* 0 on X. The pth Chow group of X is defined to be

$$CH^p(X) = \frac{Z^p(X)}{R^p(X)}$$

= group of rational equivalence classes of codimension p-cycles on X.

We will abuse notation and also use $[Z]$ to denote the class of an irreducible subvariety Z in $CH^p(X)$.

The graded abelian group

$$CH^*(X) = \bigoplus_{0 \leq p \leq \dim X} CH^p(X)$$

can be given the structure of a commutative (graded) ring via the *intersection product*. This product is characterized by the following property – if $Y \subset X$, $Z \subset X$ are irreducible of codimensions p, q respectively, and $Y \cap Z = \cup_i W_i$, where each $W_i \subset X$ is irreducible of codimension $p + q$ (we then say Y and Z *intersect properly* in X), then the intersection product of the classes $[Y]$ and $[Z]$ is

$$[Y] \cdot [Z] = \sum_i I(Y, Z; W_i)[W_i]$$

where $I(Y, Z; W_i)$ is the *intersection multiplicity* of Y and Z along W_i, defined by Serre's formula

$$I(Y, Z; W_i) = \sum_{j \geq 0} (-1)^j \ell \left(\operatorname{Tor}_j^{\mathcal{O}_{W_j, X}} (\mathcal{O}_{W_j, Y}, \mathcal{O}_{W_j, Z}) \right).$$

One of the important results proved in the book [9] is that the above procedure does give rise to a well-defined ring structure on $CH^*(X)$.

The Chow ring is an algebraic analogue for the even cohomology ring

$$\bigoplus_{i=0}^{n} H^{2i}(X, \mathbb{Z})$$

defined in algebraic topology. To illustrate this, we note the following 'cohomology-like' properties. Here, we follow the convention of [12], and use the term "vector bundle on X" to mean "(coherent) locally free sheaf of $\mathcal{O}_X$-modules", and use the term "geometric vector bundle on X", as in [12] II Ex. 5.18, to mean a Zariski locally trivial algebraic fiber bundle $V \to X$ whose fibres are affine spaces, with linear transition functions. With this convention, we can also identify vector bundles on an affine variety $X = \operatorname{Spec} A$ with finitely generated projective A-modules; as in [12], we use the notation $\widetilde{M}$ to denote the coherent sheaf corresponding to a finitely generated A-module M.

Theorem 1.1 (Properties of the Chow ring and Chern classes).

(1) $X \mapsto \bigoplus_p CH^p(X)$ *is a contravariant functor from the category of smooth varieties over k to graded rings. If $X = \coprod_i X_i$, where X_i are the irreducible (= connected) components, then $CH^*(X) = \prod_i CH^*(X_i)$.*

(2) *If X is irreducible and projective (or more generally, proper) over an algebraically closed field k and $d = \dim X$, there is a well-defined degree homomorphism*

$$\deg : CH^d(X) \to \mathbb{Z}$$

given by $\deg(\sum_i n_i[x_i]) = \sum_i n_i$. This allows one to define **intersection numbers** *of cycles of complementary dimension, in a purely algebraic way, which agree with those defined via topology when $k = \mathbb{C}$ (see (7) below).*

(3) *If $f : X \to Y$ is a proper morphism of smooth varieties, there are "Gysin" (or "push-forward") maps $f_* : CH^p(X) \to CH^{p+r}(Y)$ for all p, where $r = \dim Y - \dim X$; here if $p + r < 0$, we define f_* to be 0. The induced map $CH^*(X) \to CH^*(Y)$ is $CH^*(Y)$-linear (**projection formula**), where $CH^*(X)$ is regarded as a $CH^*(Y)$-module via the (contravariant) ring homomorphism $f^* : CH^*(Y) \to CH^*(X)$. If $f : X \hookrightarrow Y$ is the inclusion of a closed subvariety, then f_* is induced by the natural inclusions $Z^p(X) \hookrightarrow Z^{p+r}(Y)$.*

(4) $f^* : CH^*(X) \xrightarrow{\cong} CH^*(V)$ *for any geometric vector bundle $f : V \to X$ (**homotopy invariance**). In particular, $CH^*(X \times \mathbb{A}^n) = CH^*(X)$, and $CH^*(\mathbb{A}^n) = \mathbb{Z}$.*

(5) *If V is a vector bundle (i.e., locally free sheaf) of rank r on X, then there are* **Chern classes** $c_p(V) \in CH^p(X)$, *such that*

  (a) $c_0(V) = 1$,

  (b) $c_p(V) = 0$ *for $p > r$, and*

  (c) *for any exact sequence of vector bundles*

$$0 \to V_1 \to V_2 \to V_3 \to 0$$

    we have $c(V_2) = c(V_1)c(V_3)$, where $c(V_i) = \sum_p c_p(V_i)$ are the corresponding **total Chern classes**

  (d) $c_p(V^\vee) = (-1)^p c_p(V)$, *where $V^\vee$ is the dual vector bundle.*

  Moreover, we also have the following properties.

(6) *If $f : \mathbb{P}(V) = \mathbf{Proj}\, S(V) \to X$ is the projective bundle associated to a vector bundle of rank r (where $S(V)$ is the symmetric algebra of the sheaf V over $\mathcal{O}_X$), then $CH^*(\mathbb{P}(V))$ is a $CH^*(X)$-algebra generated by $\xi = c_1(\mathcal{O}_{\mathbb{P}(V)}(1))$, the first Chern class of the tautological line bundle, which is subject to the relation*

$$\xi^r - c_1(V)\xi^{r-1} + \cdots + (-1)^r c_r(V) = 0;$$

in particular, $CH^*(\mathbb{P}(V))$ is a free $CH^*(X)$-module with basis

$$1, \xi, \xi^2, \ldots, \xi^{r-1}.$$

(7) *If $k = \mathbb{C}$, there are cycle class homomorphisms $CH^p(X) \to H^{2p}(X, \mathbb{Z})$ such that the intersection product corresponds to the cup product in cohomology, and for a vector bundle E, the cycle class of $c_p(E)$ is the topological pth Chern class of E.*

(8) *The first Chern class determines an isomorphism $\mathrm{Pic}\,X \to CH^1(X)$ from the Picard group of line bundles on X to the first Chow group (i.e., the divisor class group) of X. For an arbitrary vector bundle V, of rank n, we have $c_1(V) = c_1(\det V)$, where $\det V = \bigwedge^n V$.*

(9) *If $f : X \to Y$ is a morphism between non-singular varieties, V a vector bundle on Y, then the Chern classes of the pull-back vector bundle f^*V on X are given by $c(f^*V) = f^*c(V)$, where on the right, f^* is the ring homomorphism $CH^*(Y) \to CH^*(X)$ (**functoriality of Chern classes**). In particular, taking $Y = point$, we see that $c(\mathcal{O}_X) = 1 \in CH^*(X)$.*

(10) *If $i : Y \hookrightarrow X$ is the inclusion of an irreducible smooth subvariety of codimension r in a smooth variety, with normal bundle $\mathcal{N} = (\mathcal{I}_Y/\mathcal{I}_Y^2)^\vee$ (where $\mathcal{I}_Y \subset \mathcal{O}_X$ is the ideal sheaf of Y in X), then $\mathcal{N}$ is a vector bundle on Y of rank r with top Chern class*

$$c_r(\mathcal{N}) = i^* \circ i_*[Y],$$

*where $[Y] \in CH^0(Y) = \mathbb{Z}$ is the generator (**self-intersection formula**).*

Remark 1.2. If $X = \mathrm{Spec}\,A$ is affine, we will also sometimes write $CH^*(A)$ in place of $CH^*(X)$; similarly, by the Chern classes $c_i(P)$ of a finitely generated projective A-module P, we mean $c_i(\widetilde{P})$ where $\widetilde{P}$ is the associated vector bundle on X.

We remark that the total Chern class of a vector bundle on a smooth variety X is a *unit* in the Chow ring $CH^*(X)$, since it is of the form $1+$(nilpotent element). Thus the assignment $V \mapsto c(V)$ gives a homomorphism of groups from the Grothendieck group $K_0(X)$ of vector bundles (locally free sheaves) on X to the multiplicative group of those units in the graded ring $CH^*(X)$, which are expressible as $1 +$ (higher degree terms).

On a non-singular variety X, every coherent sheaf has a resolution by locally free sheaves (vector bundles) of finite rank, and the Grothendieck group $K_0(X)$ of vector bundles coincides with the Grothendieck group of coherent sheaves. There is a finite decreasing filtration $\{F^p K_0(X)\}_{p \geq 0}$ on $K_0(X)$, where $F^p K_0(X)$ is the subgroup generated by classes of sheaves supported in codimension $\geq p$. Further, $F^p K_0(X)/F^{p+1} K_0(X)$ is generated, as an abelian group, by the classes $\mathcal{O}_Z$ for irreducible subvarieties $Z \subset X$ of codimension p – for example, if $X = \mathrm{Spec}\,A$ is affine, we can see this using the fact that any finitely generated A-module M has a finite filtration whose quotients are of the form $A/\wp$ for prime ideals $\wp$, such that the minimal primes in $\mathrm{supp}\,(M)$ all occur, and their multiplicities in the filtration are independent of the choice of filtration. Thus, we have a natural surjection $Z^p(X) \to F^p K_0(X)/F^{p+1} K_0(X)$.

If $\mathcal{F}$ is any coherent sheaf on X whose support is of codimension p, recall that we can associate to it a codimension p cycle

$$|\mathcal{F}| \in Z^p(X), \qquad \text{by} \qquad |\mathcal{F}| = \sum_{W \subset X} \ell(\mathcal{F}_{\eta_W}),$$

where W ranges over the irreducible, codimension p subvarieties of X in the support of $\mathcal{F}$, and η_W is the generic point of W, so that the stalk $\mathcal{F}_{\eta_W}$ is a module of finite length over the local ring $\mathcal{O}_{W,X} = \mathcal{O}_{\eta_W,X}$. If V is a vector bundle of rank p, and s a section, then s induces a map of sheaves $V^{\vee} \to \mathcal{O}_X$, whose image is an ideal sheaf $\mathcal{I}_Y$, where Y is the zero scheme of s. Since the ideal sheaf of Y is locally generated by p elements, each irreducible component of Y_{red} has codimension $\leq p$. If Y is purely of codimension p, there is thus an associated cycle $|Y| = |\mathcal{O}_Y| \in Z^p(X)$.

Now we can state the following result, part (d) of which is sometimes called the *Riemann-Roch theorem without denominators* (see the book [9] for a proof). Note that (c) is consistent with the self-intersection formula, in the case when Y is nonsingular, since in this case, $V \otimes \mathcal{O}_Y$ is identified with the normal bundle.

Theorem 1.3. *Let X be a non-singular variety.*

(a) *If $x \in F^p K_0(X)$, then $c_i(x) = 0$ for $i < p$, and $c_p : F^p K_0(X) \to CH^p(X)$ is a group homomorphism. Let $\overline{c_p} : F^p K_0(X)/F^{p+1} K_0(X) \to CH^p(X)$ be the induced homomorphism.*

(b) *The natural surjection $Z^p(X) \twoheadrightarrow F^p K_0(X)/F^{p+1} K_0(X)$ factors through rational equivalence, yielding a map $\psi_p : CH^p(X) \twoheadrightarrow F^p K_0(X)/F^{p+1} K_0(X)$.*

(c) *Let V be a vector bundle of rank p on X, and s a section with zero scheme Y, which has codimension p. Then $|Y| \in Z^p(X)$ is a cycle representing the pth Chern class $c_p(V) \in CH^p(X)$.*

(d) *The compositions $\overline{c_p} \circ \psi_p$ and $\psi_p \circ \overline{c_p}$ both equal multiplication by the integer $(-1)^{p-1}(p-1)!$. In particular, both $\overline{c_p}$ and ψ_p are isomorphisms $\otimes \mathbb{Q}$.*

In particular, if $Z \subset X$ is an irreducible subvariety of codimension p, then $c_i([\mathcal{O}_Z]) = 0$ for $i < p$, and $c_p([\mathcal{O}_Z]) = (-1)^{p-1}(p-1)![Z] \in CH^p(X)$.

Remark 1.4. If $X = \operatorname{Spec} A$ is affine, any element $\alpha \in K_0(X)$ can be expressed as a difference $\alpha = [P] - [A^{\oplus m}]$ for some finitely generated projective A-module P and some positive integer m. Hence the total Chern class $c(\alpha)$ coincides with $c(P)$. The above theorem now implies that for any element $a \in CH^p(X)$, there is a finitely generated projective A module P with $c_p(P) = (p-1)!a$. By the Bass stability theorem, which implies that any projective A-module of rank $> d = \dim A$ has a free direct summand of positive rank, we can find a projective A-module P with $\operatorname{rank} P \leq d$ and $c_p(P) = (p-1)!a$.

Incidentally, this statement cannot be improved, in general: for any $p > 2$, there are examples of affine non-singular varieties X and elements $a \in CH^p(X)$ such that $ma \in \operatorname{image} c_p$ for some integer $m \iff (p-1)! | m$.

2. An example of a graded ring

We now discuss our first application of these constructions, due to N. Mohan Kumar (unpublished). It is a counterexample to the "principle": if a commutative algebra problem with graded data has a solution, then it also has a graded solution.

Let $k = \bar{k}$. We give an example of a 3-dimensional, regular, graded integral domain $A = \bigoplus_{n \geq 0} A_n$, with the following properties:

1. A is generated by A_1 as an A_0-algebra, where A_0 is a regular affine k-algebra of dimension 1
2. the "irrelevant graded prime ideal" $P = \bigoplus_{n > 0} A_n$ is the radical of an ideal generated by 2 elements (i.e., the subvariety of $Z = \operatorname{Spec} A$ defined by P is a set-theoretic complete intersection in Z)
3. P cannot be expressed as the radical of an ideal generated by 2 *homogeneous* elements (i.e., the subvariety is not a "homogeneous" set-theoretic complete intersection).

For the example, take A_0 to be affine coordinate ring of a non-singular curve $C \subset \mathbb{A}_k^3$ such that the canonical module $\omega_{A_0} = \Omega_{A_0/k}$ is a non-torsion element of the divisor class group of A_0 (this implies k is not the algebraic closure of a finite field). In fact, if we choose A_0 to be a non-singular affine k-algebra of dimension 1 such that ω_{A_0} is non-torsion in the class group, then $C = \operatorname{Spec} A_0$ can be realized as a curve embedded in $\mathbb{A}_k^3$, by more or less standard arguments (see [12], IV, or [26], for example).

Let $R = k[x, y, z]$ denote the polynomial algebra, and let $\varphi : R \to A_0$ be the surjection corresponding to $C \hookrightarrow \mathbb{A}_k^3$. Let $I = \ker \varphi$ be the ideal of C. Then I/I^2 is a projective A_0-module of rank 2; we let

$$A = S(I/I^2) = \bigoplus_{n \geq 0} S^n(I/I^2)$$

be its symmetric algebra over A_0. We claim this graded ring A has the properties stated above.

Consider the exact sequence of projective A_0-modules

$$0 \to I/I^2 \xrightarrow{\psi} \Omega_{R/k} \otimes A_0 \xrightarrow{\overline{\varphi}} \omega_{A_0} \to 0 \tag{2.1}$$

with $\overline{\varphi}$ induced by φ, and ψ by the derivation $d : R \to \Omega_{R/k}$. Let

$$h : \Omega_{R/k} \otimes A_0 \to I/I^2$$

be a splitting of ψ. Use h to define a homomorphism of k-algebras

$$\Phi : R \to A,$$

by setting

$$\Phi(t) = \phi(t) + h(dt) \in A_0 \oplus A_1 = A_0 \oplus I/I^2$$

for $t = x, y, z$; this uniquely specifies a k-algebra homomorphism Φ defined on the polynomial algebra R.

Clearly $\Phi(I) \subset P = \bigoplus_{n>0} A_n$, the irrelevant graded ideal, and one verifies that Φ induces isomorphisms $R/I \to A/P$ and $I/I^2 \to P/P^2$, and in fact an isomorphism between the I-adic completion of R and the P-adic completion of A.

Since $C \subset \mathbb{A}^3_k$ is a non-singular curve, it is a set-theoretic complete intersection, from a theorem of Ferrand and Szpiro (see [32], for example). If $a, b \in I$ with $\sqrt{(a,b)} = I$, then clearly we have $\sqrt{(\Phi(a), \Phi(b))} = P \cap Q$, for some (radical) ideal Q with $P + Q = A$. We can correspondingly write $(\Phi(a), \Phi(b)) = J \cap J'$ with $\sqrt{J} = P$, $\sqrt{J'} = Q$. Then $J/J^2 \cong (A/J)^{\oplus 2}$. This implies (by an old argument of Serre) that $\operatorname{Ext}^1_A(J, A) \cong \operatorname{Ext}^2_A(A/J, A) \cong A/J$ is free of rank 1, and any generator determines an extension

$$0 \to A \to V \to J \to 0$$

where V is a projective A-module of rank 2, and such that the induced surjection $V \otimes A/J \to J/J^2 \cong (A/J)^{\oplus 2}$ is an isomorphism.

We claim the projective module V is necessarily of the form $V = V_0 \otimes_{A_0} A$; this implies $V_0 = V \otimes_A A/P \cong J/PJ \cong (A/P)^{\oplus 2}$ is free, so that V is a free A-module, and J is generated by 2 elements. To prove the claim, note that I/I^2 is a direct summand of a free $A/I = A_0$-module of finite rank; hence there is an affine A-algebra $A' \cong A_0[x_1, \ldots, x_n]$, which is a polynomial algebra over A_0, such that A is an algebra retract of A'. Now it suffices to observe that any finitely generated projective A'-module is of the form $M \otimes_{A_0} A'$, for some projective A_0-module M; this is the main result of [13], the solution of the so-called Bass-Quillen Conjecture (see also [14]).

On the other hand, we claim that it is impossible to find two *homogeneous* elements $x, y \in P$ with $\sqrt{(x,y)} = P$. Indeed, let $X = \operatorname{Proj} A$, and $\pi : X \to C = \operatorname{Spec} A_0$ be the natural morphism. Then $X = \mathbb{P}(V)$ is the $\mathbb{P}^1$-bundle over C associated to the locally free sheaf $V = \widetilde{I/I^2}$ (the sheaf determined by the projective A_0-module I/I^2). Let $\xi = c_1(\mathcal{O}_X(1)) \in CH^1(X)$ be the 1st Chern class of the tautological line bundle $\mathcal{O}_X(1)$. Then by Theorem 1.1(6) above, $CH^*(X)$ is a free $CH^*(C)$-module with basis $1, \xi$, and ξ satisfies the monic relation

$$\xi^2 - c_1(V)\xi + c_2(V) = 0.$$

Since $\dim C = 1$, $CH^i(C) = 0$ for $i > 1$, and so this relation reduces to

$$\xi^2 = c_1(V)\xi.$$

From the exact sequence (2.1), we have a relation in $CH^*(C)$

$$1 = c(\mathcal{O}_C)^3 = c(\mathcal{O}_C^{\oplus 3}) = c(\Omega_{\mathbb{A}^3/k} \otimes \mathcal{O}_C) = c(V) \cdot c(\omega_C).$$

Hence $c_1(V) = -c_1(\omega_C)$, which by the choice of C is a non-torsion element of $CH^1(C)$ (which is the divisor class group of A_0). Thus $\xi^2 \in CH^2(X)$ is a non-torsion element of $CH^2(X)$.

If homogeneous elements $x, y \in P$ exist, say of degrees r and s respectively, such that $\sqrt{(x,y)} = P$, then we may regard x, y as determining global sections of the sheaves $\mathcal{O}_X(r)$ and $\mathcal{O}_X(s)$ respectively, *which have no common zeroes on*

X. Let $D_x \subset X$, $D_y \subset X$ be the divisors of zeroes of $x \in \Gamma(X, \mathcal{O}_X(r))$ and $y \in \Gamma(X, \mathcal{O}_X(s))$ respectively. Then we have equations in $CH^1(X)$

$$[D_x] = c_1(\mathcal{O}_X(r)) = rc_1(\mathcal{O}_X(1)) = r\xi, \quad [D_y] = c_1(\mathcal{O}_X(s)) = sc_1(\mathcal{O}_X(1)) = s\xi.$$

But $D_x \cap D_y = \emptyset$. Hence in $CH^2(X)$, we have a relation

$$0 = [D_x] \cdot [D_y] = rs\xi^2,$$

contradicting that $\xi^2 \in CH^2(X)$ is a non-torsion element.

Remark 2.1. The construction of the homomorphism from the polynomial ring R to the graded ring A is an algebraic analogue of the *exponential map* in Riemannian geometry, which identifies a tubular neighbourhood of a smooth submanifold of a Riemannian manifold with the normal bundle of the submanifold (see [15, Theorem 11.1], for example). The exponential map is usually constructed using geodesics on the ambient manifold; here we use the global structure of affine space, where "geodesics" are lines, to make a similar construction algebraically. This idea appears in a paper[6] of Boratynski, who uses it to argue that a smooth subvariety of $\mathbb{A}^n$ is a set-theoretic complete intersection if and only if the zero section of its normal bundle is a set-theoretic complete intersection in the total space of the normal bundle.

3. Zero cycles on non-singular proper and affine varieties

In this section, we discuss results of Mumford and Roitman, which give criteria for the non-triviality of $CH^d(X)$ where X is a non-singular variety over $\mathbb{C}$ of dimension $d \geq 2$, which is either proper, or affine.

If X is non-singular and irreducible, and $\dim X = d$, then $Z^d(X)$ is just the free abelian group on the (closed) points of X. Elements of $Z^d(X)$ are called *zero cycles* on X (since they are linear combinations of irreducible subvarieties of dimension 0). In the presentation $CH^d(X) = Z^d(X)/R^d(X)$, the group $R^d(X)$ of relations is generated by divisors of rational functions on irreducible curves in X.

The main non-triviality result for zero cycles is the following result, called the *infinite dimensionality theorem for 0-cycles*. It was originally proved (without $\otimes \mathbb{Q}$) by Mumford [17], for surfaces, and extended to higher dimensions by Roitman [22]; the statement with $\otimes \mathbb{Q}$ follows from [23].

Theorem 3.1. (Mumford, Roitman) *Let X be an irreducible, proper, non-singular variety of dimension d over $\mathbb{C}$. Suppose X supports a non-zero regular q-form (i.e., $\Gamma(X, \Omega^q_{X/\mathbb{C}}) \neq 0$), for some $q > 0$. Then for any closed algebraic subvariety $Y \subset X$ with $\dim Y < q$, we have $CH^d(X - Y) \otimes \mathbb{Q} \neq 0$.*

Corollary 3.2. *Let X be an irreducible, proper, non-singular variety of dimension d over $\mathbb{C}$, such that $\Gamma(X, \omega_X) \neq 0$. Then for any affine open subset $V \subset X$, we have $CH^d(V) \otimes \mathbb{Q} \neq 0$.*

The corollary results from the identification of ω_X with the sheaf $\Omega^d_{X/\mathbb{C}}$ of d-forms.

Bloch [3] gave another proof of the above result, using the action of algebraic correspondences on the étale cohomology, and generalized the result to arbitrary characteristics. In [27] and [28], Bloch's argument (for the case of characteristic 0) is recast in the language of differentials, extending it as well to certain singular varieties. One way of stating the infinite dimensionality results of [27] and [28], in the smooth case, is the following. The statement is technical, but it will be needed below when discussing M. Nori's construction of indecomposable projective modules.

We recall the notion of a k-generic point of an irreducible variety; we do this in a generality sufficient for our purposes. If X_0 is an irreducible k-variety, where $k \subset \mathbb{C}$ is a countable algebraically closed subfield, a point $x \in X = (X_0)_{\mathbb{C}}$ determines an irreducible subvariety $Z \subset X$, called the k-closure of X, which is the smallest subvariety of X which is defined over k (*i.e.*, of the form $(Z_0)_{\mathbb{C}}$ for some subvariety $Z_0 \subset X_0$) and contains the chosen point x. We call x a k-*generic point* if its k-closure is X itself.

In the case X_0 (and thus also X) is affine, say $X_0 = \operatorname{Spec} A$, and $X = \operatorname{Spec} A_{\mathbb{C}}$ with $A_{\mathbb{C}} = A \otimes_k \mathbb{C}$, then a point $x \in X$ corresponds to a maximal ideal $\mathfrak{m}_x \subset A_{\mathbb{C}}$. Let $\wp_x = A \cap \mathfrak{m}_x$, which is a prime ideal of A, not necessarily maximal. Then, in the earlier notation, $\wp_x$ determines an irreducible subvariety $Z_0 \subset X_0$. The k-closure $Z \subset X$ of x is the subvariety determined by the prime ideal $\wp_x A_{\mathbb{C}}$ (since k is algebraically closed, $\wp_x A_{\mathbb{C}}$ is a prime ideal). In particular, x is a k-generic point $\iff \wp_x = 0$. In this case, x determines an inclusion $A \hookrightarrow A_{\mathbb{C}}/\mathfrak{m}_x = \mathbb{C}(x) \cong \mathbb{C}$. This in turn gives an inclusion $i_x : K \hookrightarrow \mathbb{C}$ of the quotient field K of A (*i.e.*, of the function field $k(X_0)$) into the complex numbers.

In general, even if X is not affine, if we are given a k-generic point $x \in X$, we can replace X by any affine open subset defined over k, which will (because x is k-generic) automatically contain x; one verifies easily that the corresponding inclusion $K \hookrightarrow \mathbb{C}$ does not depend on the choice of this open subset. Thus we obtain an inclusion $i_x : K \hookrightarrow \mathbb{C}$ of the function field $K = k(X_0)$ into $\mathbb{C}$, associated to any k-generic point of X.

It is easy to see that the procedure is reversible: any inclusion of k-algebras $i : K \hookrightarrow \mathbb{C}$ determines a unique k-generic point of X. Indeed, choose an affine open subset $\operatorname{Spec} A = U_0 \subset X_0$, so that K is the quotient field of A. The induced inclusion $A \hookrightarrow \mathbb{C}$ induces a *surjection* of $\mathbb{C}$-algebras $A_{\mathbb{C}} \to \mathbb{C}$, whose kernel is a maximal ideal, giving the desired k-generic point.

Suppose now that X_0 is proper over k, and so X is proper over $\mathbb{C}$ (e.g., X is projective). Let $\dim X_0 = \dim X = d$. Then by the Serre duality theorem, the sheaf cohomology group $H^d(X, \mathcal{O}_X)$ is the dual $\mathbb{C}$-vector space to

$$\Gamma(X, \Omega^d_{X/\mathbb{C}}) = \Gamma(X, \omega_X) = \Gamma(X_0, \omega_{X_0}) \otimes_k \mathbb{C}.$$

194   V. Srinivas

Hence we may identify $H^d(X, \mathcal{O}_X) \otimes_{\mathbb{C}} \Omega^d_{\mathbb{C}/k}$ with

$$\mathrm{Hom}_{\mathbb{C}}(\Gamma(X, \omega_X), \Omega^d_{\mathbb{C}/k}) = \mathrm{Hom}_k(\Gamma(X_0, \omega_{X_0}), \Omega^d_{\mathbb{C}/k}).$$

Note that a k-generic point x determines, via the inclusion $i_x : K \hookrightarrow \mathbb{C}$, a k-linear inclusion $\Omega^n_{K/k} \hookrightarrow \Omega^n_{\mathbb{C}/k}$, and hence, via the obvious inclusion

$$\Gamma(X_0, \omega_{X_0}) = \Gamma(X_0, \Omega^n_{X_0/k}) \hookrightarrow \Omega^n_{K/k},$$

a canonical element

$$di_x \in \mathrm{Hom}_k(\Gamma(X_0, \omega_{X_0}), \Omega^d_{\mathbb{C}/k}) = H^d(X, \mathcal{O}_X) \otimes_{\mathbb{C}} \Omega^d_{\mathbb{C}/k}.$$

Theorem 3.3. *Let $k \subset \mathbb{C}$ be a countable algebraically closed subfield, and X_0 an irreducible non-singular proper k-variety of dimension d, with $\Gamma(X_0, \omega_{X_0}) \neq 0$. Let $U_0 \subset X_0$ be any Zariski open subset. Let $X = (X_0)_{\mathbb{C}}$, $U = (U_0)_{\mathbb{C}}$ be the corresponding complex varieties. Then there is a homomorphism of graded rings*

$$CH^*(U) \to \bigoplus_{p \geq 0} H^p(X, \mathcal{O}_X) \otimes_{\mathbb{C}} \Omega^p_{\mathbb{C}/k},$$

with the following properties.

(i) *If $x \in U$ is a point, which is not k-generic, then the image in $H^d(X, \mathcal{O}_X) \otimes \Omega^d_{\mathbb{C}/k}$ of $[x] \in CH^d(U)$ is zero.*

(ii) *If $x \in U$ is a k-generic point, then the image in $H^d(X, \mathcal{O}_X) \otimes \Omega^d_{\mathbb{C}/k}$ of $[x] \in CH^d(U)$ is (up to sign) the canonical element di_x described above.*

As stated earlier, the above more explicit form of the infinite dimensionality theorem follows from results proved in [27] and [28].

4. Zero cycle obstructions to embedding and immersing affine varieties

We now consider the following two problems, which turn out to have some similarities. We will show how, in each case, the problem reduces to finding an example for which the Chern classes of the cotangent bundle (*i.e.*, the sheaf of Kähler differentials) have appropriate properties. We will then see, in Example 4.4, how to construct examples with these properties. The discussion is based on the article [4] of Bloch, Murthy and Szpiro.

Problem 4.1. Find examples of n-dimensional, non-singular affine algebras A over (say) the complex number field $\mathbb{C}$, for each $n \geq 1$, such that A cannot be generated by $2n$ elements as a $\mathbb{C}$-algebra, or such that the module of Kähler differentials cannot be generated by $2n - 1$ elements $da_1, \ldots, da_{2n-1}$ (in contrast, it is a "classical" result that such an algebra A can always be generated by $2n + 1$ elements, and its Kähler differentials can always be generated by $2n$ exact 1-forms; see, for example, [26]).

Problem 4.2. Find examples of prime ideals I of height $< N$ in a polynomial ring $\mathbb{C}[x_1, \ldots, x_N]$ such that $\mathbb{C}[x_1, \ldots, x_N]/I$ is regular, but I cannot be generated by $N - 1$ elements (a theorem of Mohan Kumar [16] implies that such an ideal I can always be generated by N elements).

First we discuss Problem 4.1. Suppose A is an affine smooth $\mathbb{C}$-algebra which is an integral domain of dimension n. Assume $X = \operatorname{Spec} A$ can be generated by $2n$ elements, $i.e.$, that there is a surjection $f : \mathbb{C}[x_1, \ldots, x_{2n}] \to A$ from a polynomial ring. Let $I = \ker f$. If $i : X \hookrightarrow \mathbb{A}^{2n}_{\mathbb{C}}$ is the embedding corresponding to the surjection f, then the normal bundle to i is the sheaf $V^{\vee}$, where $V = \widetilde{I/I^2}$.

From the self-intersection formula, and the formula for the Chern class of the dual of a vector bundle, we see that

$$(-1)^n c_n(V) = c_n(V^{\vee}) = i^* i_*[X] = 0, \tag{4.1}$$

since $CH^n(\mathbb{A}^{2n}_{\mathbb{C}}) = 0$.

On the other hand, suppose $j : X \hookrightarrow Y$ is any embedding as a closed subvariety of a non-singular affine variety Y whose cotangent bundle ($i.e.$, sheaf of Kähler differentials) $\Omega_{Y/\mathbb{C}}$ is a trivial bundle. For example, we could take $Y = \mathbb{A}^{2n}_{\mathbb{C}}$, and $j = i$, but below we will consider a different example as well.

Let W be the conormal bundle of X in Y (if $Y = \operatorname{Spec} B$, and $J = \ker j^* : B \twoheadrightarrow A$, then $W = \widetilde{J/J^2}$). We then have an exact sequence of vector bundles on X

$$0 \to W \to j^* \Omega_{Y/\mathbb{C}} \to \Omega^1_{X/\mathbb{C}} \to 0.$$

Since $\Omega_{Y/\mathbb{C}}$ is a trivial vector bundle, we get that

$$c(W) = c(\Omega_{X/\mathbb{C}})^{-1} \in CH^*(X). \tag{4.2}$$

Note that this expression for $c(W)$, and hence the resulting formula for $c_n(W)$ as a polynomial in the Chern classes of $\Omega_{X/\mathbb{C}}$, is in fact *independent* of the embedding j. In particular, from (4.1), we see that $c_n(W) = 0$ for *any* such embedding $j : X \hookrightarrow Y$.

Remark 4.3. In fact, the stability and cancellation theorems of Bass and Suslin imply that in the above situation, the vector bundle W itself is, up to isomorphism, independent of j, and is thus an invariant of the variety X. We call it the *stable normal bundle* of X; this is similar to the case of embeddings of smooth manifolds into Euclidean spaces. We will not need this fact in our computations below.

Returning to our discussion, we see that to find a $\mathbb{C}$-algebra A with $\dim A = n$, and which cannot be generated by $2n$ elements as a $\mathbb{C}$-algebra, it suffices to produce an embedding $j : X \hookrightarrow Y$ of $X = \operatorname{Spec} A$ into a smooth variety Y of dimension $2n$, such that

(i) $\Omega_{Y/\mathbb{C}}$ is a trivial bundle, and
(ii) if W is the conormal bundle of j, then $c_n(W) \neq 0$; in fact it suffices to produce such an embedding such that $j_* c_n(W) \in CH^{2n}(Y)$ is non-zero.

We see easily that the same example $X = \operatorname{Spec} A$ will have the property that $\Omega_{A/\mathbb{C}}$ is not generated by $2n - 1$ elements; in fact if $P = \ker(f : A^{\oplus 2n-1} \twoheadrightarrow \Omega_{A/\mathbb{C}})$ for some surjection f, then $\widetilde{P}$ is a vector bundle of rank $n - 1$, so that $c_n(\widetilde{P}) = 0$, while on the other hand, the exact sequence

$$0 \to P \to A^{\oplus 2n-1} \xrightarrow{f} \Omega_{A/\mathbb{C}} \to 0$$

implies that

$$c(\widetilde{P}) = c(\Omega_{X/\mathbb{C}})^{-1},$$

so that we would have

$$0 = c_n(\widetilde{P}) = c_n(W) \neq 0,$$

a contradiction.

Next we discuss the Problem 4.2 of finding an example of a "non-trivial" prime ideal $I \subset \mathbb{C}[x_1, \ldots, x_N]$ in a polynomial ring such that the quotient ring $A = \mathbb{C}[x_1, \ldots, x_N]/I$ is smooth of dimension > 0, while I cannot be generated by $N - 1$ elements (by the Eisenbud-Evans conjectures, proved by Sathaye and Mohan Kumar, I can always be generated by N elements).

Suppose I can be generated by $N - 1$ elements, and $\dim A/I = n > 0$. Then $I/I^2 \oplus Q = A^{\oplus N-1}$ for some projective A-module Q of rank $n - 1$; hence

$$(I/I^2 \oplus Q \oplus A) \cong A^{\oplus N} \cong (I/I^2 \oplus \Omega_{A/\mathbb{C}}).$$

Hence we have an equality between total Chern classes

$$c(\Omega_{X/\mathbb{C}}) = c(\widetilde{Q}),$$

and in particular, $c_n(\Omega_{X/\mathbb{C}}) = 0$.

So if $X = \operatorname{Spec} A$ is such that $c_n(\Omega_{X/\mathbb{C}}) \in CH^n(X)$ is non-zero, then for any embedding $X \hookrightarrow \mathbb{A}_{\mathbb{C}}^N$, the corresponding prime ideal I cannot be generated by $N - 1$ elements.

Example 4.4. We now show how to construct an example of an n-dimensional affine variety $X = \operatorname{Spec} A$ over $\mathbb{C}$, for any $n \geq 1$, such that, for some embedding $X \hookrightarrow Y = \operatorname{Spec} B$ with $\dim Y = 2n$, and ideal $I \subset B$, the projective module $P = I/I^2$ has the following properties:

(i) $c_n(P) \neq 0$ in $CH^n(X) \otimes \mathbb{Q}$
(ii) if $c(P) \in CH^*(X)$ is the total Chern class, then $c(P)^{-1}$ has a non-torsion component in $CH^n(X) \otimes \mathbb{Q}$.

Then, by the discussion earlier, the affine ring A will have the properties that

(a) A cannot be generated by $2n$ elements as a $\mathbb{C}$-algebra
(b) $\Omega_{A/\mathbb{C}}$ is not generated by $2n - 1$ elements
(c) for any way of writing $A = \mathbb{C}[x_1, \ldots, x_N]/J$ as a quotient of a polynomial ring (with n necessarily at least $2n + 1$), the ideal J requires N generators (use the formula (4.2)).

The technique is that given in [4]. Let E be an elliptic curve (*i.e.*, a non-singular, projective plane cubic curve over $\mathbb{C}$), for example,

$$E = \operatorname{Proj} \mathbb{C}[x, y, z]/(x^3 + y^3 + z^3).$$

Let $E^{2n} = E \times \cdots \times E$, the product of $2n$ copies of E. Let $Y = \operatorname{Spec} B \subset E^{2n}$ be any affine open subset. By the Mumford-Roitman infinite dimensionality theorem (Theorem 3.1 above), $CH^{2n}(Y) \otimes \mathbb{Q} \neq 0$. Also, since $Y \subset E^{2n}$, clearly the $2n$-fold intersection product

$$CH^1(Y)^{\otimes 2n} \to CH^{2n}(Y)$$

is surjective. Hence we can find an element $\alpha \in CH^1(Y)$ with $\alpha^{2n} \neq 0$ in $CH^{2n}(Y) \otimes \mathbb{Q}$. Let P be the projective B-module of rank 1 corresponding to α. Since Y is affine, by Bertini's theorem, we can find elements $a_1, \ldots, a_n \in P$ such that the corresponding divisors $H_i = \{a_i = 0\} \subset Y$ are non-singular, and intersect transversally; take $X = H_1 \cap \cdots \cap H_n$. Then $X = \operatorname{Spec} A$ is non-singular of dimension n, and the ideal $I \subset B$ of $X \subset Y$ is such that $I/I^2 \cong (P \otimes_B A)^{\oplus n}$. Thus, if $j : X \hookrightarrow Y$ is the inclusion, then we have a formula between total Chern classes

$$c(I/I^2) = j^* c(P)^n = (1 + j^* c_1(P))^n = (1 + j^* \alpha)^n.$$

Hence

$$c_n(I/I^2) = j^*(\alpha)^n,$$

and so by the projection formula,

$$j_* c_n(I/I^2) = j_*(1)\alpha^n = \alpha^{2n},$$

since

$$j_*(1) = [X] = [H_1] \cdot [H_2] \cdots \cdots [H_n] = \alpha^n \in CH^n(Y),$$

as X is the complete intersection of divisors H_i, each corresponding to the class $\alpha \in CH^1(Y)$. By construction, $j_* c_n(I/I^2) \neq 0$ in $CH^{2n}(Y) \otimes \mathbb{Q}$, and so we have that $c_n(I/I^2) \neq 0$ in $CH^n(X) \otimes \mathbb{Q}$, as desired.

Similarly

$$c(I/I^2)^{-1} = (1 + j^* \alpha)^{-n}$$

has a non-zero component of degree n, which is a non-zero integral multiple of $j^* \alpha^n$.

Remark 4.5. The existence of n-dimensional non-singular affine varieties X which do not admit closed embeddings into affine $2n$-space is in contrast to the situation of differentiable manifolds – the "hard embedding theorem" of Whitney states that any smooth n-manifold has a smooth embedding in the Euclidean space $\mathbb{R}^{2n}$.

5. Indecomposable projective modules, using 0-cycles

Now we discuss M. Nori's (unpublished) construction of indecomposable projective modules of rank d over any affine $\mathbb{C}$-algebra $A_\mathbb{C}$ of dimension d, such that $U = \operatorname{Spec} A_\mathbb{C}$ is an open subset of a non-singular projective (or proper) $\mathbb{C}$-variety X with $H^0(X, \omega_X) = H^0(X, \Omega^d_{X/\mathbb{C}}) \neq 0$.

The idea is as follows. Fix a countable, algebraically closed subfield $k \subset \mathbb{C}$ such that X and U are defined over k; in particular, we are given an affine k-subalgebra $A \subset A_\mathbb{C}$ such that $A_\mathbb{C} = A \otimes_k \mathbb{C}$. We also have a k-variety X_0 containing $U_0 = \operatorname{Spec} A$ as an affine open subset, such that $X = (X_0)_\mathbb{C}$.

Let K_n be the function field of $X_0^n = X_0 \times_k \cdots \times_k X_0$ (equivalently, K_n is the quotient field of $A^{\otimes n} = A \otimes_k \cdots \otimes_k A$). We have n induced embeddings $\varphi_i : K \hookrightarrow K_n$, where $K = K_1$ is the quotient field of A, given by $\varphi_i(a) = 1 \otimes \cdots \otimes 1 \otimes a \otimes 1 \otimes \cdots \otimes 1$ with a in the ith position.

Choose an embedding $K_n \hookrightarrow \mathbb{C}$ as a k-subalgebra. The inclusions φ_i then determine n inclusions $K \hookrightarrow \mathbb{C}$, or equivalently, k-generic points $x_1, \ldots, x_n \in X$ (in algebraic geometry, these are called "n independent generic points of X"). Let $\mathfrak{m}_i$ be the maximal ideal of $A_\mathbb{C}$ determined by x_i, and let $I = \cap_{i=1}^n \mathfrak{m}_i$. Clearly I is a local complete intersection ideal of height d in the d-dimensional regular ring $A_\mathbb{C}$. Thus we can find a projective resolution of I

$$0 \to P \to F_{d-1} \to \cdots \to F_1 \to I \to 0,$$

where F_i are free. By construction, $c(P) = c(A/I)^{(-1)^d}$. By Theorem 1.3, we have

$$c(A/I) = 1 + (-1)^{d-1}(d-1)!(\sum_{i=1}^n [x_i]) \in CH^*(U).$$

Hence $c_i(P) = 0$ for $i < d$, while $c_d(P)$ is a non-zero integral multiple of the class $\sum_i [x_i] \in CH^d(U)$. This class is non-zero, from Theorem 3.3 (we will get a stronger conclusion below). Hence rank $P \geq d$.

By Bass' stability theorem, if rank $P = d + r$, we may write $P = Q \oplus A^{\oplus r}$, where Q is projective of rank d. Then P and Q have the same Chern classes. So we can find a projective module Q of rank d with $c(Q) = 1 + m(\sum_i [x_i]) \in CH^*(U)$, for some non-zero integer m.

Suppose $Q = Q_1 \oplus Q_2$ with rank $Q_1 = p$, rank $Q_2 = d - p$, and $1 \leq p < d$. Then in $CH^*(U) \otimes \mathbb{Q}$, the class $\sum_i [x_i]$ is expressible as

$$\sum_i [x_i] = \alpha \cdot \beta, \quad \alpha \in CH^p(U) \otimes \mathbb{Q}, \ \beta \in CH^{d-p}(U) \otimes \mathbb{Q}.$$

Using the homomorphism of graded rings of Theorem 3.3,

$$CH^*(U) \otimes \mathbb{Q} \to \bigoplus_{j \geq 0} H^j(X, \mathcal{O}_X) \otimes_\mathbb{C} \Omega^j_{\mathbb{C}/k},$$

we see that the element

$$\xi = \sum_{i=1}^{n} di_{x_i} \in H^d(X, \mathcal{O}_X) \otimes \Omega^d_{\mathbb{C}/k}$$

is expressible as a product

$$\xi = \sum_{i=1}^{n} di_{x_i} = \alpha \cdot \beta, \quad \alpha \in H^p(X, \mathcal{O}_X) \otimes_{\mathbb{C}} \Omega^p_{\mathbb{C}/k}, \quad \beta \in H^{d-p}(X, \mathcal{O}_X) \otimes_{\mathbb{C}} \Omega^{d-p}_{\mathbb{C}/k}.$$

Let L be the algebraic closure of K_n in $\mathbb{C}$. The graded ring

$$\bigoplus_{j=0}^{d} H^j(X, \mathcal{O}_X) \otimes_{\mathbb{C}} \Omega^j_{\mathbb{C}/k} = \bigoplus_{j=0}^{d} H^j(X_0, \mathcal{O}_{X_0}) \otimes_k \Omega^j_{\mathbb{C}/k}$$

has a graded subring

$$\bigoplus_{j=0}^{d} H^j(X_0, \mathcal{O}_{X_0}) \otimes_k \Omega^j_{L/k}$$

which contains the above element ξ. We claim that ξ is then expressible as a product $\alpha \cdot \beta$ of homogeneous elements of degrees $p, d - p$ with α, β lying in this subring. Indeed, since $\mathbb{C}$ is the direct limit of its subrings B which are finitely generated L-subalgebras, we can find such a subring B, and homogeneous elements $\widetilde{\alpha}, \widetilde{\beta}$ of degrees $p, d-p$ in $\bigoplus_{j=0}^{d} H^j(X_0, \mathcal{O}_{X_0}) \otimes_k \Omega^j_{B/k}$ such that $\xi = \widetilde{\alpha} \cdot \widetilde{\beta}$. Choosing a maximal ideal in B, we can find an L-algebra homomorphism $B \twoheadrightarrow L$, giving rise to a graded ring homomorphism

$$f : \bigoplus_{j=0}^{d} H^j(X_0, \mathcal{O}_{X_0}) \otimes_k \Omega^j_{B/k} \to \bigoplus_{j=0}^{d} H^j(X_0, \mathcal{O}_{X_0}) \otimes_k \Omega^j_{L/k}.$$

Then $\xi = f(\widetilde{\alpha}) \cdot f(\widetilde{\beta})$ holds in $\bigoplus_{j=0}^{d} H^j(X_0, \mathcal{O}_{X_0}) \otimes_k \Omega^j_{L/k}$ itself.

Now

$$\Omega^1_{L/k} = \Omega^1_{K_n/k} \otimes_{K_n} L = \bigoplus_{j=1}^{n} \Omega^1_{K/k} \otimes_K L,$$

where the jth summand corresponds to the jth inclusion $K \hookrightarrow K_n$. We may write this as

$$\Omega^1_{L/k} = \Omega^1_{K/k} \otimes_K W,$$

where $W \cong L^{\oplus n}$ is an n-dimensional L-vector space with a distinguished basis. Then there are natural surjections

$$\Omega^r_{L/k} = \bigwedge_L^r (\Omega^1_{K/k} \otimes_K W) \twoheadrightarrow \Omega^r_{K/k} \otimes_K S^r(W),$$

where $S^r(W)$ is the rth symmetric power of W as an L-vector space. In particular, since $\Omega^d_{K/k}$ is 1-dimensional over K, we get a surjection $\Omega^d_{L/k} \twoheadrightarrow S^d(W)$. This determines the component of degree d of a graded ring homomorphism

$$\Phi : \bigoplus_{j=0}^{d} H^j(X_0, \mathcal{O}_{X_0}) \otimes_k \Omega^j_{L/k} \to \bigoplus_{j=0}^{d} H^j(X_0, \mathcal{O}_{X_0}) \otimes_k \Omega^j_{K/k} \otimes_K S^j(W).$$

As in the discussion preceding Theorem 3.3, by Serre duality on X_0, the natural inclusion $H^0(X_0, \Omega^d_{X_0/k}) \hookrightarrow \Omega^d_{K/k}$ determines a canonical element $\theta \in H^d(X_0, \mathcal{O}_{X_0}) \otimes_k \Omega^d_{K/k}$. Identifying the symmetric algebra $S^\bullet(W) = S^\bullet(L^{\oplus n})$ with the polynomial algebra $L[t_1, \ldots, t_n]$, we have that $\Phi(\xi) = \theta \cdot (t_1^d + \cdots + t_n^d)$. Hence, in the graded ring

$$\bigoplus_{j=0}^{d} H^j(X_0, \mathcal{O}_{X_0}) \otimes_k \Omega^j_{K/k} \otimes_K S^j(W),$$

the element $\theta \cdot (t_1^d + \cdots + t_n^d)$ is expressible as a product of homogeneous elements α, β of degrees p and $d - p$. Hence, by expressing

$$\alpha \in H^p(X_0, \mathcal{O}_{X_0}) \otimes_k \Omega^p_{K/k} \otimes_K S^p(W), \quad \beta \in H^{d-p}(X_0, \mathcal{O}_{X_0}) \otimes_k \Omega^{d-p}_{K/k} \otimes_K S^{d-p}(W)$$

in terms of K-bases of $H^p(X_0, \mathcal{O}_{X_0}) \otimes_k \Omega^p_{K/k}$ and $H^{d-p}(X_0, \mathcal{O}_{X_0}) \otimes_k \Omega^{d-p}_{K/k}$, we deduce that in the polynomial ring $S^\bullet(W) = L[t_1, \ldots, t_n]$, the "Fermat polynomial" $t_1^d + \cdots + t_n^d$ is expressible as a sum of pairwise products of homogeneous polynomials

$$t_1^d + \cdots + t_n^d = \sum_{m=1}^{N} a_m(t_1, \ldots, t_n) b_m(t_1, \ldots, t_n)$$

with

$$N = \binom{d}{p}\binom{d}{d-p}(\dim_k H^p(X_0, \mathcal{O}_{X_0}))(\dim_k H^{d-p}(X_0, \mathcal{O}_{X_0})).$$

If $n > 2N$, the system of homogeneous polynomial equations $a_1 = b_1 = \cdots = a_N = b_N = 0$ defines a non-empty subset of the projective variety $t_1^d + \cdots + t_n^d = 0$ in $\mathbb{P}_L^{n-1}$, along which this Fermat hypersurface is clearly singular – and this is a contradiction!

6. Stably trivial vector bundles on affine varieties, using cycles

In this section, we give a construction of stably trivial non-trivial vector bundles on affine varieties, following an argument of Mohan Kumar and Nori. Here, instead of topology, the properties of Chow rings and Chern classes provide the invariants to prove non-triviality of the vector bundles; thus the arguments are valid over any base field.

Theorem 6.1. *Let k be a field, and let*

$$A = \frac{k[x_1, \ldots, x_n, y_1, \ldots, y_n]}{(x_1 y_1 + \cdots + x_n y_n - 1)}.$$

Let $\mathbf{m} = (m_1, \ldots, m_n)$ be an n-tuple of positive integers, and set

$$P(\mathbf{m}) = \ker \left(\psi(\mathbf{m}) : A^{\oplus n} \to A \right),$$

$$(a_1, \ldots, a_n) \mapsto a_1 x_1^{m_1} + \cdots + a_n x_n^{m_n}.$$

Then $P(\mathbf{m})$ is a stably free projective A-module of rank $n - 1$. If $\prod_{i=1}^{n} m_i$ is not divisible by $(n - 1)!$, then $P(\mathbf{m})$ is not free.

Note that in particular, if all the m_i are 1, and $n \geq 3$, the corresponding projective module is not free.

Conversely, a theorem of Suslin [31] implies that, if $\prod_{i=1}^{n} m_i$ is divisible by $(n - 1)!$, then P is free, since we can find an invertible $n \times n$ matrix over A whose first row has entries $x_1^{m_1}, \ldots, x_n^{m_n}$.

The theorem above is proved by relating the freeness of $P(\mathbf{m})$ to another property. Let

$$R = \frac{k[z, x_1, \ldots, x_n, y_1, \ldots, y_n]}{(z(1 - z) - x_1 y_1 - \cdots - x_n y_n)},$$

and consider the ideals

$$I = (x_1, \ldots, x_n, z) \subset R, \quad I' = (x_1, \ldots, x_n, 1 - z).$$

Note that I, I' are prime ideals in R, and $I + I' = R$ (i.e., the corresponding subvarieties of $\operatorname{Spec} R$ are disjoint); further, $I \cap I' = II' = (x_1, \ldots, x_n)$ is a complete intersection.

Consider the ideal $(x_1^{m_1}, \ldots, x_n^{m_n}) \subset R$. Clearly its radical is the complete intersection $(x_1, \ldots, x_n) = II'$. Hence we may uniquely write

$$(x_1^{m_1}, \ldots, x_n^{m_n}) = J(\mathbf{m}) \cap J'(\mathbf{m})$$

where $J(\mathbf{m})$, $J'(\mathbf{m})$ have radicals I, I' respectively. In fact $J(\mathbf{m})$, $J'(\mathbf{m})$ are the contractions to R of the complete intersection $(x_1^{m_1}, \ldots, x_n^{m_n})$ from the overrings $R[\frac{1}{1-z}]$, $R[\frac{1}{z}]$ respectively.

Lemma 6.2. *Let $X = \operatorname{Spec} R$, and $W(\mathbf{m}) \subset X$ the subscheme with ideal $J(\mathbf{m})$. If $P(\mathbf{m})$ is a free A-module, then there exists a vector bundle V on X of rank n and a section, whose zero scheme on X is $W(\mathbf{m})$.*

Proof. We write J, J' instead of $J(\mathbf{m})$, $J'(\mathbf{m})$ to simplify notation. Consider the surjection

$$\alpha : R^{\oplus n} \to J \cap J' = (x_1^{m_1}, \ldots, x_n^{m_n}).$$

If we localize to $R[\frac{1}{z(1-z)}]$, we obtain a surjection

$$\widetilde{\alpha} : R_{z(1-z)}^{\oplus n} \to R_{z(1-z)},$$

whose kernel Q is a stably free projective $R_{z(1-z)}$-module, fitting into a split exact sequence of projective modules

$$0 \to Q \to R_{z(1-z)}^{\oplus n} \xrightarrow{\widetilde{\alpha}} R_{z(1-z)} \to 0.$$

It is easy to see that the projective module Q is free if and only if there is some $\varphi \in \operatorname{GL}_n(R_{z(1-z)})$ so that $\widetilde{\alpha} \circ \varphi$ is projection onto the first coordinate.

If this is the case, the surjection

$$R[\frac{1}{1-z}]^{\oplus n} \to JJ'[\frac{1}{1-z}] = J[\frac{1}{1-z}]$$

clearly "patches" over $\operatorname{Spec} R_{z(1-z)}$ with the "trivial" surjection

$$R[\frac{1}{z}]^{\oplus n} \to R[\frac{1}{z}] = J[\frac{1}{z}]$$

given by projection on the first coordinate, to yield a projective R-module $\widetilde{P}$ of rank n, with a surjection

$$\widetilde{P} \to J.$$

We then take V to be the vector bundle on X associated to the dual projective module $\widetilde{P}^{\vee}$, and the section of V to be dual to the map $\widetilde{P} \to J$.

Finally, we observe that there is a homomorphism $A \to R[\frac{1}{z(1-z)}]$ given by

$$x_i \mapsto x_i,$$

$$y_i \mapsto \frac{y_i}{z(1-z)},$$

which gives an identification

$$Q \cong P(\mathbf{m}) \otimes_A R_{z(1-z)}.$$

Hence if $P(\mathbf{m})$ is free, so is the projective $R_{z(1-z)}$-module Q, and we can thus construct V as above. $\square$

Lemma 6.3. $CH^i(X) = 0$ *for* $1 \leq i \leq n-1$, *and* $CH^n(X) \cong \mathbb{Z}$ *is generated by the class of* W, *the irreducible subvariety defined by the ideal* I.

Proof. Consider the smooth $2n$-dimensional projective quadric hypersurface $\overline{X}$ in $\mathbb{P}_k^{2n+1}$ given by the equation

$$\sum_{i=1}^{n} X_i Y_i = Z_0(Z_1 - Z_0).$$

We may identify $X = \operatorname{Spec} R$ with the affine open subset of $\overline{X}$ obtained by setting $Z_1 = 1$ in the above homogeneous polynomial equation. Hence the complement of X is the smooth hyperplane section $Z_1 = 0$. From the known structure of the Chow ring of a smooth "split" quadric hypersurface in any dimension, we know that

  (i) $CH^i(\overline{X})$ is spanned by the complete intersection with any projective linear subspace of codimension i, for any $i < n$, while

  (ii) $CH^n(\overline{X})$ is free abelian of rank 2, spanned by the classes of 2 linear (projective) subvarieties, which are the irreducible components of a suitable complete intersection with a projective linear subspace.

For $i < n$, we may choose a complete intersection generator of $CH^i(\overline{X})$ as in (i) to be contained in the linear subspace intersection $\overline{X} \cap \{Z_1 = 0\}$. From the exact localization sequence

$$CH^{i-1}(\overline{X} \cap \{Z_1 = 0\}) \to CH^i(\overline{X}) \to CH^i(X) \to 0$$

it follows that $CH^i(X) = 0$ for $1 \leq i \leq n-1$. Also, one sees easily that if W' is the subscheme defined by I', then the closures of W, W' in $\overline{X}$ are generators for $CH^n(\overline{X}) = \mathbb{Z}^{\oplus 2}$ as in (ii), and $W + W'$ is rationally equivalent to 0 on X, so that $CH^n(X)$ is generated by the class of W. Finally, $CH^{n-1}(\overline{X} \cap \{Z_1 = 0\}) = \mathbb{Z}$, so $CH^n(X)$ must have rank 1, again from the above exact localization sequence of Chow groups. $\qquad\square$

Thus, from Theorem 1.3 we must also have $F^1 K_0(X) = F^n K_0(X)$ where $F^i K_0(X)$ is the subgroup generated by the classes of modules supported in codimension $\geq i$; also, $F^n K_0(X)/F^{n+1} K_0(X)$ is generated by the class of $\mathcal{O}_W$.

The vector bundle V constructed in Lemma 6.2 above thus has the following properties:

   (i) $c_n(V) = [W(\mathbf{m})] = (\prod_{i=1}^n m_i)[W]$ in $CH^n(X)$.
  (ii) $[V] - [\mathcal{O}_X^{\oplus n}] \in F^1 K_0(X) = F^n K_0(X)$, so that for some $r \in \mathbb{Z}$, we must have

$$[V] - [\mathcal{O}_X^{\oplus n}] = r[\mathcal{O}_W] \pmod{F^{n+1} K_0(X)},$$

     which yields

$$c_n(V) = r c_n([\mathcal{O}_W]) \in CH^n(X)$$

(iii) from the Riemann-Roch theorem without denominators,

$$c_n(\mathcal{O}_W) = (-1)^{n-1}(n-1)![W] \in CH^n(X) = \mathbb{Z} \cdot [W].$$

Hence we must have

$$\prod_{i=1}^n m_i = r(-1)^{n-1}(n-1)!,$$

which proves Theorem 6.1.

7. 0-cycles and the complete intersection property for affine varieties

Let A be a finitely generated reduced algebra over a field k, which we assume to be algebraically closed, for simplicity. Let $d = \dim A$. A point $x \in X = \operatorname{Spec} A$ is called a *complete intersection point* if the corresponding maximal ideal $\mathfrak{M} \subset A$ has height d, and is generated by d elements of A. Any such point is necessarily a smooth point (of codimension d) in X (we will take points of codimension $< d$ to be singular, by definition, in this context). We will refer to the corresponding maximal ideals as smooth maximal ideals.

In this section, we want to discuss the following problem:

Problem 7.1. Characterize reduced affine k-varieties X such that all smooth points are complete intersections.

It is easy to show, using the theory of the Jacobian (suitably generalized in the singular case), that for $d = \dim X = 1$, we have a complete answer, as follows. Any such curve X can be written as $X = \overline{X} \setminus S$ where $\overline{X}$ is a reduced projective curve over k, and S a finite set of non-singular points of X, in a unique way. Then: all smooth points of X are complete intersections $\Leftrightarrow H^1(\overline{X}, \mathcal{O}_{\overline{X}}) = 0$.

So the interesting case of the problem is in dimensions $d > 1$. There are several conjectures and results related to this problem. We first state a general "positive" result.

Theorem 7.2. *Let $k = \overline{\mathbb{F}}_p$ be the algebraic closure of the finite field $\mathbb{F}_p$. Then for any reduced finitely generated k-algebra A of dimension $d > 1$, every smooth maximal ideal is a complete intersection.*

In the case when $\dim A \geq 3$, or A is smooth of dimension 2, this is a result essentially due to M. P. Murthy. The higher-dimensional case is reduced to the 2-dimensional case by showing that any smooth point of $V = \operatorname{Spec} A$ lies on a smooth affine surface $W \subset V$ such that the ideal of W in A is generated by $d - 2$ elements (i.e., W is a complete intersection surface in V). This argument (see [20] for details) depends on the fact that we are dealing here with *affine* algebraic varieties.

The case of an arbitrary 2-dimensional algebra is a corollary of results of Amalendu Krishna and mine [1]; the details are worked out in [2]. I will make a few remarks about this later in this paper.

Next, we state two conjectures, which are "affine versions" of famous conjectures on 0-cycles.

Conjecture 7.3 (Affine Bloch Conjecture). Let $k = \mathbb{C}$, the complex numbers. Let $V = \operatorname{Spec} A$ be a non-singular affine $\mathbb{C}$-variety of dimension $d > 1$, and let $X \supset V$ be a smooth proper (or projective) $\mathbb{C}$-variety containing V as a dense open subset. Then:
all maximal ideals of A are complete intersections
$\Leftrightarrow X$ does not support any global regular (or holomorphic) differential d-forms
$\Leftrightarrow H^d(X, \mathcal{O}_X) = 0$.

Here, $\mathcal{O}_X$ is the sheaf of algebraic regular functions on X. The non-existence of d-forms is equivalent to the cohomology vanishing condition, by Serre duality; the open question is the equivalence of either of these properties with the complete intersection property for maximal ideals.

This conjecture has been verified in several "non-trivial" examples (for example, if $V = \operatorname{Spec} A$ is a "small enough" Zariski open subset of the Kummer variety of an odd (> 1) dimensional abelian variety over $\mathbb{C}$, all smooth maximal ideals of A are complete intersections).

One consequence of the conjecture is that, for smooth affine $\mathbb{C}$-varieties, the property that all maximal ideals are complete intersections is a *birational* invariant (that is, it depends only on the quotient field of A, as a $\mathbb{C}$-algebra). This birational invariance can be proved to hold in dimension 2, using a result of Roitman; in dimensions ≥ 3, it is unknown in general.

Conjecture 7.4 (Affine Bloch-Beilinson Conjecture). Let $k = \overline{\mathbb{Q}}$ be the field of algebraic numbers (algebraic closure of the field of rational numbers). Then for any finitely generated *smooth k-algebra* of dimension $d > 1$, every maximal ideal is a complete intersection.

This very deep conjecture has not yet been verified in any "nontrivial" example (i.e., one where there do exist smooth maximal ideals of $A \otimes_{\overline{\mathbb{Q}}} \mathbb{C}$ which are not complete intersections).

However, it is part of a more extensive set of interrelated conjectures relating *K-groups of motives over algebraic number fields* and *special values of L-functions,* and there are nontrivial examples where some other parts of this system of conjectures can be verified. This is viewed as indirect evidence for the above conjecture.

I will now relate these "affine" conjectures to the more standard forms of these, in terms of algebraic cycles and K-theory. The first step is a fundamental result of Murthy, giving a K-theoretic interpretation of the complete intersection property.

Recall that $K_0(A)$ denotes the Grothendieck group of finitely generated projective A-modules. It coincides with the Grothendieck group of finitely generated A-modules of finite projective dimension: recall that M has *finite projective dimension* if there exists a finite projective resolution of M, i.e., an exact sequence

$$0 \to P_r \to P_{r-1} \to \cdots \to P_0 \to M \to 0$$

where the P_i are finitely generated projective A-modules. Then such a module M has a well-defined class $[M] \in K_0(A)$, obtained by choosing any such resolution, and defining

$$[M] = \sum_{i=0}^{r} (-1)^i [P_i] \in K_0(A).$$

Recall also that a maximal ideal $\mathfrak{M}$ has finite projective dimension precisely when the local ring $A_{\mathfrak{M}}$ is a regular local ring.

Theorem 7.5 (M.P. Murthy). *Let A be a reduced finitely generated algebra over an algebraically closed field, of dimension d. Assume $F^d K_0(A)$ has no torsion of exponent $(d-1)!$. Then a smooth maximal ideal $\mathfrak{M}$ of A is a complete intersection $\Leftrightarrow [\mathfrak{M}] = [A]$ in $K_0(A)$.*

The main point in the proof of Murthy's theorem is to show that any smooth maximal ideal $\mathfrak{M}$ is a quotient of a projective A-module P of rank d, satisfying the additional condition that $[P] - [A^{\oplus d}] \in F^d K_0(A)$. Then $[\mathfrak{M}] = [A]$ in $K_0(A)$ implies that the above projective module P has $c_d(P) = 0$, so that P is in fact stably trivial, from Riemann-Roch (Theorem 1.3), since we assumed $F^d K_0(A)$ has no $(d-1)!$-torsion. Now Suslin's cancellation theorem for projective modules implies that P is a free module, so that $\mathfrak{M}$ is a complete intersection.

Let A be a reduced, finitely generated algebra, of Krull dimension d, over an algebraically closed field k. We can associate to it the group

$$F^d K_0(A) = \text{ subgroup of } K_0(A) \text{ generated by}$$

$$[A] - [\mathfrak{M}] \text{ for all smooth maximal ideals } \mathfrak{M}.$$

If $V = \operatorname{Spec} A$, then $F^d K_0(A)$ is a quotient of the free abelian group on smooth points of V, modulo a suitable equivalence relation. When V is nonsingular, one can identify this equivalence relation with *rational equivalence*, up to torsion, using the "Riemann-Roch Theorem without denominators", using the dth Chern class map.

This also suggests a good definition of rational equivalence for 0-cycles for singular V; this was given by Levine and Weibel [11], and Levine (unpublished) has defined a suitable dth Chern class, for which the Riemann-Roch without denominators is valid.

Now assume $V = \operatorname{Spec} A$ is an affine open subset of a nonsingular projective k-variety X of dimension d. Clearly

$$CH^d(V) = \frac{CH^d(X)}{\text{subgroup generated by points of } X \setminus V}.$$

We saw earlier that *Roitman's Theorem* on torsion 0-cycles (extended by Milne to arbitrary characteristic) gives a description of the torsion in $CH^d(X)$. Using this, it can be shown that $CH^d(V)$ is a torsion free, divisible abelian group (i.e., a vector space over $\mathbb{Q}$). In particular, we see that the map $\psi_d : CH^d(V) \to F^d K_0(V)$ is an *isomorphism*.

Thus, by Murthy's theorem, for nonsingular A, all maximal ideals of A are complete intersections $\Leftrightarrow CH^d(X)$ is generated by points of $X \setminus V$.

We now restate the Bloch and Bloch-Beilinson Conjectures in something resembling their "original" forms.

Conjecture 7.6 (Bloch Conjecture). Let X be a projective smooth variety over $\mathbb{C}$. Suppose that, for some integer $r > 0$, X has no nonzero regular (or holomorphic) s-forms for any $s > r$. Then for any "sufficiently large" subvariety $Z \subset X$ of dimension r, we have $CH^d(X \setminus Z) = 0$.

For a smooth projective complex surface X, this conjecture states that if X has no holomorphic 2-forms, then $CH^2(X \setminus C) = 0$ for some curve C in X. This has been verified in several situations, for example, for surfaces of Kodaira dimension ≤ 1 (Bloch, Kas, Lieberman), for general Godeaux surfaces (Voisin), and in some other cases.

In higher dimensions, Roitman proved it for complete intersections in projective space, and there are a few other isolated examples, like the Kummer variety associated to an odd-dimensional abelian variety (see [5]).

Conjecture 7.7 (Bloch-Beilinson Conjecture). Let X be a smooth projective variety of dimension d over $\overline{\mathbb{Q}}$. Then $CH^d(X)$ is "finite dimensional"; in particular, there is a curve $C \subset X$ so that $CH^d(X \setminus C) = 0$.

As remarked earlier, there is only indirect evidence for this conjecture: *it has not been verified for any smooth projective surface over $\overline{\mathbb{Q}}$ which supports a non-zero 2-form* (e.g., any hypersurface in projective 3-space of degree ≥ 4).

To exhibit one such nontrivial example is already an interesting open question.

From the algebraic viewpoint, it seems restrictive to work only with smooth varieties. In any case, it is unknown in characteristic $p > 0$ that a smooth affine variety V can be realized as an open subset of a smooth proper variety X (in characteristic 0, this follows from Hironaka's theorem on *resolution of singularities*).

In spite of this, it is possible to make a systematic study of the singular case, and to try to extend the above conjectures, using the Levine-Weibel Chow group of 0-cycles; see [28] for further discussion.

For our purposes, let me focus on one very special situation. Let

$$Z \subset \mathbb{P}_k^N$$

be a non-singular projective algebraic k-variety, and

$$\begin{aligned} A &= \quad \oplus_{n \geq 0} A_n \\ &= \quad \textit{homogeneous coordinate ring of } Z. \end{aligned}$$

The affine variety $V = \operatorname{Spec} A$ is the *affine cone* over Z with *vertex* corresponding to the unique graded maximal ideal $\mathfrak{M} = \oplus_{n > 0} A_n$, and the vertex is the unique singular point of V.

The *projective cone* $C(Z)$ over Z with the same vertex naturally contains V as an open subset, whose complement is a divisor isomorphic to Z, and the vertex is again the only singular point of $C(Z)$.

The following theorem is obtained using results from my paper with Amalendu Krishna [1], in the 2-dimensional case, and a preprint of Krishna's in the higher-dimensional case; see [2] for more details.

Theorem 7.8. (i) *Let $k = \overline{\mathbb{Q}}$. Then every smooth maximal ideal of A is a complete intersection.*

(ii) *Let $k = \mathbb{C}$. Then every smooth maximal ideal of A is a complete intersection $\Rightarrow H^{d-1}(Z, \mathcal{O}_Z(1)) = 0$ $(\Leftrightarrow H^d(C(Z), \mathcal{O}_{C(Z)}) = 0)$. If V is Cohen-Macaulay of dimension ≤ 3, then the converse holds: if $H^{d-1}(Z, \mathcal{O}_Z(1)) = 0$, then every smooth maximal ideal of A is a complete intersection.*

Here, (i) is analogous to the Bloch-Beilinson Conjecture, while (ii) is analogous to the Bloch Conjecture.

Here are two examples, which shed some light on the content of the above theorems.

Example 7.9.   (Amalendu Krishna + V. S.)

$$A = \frac{\overline{\mathbb{Q}}[x, y, z]}{(x^4 + y^4 + z^4)}.$$

The following properties hold.

(i) All smooth maximal ideals of A are complete intersections.

(ii) "Most" smooth maximal ideals of $A \otimes_{\overline{\mathbb{Q}}} \mathbb{C}$ are *not* complete intersections.

(iii) The complete intersection smooth points on $V_{\mathbb{C}}$ are those lying on rulings of the cone over $\overline{\mathbb{Q}}$-*rational points* of the Fermat Quartic curve.

This is a consequence of Theorem 7.8.

Example 7.10.

$$A = \frac{\overline{\mathbb{Q}}[x, y, z]}{(xyz(1 - x - y - z))}.$$

Again, all smooth maximal ideals of A are complete intersections, while "most" smooth maximal ideals of $A \otimes_{\overline{\mathbb{Q}}} \mathbb{C}$ are *not* complete intersections.

In fact, there is an identification

$$F^2 K_0(A \otimes_{\overline{\mathbb{Q}}} k) = K_2(k),$$

where K_2 denotes the Milnor K_2 functor.

Now one has the result of Garland (vastly generalized by Borel) that $K_2(\overline{\mathbb{Q}}) = 0$, while $K_2(\mathbb{C})$ is "very large".

8. 0-cycles on normal surfaces

Here, we give some idea of the proof of Theorem 7.8, for the case of surfaces.

Let X be a normal, quasiprojective surface over a field k. Then $CH^2(X)$ is identified with $F^2 K_0(X)$, which in turn is identified with the subgroup of the Grothendieck group $K_0(X)$ of vector bundles consisting of elements (virtual bundles) of trivial rank and determinant.

Let $\pi : Y \to X$ be a resolution of singularities, and let E be the exceptional set, with reduced structure. Let nE denote the subscheme of Y with ideal sheaf $\mathcal{O}_Y(-nE)$; this is in fact an effective Cartier divisor. One can define *relative algebraic K-groups* $K_i(Y, nE)$ for any $n \geq 0$, along the following lines.

The algebraic K-groups of a scheme T are defined (by Quillen) by

$$K_i(T) = \pi_{i+1}(\mathbb{K}(T)),$$

where $\mathbb{K}(T)$ is a certain connected CW complex associated to the category of vector bundles on T, and π_n denotes the nth homotopy group. Quillen shows that $K_0(T)$ defined in this way coincides with the usual Grothendieck group, and if $T = \operatorname{Spec} A$ is affine, then $K_1(T)$, $K_2(T)$ coincide with the groups $K_1(A)$ of Bass, and $K_2(A)$ of Milnor, respectively. For an introduction to these ideas, see [30].

Given a morphism $f : S \to T$ of schemes, there is an induced continuous map $f^* : \mathbb{K}(T) \to \mathbb{K}(S)$, and hence induced homomorphisms $K_i(T) \to K_i(S)$, for all i. Let $\mathbb{K}(f)$ denote the *homotopy fibre* of the continuous map $f^* : \mathbb{K}(T) \to \mathbb{K}(S)$. Then there is an associated long exact sequence of homotopy groups

$$\cdots \to \pi_{i+1}(\mathbb{K}(S)) \to \pi_i(\mathbb{K}(f)) \to \pi_i(\mathbb{K}(T)) \to \pi_i(\mathbb{K}(S)) \to \cdots$$

If we define $K_i(f) = \pi_{i+1}(\mathbb{K}(f))$, this exact sequence may be rewritten as

$$\cdots \to K_{i+1}(S) \to K_i(f) \to K_i(T) \to K_i(S) \to \cdots$$

In particular, if T is a scheme, and $f : S \to T$ is the inclusion of a closed subscheme, we write $K_i(T, S)$ instead of $K_i(f)$, and call it the ith relative K group of the pair (T, S). We also write $\mathbb{K}(T, S)$ to mean $\mathbb{K}(f)$, in this situation, so that $K_i(T, S) = \pi_{i+1}(\mathbb{K}(T, S))$.

The relative K-groups have certain functoriality properties. In particular, if S is the singular locus (with reduced scheme structure) of our normal surface X, and nS is the subscheme of X defined by the nth power of the ideal sheaf of S, then there is a commutative diagram of schemes and morphisms

$$
\begin{array}{ccc}
nE & \to & Y \\
\pi\,|_{nE}\downarrow & & \downarrow \pi \\
nS & \to & X
\end{array}
$$

which gives rise to a commutative diagram of relative K-groups

$$
\begin{array}{ccccccccc}
\to K_1(Y) \to & K_1(nE) & \to K_0(Y, nE) \to & K_0(Y) & \to K_0(nE) \\
\uparrow & \uparrow & \uparrow & \uparrow & \uparrow \\
\to K_1(X) \to & K_1(nS) & \to K_0(X, nS) \to & K_0(X) & \to K_0(nS)
\end{array}
$$

Another functoriality property of relative K-groups implies the following. Let (T, S) be a pair consisting of a scheme and a closed subscheme, and $i : T' \hookrightarrow T$ another closed subscheme such that (a) $T' \cap S = \emptyset$, and (b) all vector bundles on T' have finite $\mathcal{O}_T$-homological dimension (e.g., $T \setminus S$ is a Noetherian regular scheme). There is a well-defined homotopy class of maps $i_* : \mathbb{K}(T') \to \mathbb{K}(T)$, together with a lifting $\mathbb{K}(T') \to \mathbb{K}(T, S)$, which is again well defined up to homotopy. Thus the natural "Gysin" maps $K_i(T') \to K_i(T)$ lift to maps $K_i(T') \to K_i(T, S)$.

In our context, in particular, any point $x \in X \setminus S = Y \setminus E$ defines homomorphisms $K_0(x) \to K_0(X, nS)$ and $K_0(x) \to K_0(Y, nE)$, giving a commutative triangle. Identifying $K_0(k(x)) = \mathbb{Z}$, we see that any point in $X \setminus S$ has a class in $K_0(X, nS)$ as well as in $K_0(Y, nE)$, compatibly with the map $\pi^* : K_0(X, nS) \to K_0(Y, nE)$. Define

$$F^2 K_0(Y, nE) \subset K_0(Y, nE), \quad F^2 K_0(X, nS) \subset K_0(X, nS)$$

to be the subgroups generated by the classes of points of $X \setminus S$. We then have a commutative square

$$
\begin{array}{ccc}
F^2 K_0(Y, nE) & \to & F^2 K_0(Y) \\
\uparrow & & \uparrow \\
F^2 K_0(X, nS) & \to & F^2 K_0(X)
\end{array}
$$

In fact, it is easy to see that all the four maps in this square are *surjective*. Indeed, this is clear for the maps with domain $F^2 K_0(X, nS)$, since both the target groups are also generated by points of $X \setminus S$, by definition. This same set of points also generates $F^2 K_0(Y)$, from an easy moving lemma, since the class group of any curve is generated by the classes of points in any nonempty Zariski open subset.

For any scheme T, one has a decomposition

$$K_1(T) = \Gamma(T, \mathcal{O}_T^*) \oplus SK_1(T),$$

which is functorial for arbitrary morphisms. For affine T, this is defined using the determinant. In fact for any T there is a functorial morphism $T \to \operatorname{Spec}\Gamma(T, \mathcal{O}_T)$, inducing a map on $K_1(\Gamma(T, \mathcal{O}_T)) \to K_1(T)$, as well as a map $K_1(T) \to \Gamma(T, \mathcal{O}_T^*)$, where we identify the sheaf of units $\mathcal{O}_T^*$ with the sheaf (for the Zariski topology) associated to the presheaf $V \mapsto K_1(V)$.

Next, one sees that

$$\ker\left(F^2 K_0(Y, nE) \to F^2 K_0(Y)\right) \subset \operatorname{image} SK_1(nE),$$

and similarly

$$\ker\left(F^2 K_0(X, nS) \to F^2 K_0(X)\right) \subset \operatorname{image} SK_1(nS).$$

This follows easily from the fact that any invertible function on the complement of a finite subset of X extends to one on all of X, and similarly for Y. However, $SK_1(nS) = 0$ since nS is a 0-dimensional affine scheme. Hence we see that $F^2 K_0(X, nS) = F^2 K_0(X)$. Thus we obtain an induced surjective map

$$CH^2(X) = F^2 K_0(X) \to F^2 K_0(Y, n),$$

for each n, compatible with the natural restriction maps

$$F^2 K_0(Y, nE) \to F^2 K_0(Y, (n-1)E)) \to F^2 K_0(Y).$$

The following theorem, which is the main new ingredient in the proof of Theorem 7.8, proves a conjecture of Bloch and myself, first stated in my Chicago thesis (1982) (see also [24], page 6).

Theorem 8.1. *Let $\pi : Y \to X$ be a resolution of singularities of a normal, quasi-projective surface, and let E be the exceptional locus, with its reduced structure. Then for all large $n > 0$, the maps*

$$F^2 K_0(X) \to F^2 K_0(Y, nE), \quad F^2 K_0(Y, nE) \to F^2 K_0(Y, (n-1)E)$$

are isomorphisms.

The proof of this theorem is in two steps, and is motivated by a paper [34] of Weibel, which studied negative K-groups of surfaces, and proved two old conjectures of mine from the paper [25].

First, one shows that the resolution $\pi : Y \to X$ can be factorized as a composition of two maps $f : Y \to Z$, $g : Z \to X$, where $g : Z \to X$ is the blow up of a local complete intersection subscheme supported on S (the singular locus of X), and $f : Y \to Z$ is the normalization map.

Next, one has that the maps on algebraic k-groups $g^* : K_i(X) \to K_i(Z)$ are split inclusions for all $i \geq 0$, since g is a proper, birational morphism of finite Tor dimension: indeed, these conditions imply that there is a well-defined *push-forward map* $g_* : K_i(Z) \to K_i(X)$, satisfying the projection formula, which implies that $g_* \circ g^*$ equals multiplication by the class of $g_*[\mathcal{O}_Z] \in K_0(X)$. But this element of the ring $K_0(X)$ is invertible, since on an open dense subset U of X, it restricts to the unit element of $K_0(U)$; now one remarks that $\ker K_0(X) \to K_0(U))$ is a nilpotent ideal.

In particular, we see that $F^2 K_0(X) \to F^2 K_0(Z)$ is an isomorphism. Hence, for any closed subscheme T supported in $g^{-1}(S)$, we see that the two maps

$$F^2 K_0(X) \to F^2 K_0(Z,T), \quad F^2 K_0(Z,T) \to F^2 K_0(Z)$$

are isomorphism as well (since both are surjective, and their composition is an isomorphism. We also get that if $T \subset T' \subset Z$ are two such subschemes of Z, then $F^2 K_0(Z,T') \to F^2 K_0(Z,T)$ is an isomorphism.

Next, one applies a suitable Mayer-Vietoris technique to study the relation between K-groups of Z and its normalization Y. Let T be a conductor subscheme for $f : Y \to Z$, and let $\widetilde{T}$ be the corresponding subscheme of Y. There are natural maps $K_i(Z,T) \to K_i(Y,\widetilde{T})$, inducing in particular a surjection

$$F^2 K_0(Z,T) \to F^2 K_0(Y,\widetilde{T}).$$

Now, using a fundamental localization theorem of Thomason-Trobaugh [33], it is shown in [21] (Cor. A.6) that there is an exact sequence

$$H^1(\widetilde{T}, \mathcal{I}/\mathcal{I}^2 \otimes \Omega_{\widetilde{T}/T}) \to K_0(Z,T) \to K_0(Y,\widetilde{T}),$$

which is functorial in T. Here $\mathcal{I}$ is the ideal sheaf of $\widetilde{T}$ on Y (whose direct image on Z equals the ideal sheaf of T). Let $2\widetilde{T}$ be the subscheme of Z defined by the ideal sheaf $\mathcal{I}^2$, and let $2T$ denote the subscheme of Z defined similarly. There is then a commutative diagram with exact rows

$$
\begin{array}{ccccc}
H^1(\mathcal{I}^2/\mathcal{I}^4 \otimes \Omega_{2\widetilde{T}/2T}) & \to & K_0(Z,2T) \to & K_0(Y,2\widetilde{T}) \\
\downarrow & & \downarrow & \downarrow \\
H^1(\mathcal{I}/\mathcal{I}^2 \otimes \Omega_{\widetilde{T}/T}) & \to & K_0(Z,T) \to & K_0(Y,2\widetilde{T})
\end{array}
$$

But the left-hand vertical arrow is 0, since the sheaf map $\mathcal{I}^2/\mathcal{I}^4 \to \mathcal{I}/\mathcal{I}^2$ is 0! Hence

$$\ker\left(F^2 K_0(Z,2T) \to F^2 K_0(Y,2\widetilde{T})\right) \subset \ker\left(F^2 K_0(Z,2T) \to F^2 K_0(Z,T)\right).$$

But we've seen already that $F^2 K_0(Z,2T) \to F^2 K_0(Z,T)$ is an isomorphism, and in fact both of these groups are isomorphic to $F^2 K_0(X)$ (as well as to $F^2 K_0(Z)$). Hence the surjective map

$$F^2 K_0(Z,2T) \to F^2 K_0(Y,2\widetilde{T})$$

must in fact be an isomorphism, and so $F^2 K_0(X) \to F^2 K_0(Y,2\widetilde{T})$ is an isomorphism. Finally, if $n > 0$ is large enough so that $2\widetilde{T}$ is a subscheme of nE, then we have that the two maps

$$F^2 K_0(X) \to F^2 K_0(Y,nE), \quad F^2 K_0(Y,E) \to F^2 K_0(Y,2\widetilde{T})$$

are isomorphisms, since both maps are surjective, and their composition is an isomorphism. This proves Theorem 8.1.

As a consequence of this theorem, we see that $\ker(F^2 K_0(X) \to F^2 K_0(Y))$ is identified with a subgroup of $\mathrm{coker}\,(SK_1(Y) \to SK_1(nE))$, for sufficiently large n. In fact, it is shown in [1] that equality holds, i.e., that there is an exact sequence (for any large enough n)

$$SK_1(Y) \to SK_1(nE) \to F^2 K_0(X) \to F^2 K_0(Y) \to 0.$$

Assume now that k has characteristic 0, and E is a divisor with simple normal crossings. The groups $SK_1(nE)$ is then shown to fit into an exact sequence

$$H^1(Y, \mathcal{I}_E / \mathcal{I}_E^n) \otimes_k \Omega_{k/\mathbb{Z}} \to SK_1(nE) \to SK_1(E) \to 0. \tag{8.1}$$

Without going into technical details, let me say that this follows from a combination of several ingredients.

The first is a formula $SK_1(W) = H^1(W, \mathcal{K}_{2,W})$ for any scheme W of dimension ≤ 1, where $\mathcal{K}_{2,W}$ is the sheaf associated to the presheaf $U \mapsto K_2(\Gamma(\mathcal{O}_U))$ of Milnor K-groups.

In our case, since E is a reduced divisor on a smooth surface, the local ring of nE at a smooth point $x \in E$ has the form $\mathcal{O}_{x,E}[t]/(t^n)$, where t is a local generator for the ideal sheaf $\mathcal{I}_E$ in the local ring $\mathcal{O}_{x,Y}$. A result of Bloch gives a formula for any local $\mathbb{Q}$-algebra of the form $A = R[t_1, \ldots, t_r]$, where the ideal I in A generated by $t_1, \ldots, t_r$ is nilpotent, with quotient ring $A/I = R$. His result is that

$$K_2(A) = K_2(R) \oplus \frac{\ker \Omega_{A/\mathbb{Z}} \to \Omega_{R/\mathbb{Z}}}{d(I)}.$$

In particular, this is applicable when $A = R[t]/(t^n)$ is a truncated polynomial algebra, and thus gives a local description of $\mathcal{K}_{2,nE}$ at smooth points $x \in E$.

Since E has simple normal crossings, at a singular point of E, the local ring of nE has the form $\mathcal{O}_{x,Y}/(s^n t^n)$ where s, t are a regular system of parameters. This local ring is not of the type covered by the Bloch formula, but one has a Mayer-Vietoris sequence relating the K_2 of such a ring to the K_2 groups of the two (non-reduced) branches, and to K_2 of their intersection, and the Bloch formula is applicable to compute these K_2 groups. The local identifications with truncated polynomial algebras depend on choices of local generators for the ideal sheaves in Y of components of E, so when the above descriptions of the stalks of the $\mathcal{K}_2$ sheaf are globalized, one has appropriate terms involving the ideal sheaves of components of E.

Finally, one manages to reduce to the above sequence (8.1), by showing that certain additional terms appearing at the sheaf level have no contribution to the cohomology of the $\mathcal{K}_2$ sheaf, using the Grauert-Riemenschneider vanishing theorem for $\pi : Y \to X$, that $R^1 \pi_* \omega_Y = 0$.

Now if $k = \overline{\mathbb{Q}}$, we see that since $\Omega_{k/\mathbb{Z}} = 0$ in this case, we get that for a resolution $\pi : Y \to X$ of a normal surface X over $\overline{\mathbb{Q}}$, with a normal crossing exceptional locus E, we have a formula

$$CH^2(X) = F^2 K_0(X) \cong F^2 K_0(Y, E).$$

If X is the affine cone over a smooth projective curve C over $\overline{\mathbb{Q}}$, then Y may be taken to be the blow up of the vertex of the cone. Then Y is a geometric line bundle over the original curve C, and E is its 0-section. Hence $K_i(Y) \to K_i(E)$ is an isomorphism for all i, by the homotopy invariance of algebraic K-theory for regular schemes, and so $K_i(Y, E) = 0$ for all i. This gives in particular, from Theorem 8.1, that $CH^2(X) = 0$. This was the conclusion of Theorem 7.8 for $k = \overline{\mathbb{Q}}$.

On the other hand, suppose $k = \mathbb{C}$, the complex numbers, and X is the cone over a smooth projective complex curve C. Again using the blow-up $Y \to X$ of the vertex, and the line bundle structure on Y with 0-section E, we first see that (8.1) simplifies in this case to a formula

$$SK_1(nE) = SK_1(E) \oplus \oplus_{j=1}^{n-1} H^1(C, \mathcal{O}_C(j)) \otimes_{\mathbb{C}} \Omega_{\mathbb{C}/\mathbb{Z}}.$$

This uses that the inclusion $E \hookrightarrow nE$ has a retraction, so (8.1) becomes a split exact sequence; then one uses that $\mathcal{I}/\mathcal{I}^n \cong \oplus_{j=1}^{n-1} \mathcal{O}_C(j))$ as sheaves of $\mathbb{C}$-vector spaces, using the underlying graded structure.

Theorem 8.1 now implies that

$$CH^2(X) = \oplus_{j=1}^{\infty} H^1(C, \mathcal{O}_C(1)) \otimes_{\mathbb{C}} \Omega_{\mathbb{C}/\mathbb{Z}},$$

where the direct sum is of course finite, since $H^1(C, \mathcal{O}_C(j))$ vanishes for all large j. Now for the curve C,

$$H^1(C, \mathcal{O}_C(j)) = 0 \implies H^1(C, \mathcal{O}_C(j+1)) = 0 \quad \forall\, j,$$

since multiplication by a suitable section of $\mathcal{O}_C(1)$ gives an inclusion $\mathcal{O}_C(j) \to \mathcal{O}_C(j+1)$ with a cokernel supported at points, and so this induces a surjection on H^1. Hence, we see that

$$CH^2(X) = 0 \Leftrightarrow H^1(C, \mathcal{O}_C(1)) = 0.$$

This is the other conclusion of Theorem 7.8 for the case $k = \mathbb{C}$.

It was known earlier that Theorem 7.2 follows from the special case of normal affine surfaces, as discussed earlier. In this case, it can be deduced using Theorem 8.1. Again, one first takes a resolution $Y \to X$ with normal crossing exceptional locus E, and makes the analysis of $SK_1(nE)$ as in characteristic 0. Then, instead of using the Bloch formula, which is in fact not valid in characteristic p, one shows that $\ker(SK_1(nE) \to SK_1(E))$ is p^N torsion, for some N, using certain local descriptions of the $\mathcal{K}_2$ sheaves, found for example in work of Bloch and Kato.

On the other hand, one sees also that $\ker(F^2 K_0(X) \to F^2 K_0(Y))$ is a divisible abelian group, and further, that

$$\operatorname{coker} SK_1(Y) \to SK_1(E)$$

is a torsion-free divisible group (one ingredient in the proof of the latter is the negative definiteness of the intersection pairing on components of E). This implies that $F^2 K_0(Y, E) \to F^2 K_0(Y)$ is an isomorphism on torsion, and so $\ker F^2 K_0(X) \to F^2 K_0(Y, E)$ is also divisible, by a diagram chase. But we have also seen that it is

214   V. Srinivas

p^N-torsion for some large enough N, so it is 0, i.e., $F^2 K_0(X) \cong F^2 K_0(Y, E)$, and also $F^2 K_0(X) \to F^2 K_0(Y)$ is an isomorphism on torsion subgroups.

This analysis is valid over an arbitrary algebraically closed ground field of characteristic p. Further, if X is affine, one sees that $CH^2(Y)$ is in fact torsion-free, from the theorems of Roitman and Milne, cited earlier. Hence $CH^2(X)$ is torsion-free. But if now $k = \overline{\mathbb{F}}_p$, clearly $CH^2(X)$ is torsion, since the Picard group of any affine curve over $\overline{\mathbb{F}}_p$ is torsion. Hence we get that $CH^2(X) = 0$ in this case, as claimed in Theorem 7.2.

References

[1] Amalendu Krishna, V. Srinivas, *Zero cycles and K-theory on normal surfaces*, Annals of Math. 156 (2002) 155–195.

[2] Amalendu Krishna, V. Srinivas, *Zero cycles on singular varieties*, to appear in Proceedings of the EAGER Conference *Workshop: Algebraic Cycles and Motives, Leiden,* 2004.

[3] Bloch, S., *Lectures on Algebraic Cycles*, Duke Univ. Math. Ser. IV, Durham, 1979.

[4] Bloch, S., Murthy, M.P., Szpiro, L., *Zero cycles and the number of generators of an ideal*, Mémoire No. 38 (nouvelle série), Supplément au Bulletin de la Soc. Math. de France, Tome 117 (1989) 51–74.

[5] Bloch, S. Srinivas, V., *Remarks on correspondences and algebraic cycles*, Amer. J. Math. 105 (1983), no. 5, 1235–1253.

[6] Boratynski, M., *On a conormal module of smooth set-theoretic complete intersections*, Trans. Amer. Math. Soc. 296 (1986).

[7] Deligne, P., *Théorie de Hodge II*, Publ. Math. I.H.E.S. 40 (1972) 5–57.

[8] Eisenbud, D., *Commutative Algebra with a view toward Algebraic Geometry*, Grad. Texts in Math. 150, Springer-Verlag (1995).

[9] Fulton, W., *Intersection Theory*, Ergeb. Math. Folge 3, Band 2, Springer-Verlag (1984).

[10] Grothendieck, A., Berthellot, P. Illusie, L., *SGA6, Théorie des Intersections et Théorème de Riemann-Roch*, Lect. Notes in Math. 225, Springer-Verlag (1971).

[11] Levine, M., Weibel, C., *Zero cycles and complete intersections on singular varieties*, J. Reine Angew. Math. 359 (1985), 106–120.

[12] Hartshorne, R., *Algebraic Geometry*, Grad texts in Math. 52, Springer-Verlag (1977).

[13] Lindel, H., *On the Bass-Quillen conjecture concerning projective modules over polynomial rings*, Inventiones Math. 65 (1981/82) 319–323.

[14] Lindel, H., *On projective modules over polynomial rings over regular rings*, in *Algebraic K-Theory, Part I (Oberwolfach, 1980)*, Lecture Notes in Math. 966, Springer-Verlag (1982) 169–179

[15] Milnor, J.W., Stasheff, J.D., *Characteristic Classes*, Ann. Math. Studies 76, Princeton (1974).

[16] Mohan Kumar, N., *On two conjectures about polynomial rings*, Invent. Math. 46 (1978) 225–236.

[17] Mumford, D., *Rational equivalence of 0-cycles on surfaces*, J. Math. Kyoto Univ. 9 (1968) 195–204.

[18] Mumford, D., *Lectures on Curves on an Algebraic Surface*, Ann. Math. Studies 59, Princeton (1966).

[19] Murthy, M.P., *Zero cycles and projective modules*, Annals of Math. 140 (1994) 405–434.

[20] Murthy, M.P., Mohan Kumar, N., Roy, A., in *Algebraic Geometry and Commutative Algebra, Vol. I (in honour of Masayoshi Nagata)*, Kinokuniya, Tokyo (1988), 281–287.

[21] Pedrini, C., Weibel, C.A., *Divisibility in the Chow group of 0-cycles on a singular surface*, Astérisque 226 (1994) 371–409.

[22] Roitman, A.A., *Rational equivalence of 0-cycles*, Math. USSR Sbornik, 18 (1972) 571–588.

[23] Roitman, A.A., *The torsion of the group of 0-cycles modulo rational equivalence*, Ann. Math. 111 (1980) 553–569.

[24] V. Srinivas, *Zero cycles on a singular surface II*, J. Reine Ang. Math. 362 (1985) 3–27.

[25] Srinivas, V., *Grothendieck groups of polynomial and Laurent polynomial rings*, Duke Math. J. 53 (1986) 595–633.

[26] Srinivas, V., *The embedding dimension of an affine variety*, Math. Ann. 289 (1991) 125–132.

[27] Srinivas, V., *Gysin maps and cycle classes for Hodge cohomology*, Proc. Indian Acad. Sci. (Math. Sci.) 103 (1993) 209–247.

[28] Srinivas, V., *Zero cycles on singular varieties*, in *Arithmetic and Geometry of Algebraic Cycles*, NATO Science Series C, Vol. 548, Kluwer (2000) 347–382.

[29] Srinivas, V., *Some Geometric Methods in Commutative Algebra*, in *Computational Commutative Algebra and Combinatorics (Osaka, 1999)*, Advanced Studies in Pure Math. 33 (2002) 231–276.

[30] Srinivas, V., *Algebraic K-theory*. Second edition. Progress in Math., **90**, Birkhäuser, Boston, MA, 1996.

[31] Suslin, A.A., *Stably free modules*. (Russian) Mat. Sb. (N.S.) 102(144) (1977), no. 4, 537–550, 632.

[32] Szpiro, L., *Lectures on Equations defining Space Curves*, Tata Institute of Fundamental Research Lecture Notes, No. 62, Springer-Verlag (1979).

[33] Thomason, R., Trobaugh, T, *Higher algebraic K-theory of schemes and of derived categories*, in *The Grothendieck Festschrift III*, Progress in Math. 88, Birkhäuser (1990).

[34] Weibel, C.A., *The negative K-theory of normal surfaces*, Duke Math. J. 108 (2001) 1–35.

Vasudevan Srinivas
School of Mathematics, Tata Institute of Fundamental Research
Homi Bhabha Road, Mumbai-400005, India
e-mail: `srinivas@math.tifr.res.in`

Algebraic Cycles, Sheaves, Shtukas, and Moduli
Trends in Mathematics, 217–236
© 2007 Birkhäuser Verlag Basel/Switzerland

Introduction to the Stacks of Shtukas

Ngo Dac Tuan

These notes are an attempt to provide an introduction to the stacks of shtukas and their compactifications. The notion of shtukas (or F-bundles) was first defined by Drinfeld [Dri87, Dri89] in his proof of the Langlands correspondence for the general linear group GL_2 over function fields. It is recently used in the Lafforgue's proof of the Langlands correspondence for the general linear group of higher rank GL_r over function fields, *cf.* [Laf02].

These notes grew out of my lectures at the mini-school "Moduli spaces" (Warsaw, Poland) in May 2005. I would like to thank the organizers of this school, Adrian Langer, Piotr Pragacz and Halszka Tutaj-Gasinska, for the invitation and hospitality. These notes have been written during my visit to the Institute for Advanced Study in Fall 2006. I wish to thank this institution for its hospitality and excellent working conditions.

Notations

For the rest of these notes, let X be a smooth projective geometrically connected curve over a finite field $\mathbb{F}_q$ of q elements. The field F of $\mathbb{F}_q$-valued rational functions over X is called *the function field of X*. One can identify the places of F with the set $|X|$ of closed points of X.

For example, we could take the projective line $X = \mathbb{P}^1$. Then the function field of X is just the field of rational functions in one variable. Its elements are fractions $P(t)/Q(t)$, where $P(t)$ and $Q(t)$ are polynomials over $\mathbb{F}_q$ without common factors, and $Q(t) \neq 0$. The set $|X|$ corresponds to the set of irreducible polynomials over $\mathbb{F}_q$.

We will use freely the theory of schemes and stacks. For a reader unfamiliar with such matters, the book of Hartshorne [Har77] and that of Moret-Bailly and Laumon [Lau-MB99] would be good references. All schemes (or stacks) will be defined over $\operatorname{Spec}\mathbb{F}_q$ and we denote by $Y \times Z$ the fiber product $Y \times_{\operatorname{Spec}\mathbb{F}_q} Z$ for any such schemes (or stacks) Y and Z.

This work is partially supported by National Science Foundation grant number DMS-0111298.

Suppose that S is a scheme over $\operatorname{Spec} \mathbb{F}_q$. We denote by $\operatorname{Frob}_S : S \longrightarrow S$ the Frobenius morphism which is identity on points, and is the Frobenius map $t \mapsto t^q$ on functions. For any $\mathcal{O}_{X \times S}$-module $\mathcal{F}$, we denote by $\mathcal{F}^\sigma$ the pull back $(\operatorname{Id}_X \times \operatorname{Frob}_S)^* \mathcal{F}$.

We fix also an algebraic closure k of $\mathbb{F}_q$. We define

$$\overline{X} = X \times_{\operatorname{Spec} \mathbb{F}_q} \operatorname{Spec} k.$$

Finally, fix once for all an integer $r \geq 1$.

1. Modifications. Hecke stacks

1.1. Let $\overline{T}$ be a finite closed subscheme of the geometric curve $\overline{X}$. Suppose that $\mathcal{E}$ and $\mathcal{E}'$ are two vector bundles of rank r over $\overline{X}$. By definition, a $\overline{T}$-*modification* from $\mathcal{E}$ into $\mathcal{E}'$ is an isomorphism between the restrictions of $\mathcal{E}$ and $\mathcal{E}'$ to the open subscheme $\overline{X} - \overline{T}$ of $\overline{X}$:

$$\varphi : \mathcal{E}\big|_{\overline{X} - \overline{T}} \xrightarrow{\sim} \mathcal{E}'\big|_{\overline{X} - \overline{T}}.$$

Suppose that φ is a $\overline{T}$-modification from $\mathcal{E}$ into $\mathcal{E}'$ as above. If x is a geometric point in $\overline{T}$, $\mathcal{O}_x$ denotes the completion of $\mathcal{O}_{\overline{X}}$ at x, and F_x denotes the fraction field of $\mathcal{O}_x$. The isomorphism φ induces an isomorphism between the generic fibers V and V' of $\mathcal{E}$ and $\mathcal{E}'$, respectively:

$$\varphi : V \xrightarrow{\sim} V',$$

hence an isomorphism

$$\varphi : V_x \xrightarrow{\sim} V'_x.$$

The completion of $\mathcal{E}$ (resp. $\mathcal{E}'$) at x defines an $\mathcal{O}_x$-lattice $\mathcal{E}_x$ in V_x (resp. E'_x in V'_x). Then it follows from the theorem of elementary divisors that the relative position of two lattices $\mathcal{E}_x$ and $\mathcal{E}'_x$ is given by a sequence of integers

$$\lambda_x = (\lambda_1 \geq \lambda_2 \geq \cdots \geq \lambda_r) \in \mathbb{Z}^r.$$

This sequence is called *the invariant of the modification φ at x*. One remarks immediately that the invariant of the modification φ^{-1} at x is the sequence $(-\lambda_r \geq -\lambda_{r-1} \geq \cdots \geq -\lambda_1)$.

Consider the simplest example of all, the case that $\overline{T}$ contains only one geometric point x of $\overline{X}$ and $\lambda = (1, 0, \ldots, 0)$. Then an x-modification from a vector bundle $\mathcal{E}$ of rank r over $\overline{X}$ into another vector bundle $\mathcal{E}'$ of same rank whose invariant at x is λ is just an injection $\mathcal{E} \hookrightarrow \mathcal{E}'$ such that the quotient $\mathcal{E}'/\mathcal{E}$ is supported by x and of length 1. It is well known as *an (elementary) upper modification at x*.

Similarly, we keep $\overline{T} = x \in \overline{X}$ and modify $\lambda = (0, 0, \ldots, -1)$. Then an x-modification from $\mathcal{E}$ into $\mathcal{E}'$ whose invariant at x is λ is just an injection $\mathcal{E}' \hookrightarrow \mathcal{E}$ such that the quotient $\mathcal{E}/\mathcal{E}'$ is supported by x and of length 1. It is well known as *an (elementary) lower modification at x*.

1.2. We are ready to define (elementary) Hecke stacks Hecke^r.
Over $\mathrm{Spec}\,k$, $\mathrm{Hecke}^r(\mathrm{Spec}\,k)$ is the category whose objects consist of

- a vector bundle $\mathcal{E}$ of rank r over $\overline{X}$,
- two geometric points $\infty, 0 \in \overline{X}$,
- an upper modification $\mathcal{E} \hookrightarrow \mathcal{E}'$ at ∞ and a lower modification $\mathcal{E}' \hookleftarrow \mathcal{E}''$ at 0,

and whose morphisms are isomorphisms between such data.

More generally, for a scheme S over $\mathbb{F}_q$, $\mathrm{Hecke}^r(S)$ is the category whose objects are the data consisting of

- a vector bundle $\mathcal{E}$ of rank r over $X \times S$,
- two morphisms $\infty, 0 : S \longrightarrow X$,
- a modification consisting of two injections

$$\mathcal{E} \hookrightarrow \mathcal{E}' \hookleftarrow \mathcal{E}''$$

of $\mathcal{E}$ by two vector bundles $\mathcal{E}'$ and $\mathcal{E}''$ of rank r over $X \times S$ such that the quotients $\mathcal{E}'/\mathcal{E}$ and $\mathcal{E}'/\mathcal{E}''$ are supported by the graphs of ∞ and 0 respectively, and they are invertible on their support,

and whose morphisms are isomorphisms between such data.

Assume that $f : S' \longrightarrow S$ is a morphism of schemes. Then the pull-back operator induces a functor between the two categories $\mathrm{Hecke}^r(S)$ and $\mathrm{Hecke}^r(S')$:

$$f^* : \mathrm{Hecke}^r(S) \longrightarrow \mathrm{Hecke}^r(S')$$

$$(\mathcal{E}, \infty, 0, \mathcal{E} \hookrightarrow \mathcal{E}' \hookleftarrow \mathcal{E}'') \mapsto (f^*\mathcal{E}, f \circ \infty, f \circ 0, f^*\mathcal{E} \hookrightarrow f^*\mathcal{E}' \hookleftarrow f^*\mathcal{E}'')$$

Hence the collection of categories $\mathrm{Hecke}^r(S)$ when S runs through the category of schemes over $\mathbb{F}_q$ defines *the so-called (elementary) Hecke stack* Hecke^r.

We consider the stack Bun^r classifying vector bundles of rank r over X. It means that this stack associates to each scheme S the groupoid of vector bundles of rank r over the product $X \times S$. It is well known that Bun^r is an algebraic Deligne-Mumford stack, *cf.* [Lau-MB99].

Proposition 1.3. *The morphism of stacks*

$$\mathrm{Hecke}^r \longrightarrow X \times X \times \mathrm{Bun}^r$$

$$(\mathcal{E}, \infty, 0, \mathcal{E} \hookrightarrow \mathcal{E}' \hookleftarrow \mathcal{E}'') \mapsto (\infty, 0, \mathcal{E})$$

is representable, projective and smooth of relative dimension $2r - 2$.

2. Stacks of Drinfeld shtukas

2.1. We have introduced the stacks Bun^r and Hecke^r classifying vector bundles of rank r and elementary Hecke modifications of rank r, respectively. We want to use these stacks to define the main object of these notes: the stack Sht^r of Drinfeld shtukas of rank r. It is simply the fiber product:

$$
\begin{array}{ccc}
\mathrm{Sht}^r & \longrightarrow & \mathrm{Bun}^r \\
\downarrow & & \downarrow \\
\mathrm{Hecke}^r & \longrightarrow & \mathrm{Bun}^r \times \mathrm{Bun}^r,
\end{array}
$$

where the lower horizontal map is

$$
\big(\mathcal{E} \hookrightarrow \mathcal{E}' \hookleftarrow \mathcal{E}''\big) \mapsto (\mathcal{E}, \mathcal{E}''),
$$

and the right vertical map is

$$
\mathcal{E} \mapsto (\mathcal{E}, \mathcal{E}^\sigma).
$$

In other words, the stack Sht^r of Drinfeld shtukas (or F-bundles, F-sheaves) of rank r associates to any scheme S the data $\widetilde{\mathcal{E}} = (\mathcal{E} \hookrightarrow \mathcal{E}' \hookleftarrow \mathcal{E}'' \xleftarrow{\sim} \mathcal{E}^\sigma)$ consisting of

- a vector bundle $\mathcal{E}$ of rank r over $X \times S$,
- two morphisms $\infty, 0 : S \longrightarrow X$,
- a modification consisting of two injections

$$
\mathcal{E} \hookrightarrow \mathcal{E}' \hookleftarrow \mathcal{E}''
$$

  of $\mathcal{E}$ by two vector bundles $\mathcal{E}'$ and $\mathcal{E}''$ of rank r over $X \times S$ such that the quotients $\mathcal{E}'/\mathcal{E}$ and $\mathcal{E}'/\mathcal{E}''$ are supported by the graphs of ∞ and 0 respectively, and they are invertible on their support,
- an isomorphism $\mathcal{E}^\sigma \xrightarrow{\sim} \mathcal{E}''$.

The morphisms 0 and ∞ are called *the zero and the pole* of the shtuka $\widetilde{\mathcal{E}}$.

In the case $r = 1$, a shtuka of rank 1 over a scheme S consists of a line bundle $\mathcal{L}$ over the fiber product $X \times S$ together with an isomorphism

$$
\mathcal{L}^\sigma \otimes \mathcal{L}^{-1} \xrightarrow{\sim} \mathcal{O}_{X \times S}(\Gamma_\infty - \Gamma_0)
$$

where $\infty, 0 : S \longrightarrow X$ are two morphisms, and Γ_∞, Γ_0 are the graphs of ∞ and 0, respectively.

From this observation, it is easy to construct shtukas of higher rank. For higher rank $r \geq 2$, suppose that

$$
\widetilde{\mathcal{L}} = (\mathcal{L}, \mathcal{L} \xrightarrow{j} \mathcal{L}' \xleftarrow{t} \mathcal{L}'' \xleftarrow{\sim} \mathcal{L}^\sigma)
$$

is a shtuka of rank 1 over a scheme S with the pole ∞ and the zero 0, and $\mathcal{F}$ is a vector bundle of rank $r - 1$ over the curve X. Then one sees immediately that the direct sum

$$
(\mathcal{L} \oplus \mathcal{F}, \mathcal{L} \oplus \mathcal{F} \xrightarrow{j \oplus \mathrm{id}} \mathcal{L}' \oplus \mathcal{F} \xleftarrow{t \oplus \mathrm{id}} \mathcal{L}' \oplus \mathcal{F} \xleftarrow{\sim} \mathcal{L}^\sigma \oplus \mathcal{F})
$$

is a shtuka of rank r over S with the same pole ∞ and the same zero 0.

Theorem 2.2 (Drinfeld). *The stack* Sht^r *is an algebraic Deligne-Mumford stack and the characteristic map*

$$(\infty, 0) : \mathrm{Sht}^r \longrightarrow X \times X$$

is smooth of relative dimension $2r - 2$. *Furthermore, it is locally of finite type.*

Thanks to Proposition 1.3, this theorem is a direct corollary of the following lemma applying to $\mathcal{W} = \mathrm{Sht}^r$, $\mathcal{M} = \mathrm{Hecke}^r$, $\mathcal{U} = \mathrm{Bun}^r$ and $Y = X \times X$.

Lemma 2.3. *Consider a cartesian diagram of stacks*

$$
\begin{array}{ccc}
\mathcal{W} & \longrightarrow & \mathcal{U} \\
\downarrow & & \downarrow {\scriptstyle (\mathrm{Frob}_{\mathcal{U}}, \mathrm{id}_{\mathcal{U}})} \\
\mathcal{M} & \xrightarrow{(\alpha, \beta)} & \mathcal{U} \times \mathcal{U} \\
{\scriptstyle \pi} \downarrow & & \\
Y & &
\end{array}
$$

where

- Y *is a scheme,*
- $\mathcal{U}$ *is algebraic and locally of finite type,*
- $\mathcal{M}$ *is algebraic and locally of finite type over* Y,
- *the morphism* $(\pi, \alpha) : \mathcal{M} \longrightarrow Y \times \mathcal{U}$ *is representable.*

Then $\mathcal{W}$ *is algebraic and locally of finite type. Moreover, the diagonal map* $\mathcal{W} \longrightarrow \mathcal{W} \times \mathcal{W}$ *which is representable, separated and of finite type is unramified everywhere, hence quasi-finite.*

If $\mathcal{U}$ *is smooth, and the morphism* $(\pi, \alpha) : \mathcal{M} \longrightarrow Y \times \mathcal{U}$ *is smooth of relative dimension* n, *then the morphism* $\mathcal{W} \longrightarrow Y$ *is also smooth of relative dimension* n.

The stack Sht^r has an infinite number of connected components $\mathrm{Sht}^{r,d}$ indexed by the degree of the associated vector bundle $\mathcal{E}$, i.e., one requires that $\deg \mathcal{E} = d$. Then

$$\mathrm{Sht}^r = \coprod_{d \in \mathbb{Z}} \mathrm{Sht}^{r,d}.$$

2.4. We consider the stack Triv^r_X which associates to each scheme S the data consisting of a vector bundle $\mathcal{E}$ of rank r over $X \times S$ together with an isomorphism $\mathcal{E}^\sigma \xrightarrow{\sim} \mathcal{E}$. As observed by Drinfeld, this stack is not interesting:

$$\mathrm{Triv}^r_X = \coprod_E \mathrm{Spec}\, \mathbb{F}_q / \mathrm{Aut}(E).$$

Here E runs through the set of vector bundles of rank r over the curve X and $\mathrm{Aut}(E)$ denotes the automorphism group of E.

Similarly, let I be a level, i.e., a finite closed subscheme of X. We consider the stack Triv^r_I which associates to each scheme S the data consisting of a vector bundle $\mathcal{E}$ of rank r over $I \times S$ together with an isomorphism $(\mathrm{id}_I \times \mathrm{Frob}_S)^* \mathcal{E} \xrightarrow{\sim} \mathcal{E}$. One can show that

$$\mathrm{Triv}_I^r = \mathrm{Spec}\, \mathbb{F}_q / \mathrm{GL}_r(\mathcal{O}_I).$$

For the rest of this section, we fix a level I of X. Suppose that $\widetilde{\mathcal{E}} = (\mathcal{E}, \mathcal{E} \hookrightarrow \mathcal{E}' \hookleftarrow \mathcal{E}'' \xleftarrow{\sim} \mathcal{E}^\sigma)$ is a shtuka of rank r over a scheme S such that the graphs of $\infty : S \longrightarrow X$ and $0 : S \longrightarrow X$ do not meet $I \times S$. Under this hypothesis, one obtains an isomorphism of restriction to $I \times S$:

$$\psi : \mathcal{E}\big|_{I \times S} \xrightarrow{\sim} \mathcal{E}^\sigma\big|_{I \times S}.$$

By definition, a *level structure of $\widetilde{\mathcal{E}}$ on I* is an isomorphism

$$\varphi : \mathcal{E}\big|_{I \times S} \xrightarrow{\sim} \mathcal{O}_{I \times S}^r$$

such that the following diagram is commutative:

$$
\begin{array}{ccc}
\mathcal{E}\big|_{I \times S} & \xrightarrow{\ \psi\ } & \mathcal{E}^\sigma\big|_{I \times S} \\[4pt]
\varphi \downarrow & & \downarrow \varphi^\sigma \\[4pt]
\mathcal{O}_{I \times S}^r & =\!=\!= & \mathcal{O}_{I \times S}^r.
\end{array}
$$

We denote by $\mathrm{Sht}_I^r(S)$ the category whose objects are shtukas of rank r over S together with a level structure on I and whose morphisms are isomorphisms between these objects.

Then the collection of categories $\mathrm{Sht}_I^r(S)$ when S runs over the category of schemes S over $\mathbb{F}_q$ defines a stack Sht_I^r called *the stack of shtukas of rank r with level structure on I.*

It fits into the cartesian diagram

$$
\begin{array}{ccc}
\mathrm{Sht}_I^r & \longrightarrow & \mathrm{Spec}\,\mathbb{F}_q \\[4pt]
\downarrow & & \downarrow \\[4pt]
\mathrm{Sht}^r \times_{X^2} (X - I)^2 & \longrightarrow & \mathrm{Triv}_I^r
\end{array}
$$

where

- the left arrow corresponds to the morphism forgetting the level structure,
- the right arrow corresponds to the trivial object $(\mathcal{O}_I^r, \mathrm{id})$ of $\mathrm{Triv}_I^r(\mathrm{Spec}\,\mathbb{F}_q)$,
- the lower arrow is given by the restriction to the level I which with the above notations sends $(\widetilde{\mathcal{E}}, \varphi : \mathcal{E}\big|_{I \times S} \xrightarrow{\sim} \mathcal{O}_{I \times S}^r)$ to $(\mathcal{E}\big|_{I \times S}, \psi : \mathcal{E}\big|_{I \times S} \xrightarrow{\sim} \mathcal{E}^\sigma\big|_{I \times S})$.

It follows immediately:

Proposition 2.5 (Drinfeld). *The forgetful morphism*

$$\mathrm{Sht}_I^r \longrightarrow \mathrm{Sht}^r \times_{X^2} (X - I)^2$$

is representable, finite, étale and Galois with Galois group $\mathrm{GL}_r(\mathcal{O}_I)$.

3. Harder-Narasimhan polygons

We have given a simple description of shtukas of rank 1. It implies that the stack $\mathrm{Sht}^{1,d}$ is of finite type for every integer d. However, this statement is not true for higher rank $r \geq 2$: the stack $\mathrm{Sht}^{r,d}$ is only locally of finite type but no longer of finite type. In order to construct interesting open substacks of finite type of $\mathrm{Sht}^{r,d}$, we digress for a moment to review the notion of Harder-Narasimhan polygons.

For the rest of these notes, a polygon is a map

$$p : [0, r] \longrightarrow \mathbb{R}$$

such that

- $p(0) = p(r) = 0$,
- p is affine on each interval $[i - 1, i]$ for $0 < i \leq r$.

A polygon p is called

- *rational* if all the numbers $\{p(i)\}_{1 < i < r}$ are rational,
- *convex* if for every integer i with $0 < i < r$, we have

$$-p(i + 1) + 2p(i) - p(i - 1) \geq 0,$$

- *big enough with respect to a real number* μ if all the terms $2p(i) - p(i - 1) - p(i + 1)$, with $0 < i < r$, are big enough with respect to μ,
- *integral with respect to an integer* d if all the terms $p(i) + \frac{i}{r}d$, with $0 < i < r$, are integers.

Suppose that $\mathcal{E}$ is a vector bundle over the geometric curve $\overline{X}$. Then the slope $\mu(E)$ of $\mathcal{E}$ is defined as the quotient

$$\mu(E) = \frac{\deg E}{\mathrm{rk}\, E}.$$

With this definition, $\mathcal{E}$ is called *semistable* (resp. *stable*) if for any proper subbundle $\mathcal{F}$ of $\mathcal{E}$, we have

$$\mu(\mathcal{F}) \leq \mu(E) \quad (resp. <).$$

It is obvious that

a) Any stable vector bundle over $\overline{X}$ is semistable.
b) Every line bundle over the geometric curve $\overline{X}$ is stable, hence semistable.

Suppose that $X = \mathbb{P}^1$. It is well known that every vector bundle over $\mathbb{P}^1$ is totally decomposable, i.e., there exist integers $n_1 \geq n_2 \geq \cdots \geq n_r$ such that

$$\mathcal{E} = \mathcal{O}(n_1) \oplus \mathcal{O}(n_2) \oplus \cdots \oplus \mathcal{O}(n_r).$$

Then one sees immediately

a) $\mathcal{E}$ is semistable if and only if $n_1 = n_2 = \cdots = n_r$,
b) $\mathcal{E}$ is stable if and only if $r = 1$.

Harder and Narasimhan proved, *cf.* [Har-Nar75]:

Theorem 3.1 (Harder-Narasimhan). *Let $\mathcal{E}$ be a vector bundle over $\overline{X}$. Then there exists a unique filtration of maximal subbundles*

$$\mathcal{E}_0 = 0 \subsetneq \mathcal{E}_1 \subsetneq \cdots \subsetneq \mathcal{E}_n = \mathcal{E}$$

of $\mathcal{E}$ which satisfies the following properties:

- *$\mathcal{E}_i/\mathcal{E}_{i-1}$ is semistable,*
- *$\mu(\mathcal{E}_1/\mathcal{E}_0) > \mu(\mathcal{E}_2/\mathcal{E}_1) > \cdots > \mu(\mathcal{E}_n/\mathcal{E}_{n-1})$.*

This filtration is called the (canonical) Harder-Narasimhan filtration associated to $\mathcal{E}$.

We define a polygon $p : [0, r] \longrightarrow \mathbb{R}^+$ as follows:

- *$p(0) = p(r) = 0$,*
- *p is affine on each interval $[\operatorname{rk}\mathcal{E}_i, \operatorname{rk}\mathcal{E}_{i+1}]$, for $0 \leq i < r$,*
- *$p(\operatorname{rk}\mathcal{E}_i) = \deg\mathcal{E}_i - \frac{\operatorname{rk}\mathcal{E}_i}{r}\deg\mathcal{E}$.*

Then it satisfies the following properties:

i) *p is convex.*
ii) *For any subbundle $\mathcal{F}$ of $\mathcal{E}$, we have*

$$\deg\mathcal{F} \leq \frac{\operatorname{rk}\mathcal{F}}{r}\deg\mathcal{E} + p(\operatorname{rk}\mathcal{F}). \qquad (3.1.1)$$

This polygon is called the canonical Harder-Narasimhan filtration associated to $\mathcal{E}$.

We will give a full proof of this theorem in the case $r = 2$. For the general case, see [Har-Nar75]. Suppose that $\mathcal{E}$ is a vector bundle of rank 2 over $\overline{X}$.

If $\mathcal{E}$ is semistable, one verifies easily that

- $0 \subsetneq \mathcal{E}$ is the Harder-Narasimhan filtration of $\mathcal{E}$,
- the nil polygon $p = 0$ is the canonical Harder-Narasimhan polygon of $\mathcal{E}$.

If $\mathcal{E}$ is not semistable, then there exists a maximal line bundle $\mathcal{L}$ of $\mathcal{E}$ with $\deg\mathcal{L} > \deg\mathcal{E}/2$. We claim that

- $0 \subsetneq \mathcal{L} \subsetneq \mathcal{E}$ is the Harder-Narasimhan filtration of $\mathcal{E}$,
- the polygon $p(0) = 0, p(1) = \deg\mathcal{L} - \deg\mathcal{E}/2, p(2) = 0$ is the canonical Harder-Narasimhan polygon of $\mathcal{E}$.

In fact, one needs to check that if $\mathcal{L}'$ is a line bundle of $\mathcal{E}$, then $\deg\mathcal{L}' \leq \deg\mathcal{L}$. If $\mathcal{L}' \subseteq \mathcal{L}$, then it is obvious. Otherwise, the fact that $\mathcal{L}' \cap \mathcal{L} = 0$ implies an injection $\mathcal{L}' \hookrightarrow \mathcal{E}/\mathcal{L}$. Hence $\deg\mathcal{L}' \leq \deg\mathcal{E}/\mathcal{L} < \deg\mathcal{L}$. We are done.

Let us return to the projective line $X = \mathbb{P}^1$. Suppose that

$$\mathcal{E} = \mathcal{O}(n_1) \oplus \mathcal{O}(n_2) \oplus \cdots \oplus \mathcal{O}(n_r)$$

with $n_1 = \cdots = n_{i_1} > n_{i_1+1} = \cdots = n_{i_2} > \cdots > n_{i_{n-1}+1} = \cdots = n_r$. Then

$$0 \subsetneq \mathcal{O}(n_{i_1})^{\oplus n_{i_1}} \subsetneq \mathcal{O}(n_1)^{\oplus n_{i_1}} \oplus \mathcal{O}(n_{i_2})^{\oplus n_{i_2}-n_{i_1}} \subsetneq \cdots \subsetneq \mathcal{E}$$

is the canonical Harder-Narasimhan filtration of $\mathcal{E}$.

Using the notion of canonical Harder-Narasimhan polygon, Lafforgue has introduced an interesting family of open substacks of finite type of $\mathrm{Sht}^{r,d}$, cf. [Laf98, théorème II.8].

Proposition 3.2 (Lafforgue). *Let $p : [0, r] \longrightarrow \mathbb{R}$ be a convex polygon which is big enough with respect to the genus of X (or X for short) and r. Then there exists a unique open substack $\mathrm{Sht}^{r,d,p}$ of $\mathrm{Sht}^{r,d}$ such that a geometric point $\widetilde{\mathcal{E}}$ lies in this open if and only if the canonical Harder-Narasimhan polygon associated to $\mathcal{E}$ is bounded by p.*

The stack $\mathrm{Sht}^{r,d,p}$ is of finite type. Moreover, $\mathrm{Sht}^{r,d}$ is the union of these open substacks $\mathrm{Sht}^{r,d,p}$.

One can prove that these substacks $\mathrm{Sht}^{r,d,p}$ verify the valuative criterion for separatedness. However, they are not proper. This raises the compactification problem:

Problem 3.3. Find an algebraic proper stack $\mathcal{X}$ containing $\mathrm{Sht}^{r,d,p}$ as an open dense substack.

In the case $r = 2$, Drinfeld has constructed such a proper stack $\mathcal{X}$. For higher rank, Lafforgue has given a solution to this problem generalizing Drinfeld's construction. Let us pause for a moment to describe what needs to be done in order to construct such compactifications.

Step 1. *Introduce the stacks of degenerated and iterated shtukas which extends that of shtukas.*

This step is based on the well-studied scheme of complete homomorphisms of rank r which is obtained from the scheme of non-zero $n \times n$ matrices by a series of blow-ups. Roughly speaking, the last condition $\mathcal{E}^\sigma \xrightarrow{\sim} \mathcal{E}''$ will be replaced by a complete homomorphism $\mathcal{E}^\sigma \Rightarrow \mathcal{E}''$.

Step 2. *Stratify $\overline{\mathrm{Sht}^r}$ and give a modular description of each stratum.*

This step allows us to see how an iterated shtuka in each stratum can be "decomposed" as a product of shtukas.

Step 3. *Truncate $\overline{\mathrm{Sht}^{r,d}}$ to define proper compactifications $\overline{\mathrm{Sht}^{r,d,p}}$.*

This step is the hardest one: given a convex polygon p, one needs to truncate $\overline{\mathrm{Sht}^{r,d}}$ to obtain a proper substack $\overline{\mathrm{Sht}^{r,d,p}}$ containing $\mathrm{Sht}^{r,d,p}$. Lafforgue has used the semistable reduction à la Langton.

To simplify the exposition, from now on we will suppose $r = 2$ and refer any curious reader to the article of Lafforgue [Laf98] for the higher rank case.

4. Degenerated shtukas of rank 2

4.1. Drinfeld has remarked that to construct the desired compactifications, one needs to generalize the notion of shtukas. A simple way to do that is to loosen the last condition in the definition of a shtuka by replacing an isomorphism by a so-called *complete pseudo-homomorphism* or a *complete homomorphism*.

Suppose that S is a scheme over $\mathbb{F}_q$, and $\mathcal{E}$ and $\mathcal{F}$ are two vector bundles of rank 2 over S. Let $\mathcal{L}$ be a line bundle over S together with a global section $l \in \mathrm{H}^0(S, \mathcal{L})$. By definition, a *complete pseudo-homomorphism* of type $(\mathcal{L}, l)$

$$\mathcal{E} \Rightarrow \mathcal{F}$$

from $\mathcal{E}$ to $\mathcal{F}$ is a collection of two morphisms

$$u_1 : \mathcal{E} \longrightarrow \mathcal{F}$$
$$u_2 : \det \mathcal{E} \otimes \mathcal{L} \longrightarrow \det \mathcal{F}$$

verifying

i) $\det u_1 = l u_2$,
ii) u_2 is an isomorphism.

A *complete homomorphism* of type $(\mathcal{L}, l)$

$$\mathcal{E} \Rightarrow \mathcal{F}$$

from $\mathcal{E}$ to $\mathcal{F}$ is a complete pseudo-homomorphism of type $(\mathcal{L}, l)$

$$\mathcal{E} \Rightarrow \mathcal{F}$$

from $\mathcal{E}$ to $\mathcal{F}$ satisfying the following additional condition:

iii) u_1 vanishes nowhere.

Suppose that the global section l is invertible. Hence the couple $(\mathcal{L}, l)$ can be identified with $(\mathcal{O}_S, 1)$. With this identification, a complete pseudo-homomorphism $\mathcal{E} \Rightarrow \mathcal{F}$ of type $(\mathcal{L}, l)$ is just an isomorphism $\mathcal{E} \xrightarrow{\sim} \mathcal{F}$.

Now assume that $l = 0$. Then a complete pseudo-homomorphism (resp. a complete homomorphism) $\mathcal{E} \Rightarrow \mathcal{F}$ of type $(\mathcal{L}, l)$ consists of the following data:

- a maximal line bundle $\mathcal{E}_1$ of $\mathcal{E}$, a maximal line bundle $\mathcal{F}_1$ of $\mathcal{F}$ and an injection

$$\mathcal{E}/\mathcal{E}_1 \hookrightarrow \mathcal{F}_1$$

(resp. an isomorphism $\mathcal{E}/\mathcal{E}_1 \xrightarrow{\sim} \mathcal{F}_1$),
- an isomorphism

$$\det \mathcal{E} \otimes \mathcal{L} \xrightarrow{\sim} \det \mathcal{F}.$$

4.2. The stack DegSht^2 of degenerated shtukas of rank 2 associates to any scheme S the data $\widetilde{\mathcal{E}} = (\mathcal{E}, \mathcal{L}, l, \mathcal{E} \hookrightarrow \mathcal{E}' \hookleftarrow \mathcal{E}'' \Leftarrow \mathcal{E}^\sigma)$ consisting of

- a vector bundle $\mathcal{E}$ of rank 2 over $X \times S$,
- two morphisms $0, \infty : S \longrightarrow X$,
- a modification consisting of two injections

$$\mathcal{E} \hookrightarrow \mathcal{E}' \hookleftarrow \mathcal{E}''$$

of $\mathcal{E}$ by two vector bundles $\mathcal{E}'$ and $\mathcal{E}''$ of rank 2 over $X \times S$ such that the quotients $\mathcal{E}'/\mathcal{E}$ and $\mathcal{E}'/\mathcal{E}''$ are supported by the graphs of ∞ and 0 respectively, and are invertible on their support,
- a line bundle $\mathcal{L}$ over S and a global section $l \in \mathrm{H}^0(S, \mathcal{L})$,

- a complete pseudo-homomorphism of type $(\mathcal{L}, l)^{\otimes(q-1)}$

$$\mathcal{E}^{\sigma} \Rightarrow \mathcal{E}''$$

consisting of a morphism $u_1 : \mathcal{E}^{\sigma} \longrightarrow \mathcal{E}''$ and an isomorphism $u_2 : \det \mathcal{E}^{\sigma} \otimes \mathcal{L}^{q-1} \longrightarrow \det \mathcal{E}''$ such that

  i) $\det u_1 = l^{q-1} u_2$.

  ii) Generically, one can identify $\mathcal{E}''$ with $\mathcal{E}$ and with this identification, one requires that u_1 is not nilpotent.

In the above definition, the condition i) is exactly the condition appeared in the definition of a complete pseudo-homomorphism. However, the condition ii) is an additional condition.

This stack is algebraic in the sense of Artin, *cf.* [Laf98]. As before, we denote by $\mathrm{DegSht}^{2,d}$ the stack classifying degenerated shtukas of rank 2 and of degree d, i.e., one requires that $\deg \mathcal{E} = d$. Hence

$$\mathrm{DegSht}^2 = \coprod_{d \in \mathbb{Z}} \mathrm{DegSht}^{2,d}.$$

Suppose that $\widetilde{\mathcal{E}} = (\mathcal{E}, \mathcal{L}, l, \mathcal{E} \hookrightarrow \mathcal{E}' \hookleftarrow \mathcal{E}'' \Leftarrow \mathcal{E}^{\sigma})$ is a degenerated shtuka of rank 2 over k. Then we can identify $\mathcal{L}$ with the trivial line bundle $\mathcal{O}_{\mathrm{Spec}\,k}$ and l with an element of k.

a) If $l \neq 0$, then l is invertible. We have seen that the complete pseudo-homomorphism $\mathcal{E}^{\sigma} \Rightarrow \mathcal{E}''$ is just an isomorphism $\mathcal{E}^{\sigma} \xrightarrow{\sim} \mathcal{E}''$ and the condition ii) in the above definition is automatically verified. This degenerated shtuka is in fact a Drinfeld shtuka.

b) Otherwise, $l = 0$. We have seen that the complete pseudo-homomorphism $\mathcal{E}^{\sigma} \Rightarrow \mathcal{E}''$ is the data of

- a maximal line bundle $\overline{\mathcal{E}}_1$ of $\overline{\mathcal{E}} = \mathcal{E}^{\sigma}$ and a maximal line bundle $\mathcal{E}''_1$ of $\mathcal{E}''$,
- an injection $w : \overline{\mathcal{E}}/\overline{\mathcal{E}}_1 \hookrightarrow \mathcal{E}''_1$.

We denote by $\mathcal{E}_1$ the maximal line bundle of $\mathcal{E}$ induced by $\mathcal{E}''_1$ using the elementary modification $\mathcal{E} \hookrightarrow \mathcal{E}' \hookleftarrow \mathcal{E}''$. Then the condition ii) in the above definition is equivalent to the condition that $\overline{\mathcal{E}}_1 \cap \mathcal{E}^{\sigma}_1 = 0$ in $\overline{\mathcal{E}}$. Consequently, $\overline{\mathcal{E}}_1 \oplus \mathcal{E}^{\sigma}_1$ is a subbundle of same rank as $\overline{\mathcal{E}}$ and so the quotient $\overline{\mathcal{E}}/\overline{\mathcal{E}}_1 \oplus \mathcal{E}^{\sigma}_1$ is of finite length.

A geometric point $x \in \overline{X}$ is called a *degenerator* of the degenerated shtuka $\widetilde{\mathcal{E}}$ if one of the following conditions is verified:

- x is in the support of $\overline{\mathcal{E}}/\overline{\mathcal{E}}_1 \oplus \mathcal{E}^{\sigma}_1$,
- x is in the support of the injection $w : \overline{\mathcal{E}}/\overline{\mathcal{E}}_1 \hookrightarrow \mathcal{E}''_1$.

Proposition 4.3. *With the above notations, suppose that the canonical Harder-Narasimhan polygon of $\mathcal{E}$ is bounded by a polygon p_0. Then the number of degenerators of $\widetilde{\mathcal{E}}$ is bounded by a function of p_0.*

To prove this proposition, we observe that the injection $w : \overline{\mathcal{E}}/\overline{\mathcal{E}}_1 \hookrightarrow \mathcal{E}_1''$ implies that $\deg \mathcal{E} \leq \deg \overline{\mathcal{E}}_1 + \deg \mathcal{E}_1''$. Since $\deg \mathcal{E}_1'' \leq \deg \mathcal{E}_1 + 1$, then $\deg \mathcal{E} \leq \deg \overline{\mathcal{E}}_1 + \deg \mathcal{E}_1 + 1$. Hence the support of $\overline{\mathcal{E}}/\overline{\mathcal{E}}_1 \oplus \mathcal{E}_1^\sigma$ is of length at most 1.

On the other hand, we have just seen that the support of the injection w is of length bounded by $\deg \overline{\mathcal{E}}_1 + \deg \mathcal{E}_1 + 1 - \deg \mathcal{E}$, hence bounded by $2p_0(1) + 1$. We are done

5. Iterated shtukas of rank 2

Roughly speaking, iterated shtukas are degenerated shtukas satisfying several additional conditions. We will see that iterated shtukas can be "decomposed" as a product of shtukas.

The stack PreSht^2 *of pre-iterated shtukas of rank* 2 associates to any scheme S the data $\widetilde{\mathcal{E}} = (\mathcal{E}, \mathcal{L}, l, \mathcal{E} \hookrightarrow \mathcal{E}' \hookleftarrow \mathcal{E}'' \Leftarrow \mathcal{E}^\sigma)$ consisting of

- a vector bundle $\mathcal{E}$ of rank 2 over $X \times S$,
- two morphisms $\infty, 0 : S \longrightarrow X$,
- a modification consisting of two injections

$$\mathcal{E} \hookrightarrow \mathcal{E}' \hookleftarrow \mathcal{E}''$$

  of $\mathcal{E}$ by two vector bundles $\mathcal{E}'$ and $\mathcal{E}''$ of rank 2 over $X \times S$ such that the quotients $\mathcal{E}'/\mathcal{E}$ and $\mathcal{E}'/\mathcal{E}''$ are supported by the graphs of ∞ and 0 respectively, and are invertible on their support,
- a line bundle $\mathcal{L}$ over S and a global section l of $\mathcal{L}$,
- a complete homomorphism of type $(\mathcal{L}, l)^{\otimes(q-1)}$

$$\mathcal{E}^\sigma \Rightarrow \mathcal{E}''$$

  consisting of a morphism $u_1 : \mathcal{E}^\sigma \longrightarrow \mathcal{E}''$ and an isomorphism $u_2 : \det \mathcal{E}^\sigma \otimes \mathcal{L}^{q-1} \longrightarrow \det \mathcal{E}''$ such that
  i) $\det u_1 = l^{q-1} u_2$.
  i′) u_1 vanishes nowhere.
  ii) Generically, one can identify $\mathcal{E}''$ with $\mathcal{E}$ and with this identification, one requires that u_1 is not nilpotent.

As before, this stack has a stratification containing two strata as follows:

a) The open stratum $\mathrm{PreSht}^{2,\mathrm{open}}$ corresponds pre-iterated shtukas with an invertible global section l. One can then identify $(\mathcal{L}, l)$ with $(\mathcal{O}_S, 1)$ and show that the pre-iterated shtuka is in fact a shtuka. The open stratum is just the stack Sht^2 of shtukas of rank 2.

b) The closed stratum $\mathrm{PreSht}^{2,\mathrm{closed}}$ corresponds to the condition that the global section l is zero. Under this condition, the complete homomorphism $\mathcal{E}^\sigma \Rightarrow \mathcal{E}''$ consists of

- a maximal line bundle $\overline{\mathcal{E}}_1$ of $\overline{\mathcal{E}} = \mathcal{E}^\sigma$ and a maximal line bundle $\mathcal{E}_1''$ of $\mathcal{E}''$,
- an isomorphism $w : \overline{\mathcal{E}}/\overline{\mathcal{E}}_1 \xrightarrow{\sim} \mathcal{E}_1''$.

We consider the substack $\overline{\mathrm{Sht}}^{2,\mathrm{closed}}$ of the closed stratum by imposing the following conditions:

i) If we set $\mathcal{E}'_1 = \mathcal{E}''_1$, then $\mathcal{E}'/\mathcal{E}'_1$ is torsion-free, hence locally free of rank 1 over $X \times S$.

ii) The natural morphism $\mathcal{E}'_1 \longrightarrow \mathcal{E}'/\mathcal{E}$ is surjective, hence the kernel $\mathcal{E}_1$ of this morphism is locally free of rank 1 over $X \times S$.

Drinfeld proved, *cf.* [Dri89]:

Proposition 5.1 (Drinfeld). *There exists a unique substack* $\overline{\mathrm{Sht}}^2$ *of* PreSht^2 *such that* $\overline{\mathrm{Sht}}^2 \cap \mathrm{PreSht}^{2,\mathrm{open}} = \mathrm{Sht}^2$ *and* $\overline{\mathrm{Sht}}^2 \cap \mathrm{PreSht}^{2,\mathrm{closed}} = \overline{\mathrm{Sht}}^{2,\mathrm{closed}}$.

This stack is called *the stack of iterated shtukas of rank 2*. It is an algebraic stack in the sense of Artin. We denote by $\overline{\mathrm{Sht}}^{2,d}$ the stack classifying iterated shtukas of rank 2 and of degree d, i.e., $\deg \mathcal{E} = d$. Hence

$$\overline{\mathrm{Sht}}^2 = \coprod_{d \in \mathbb{Z}} \overline{\mathrm{Sht}}^{2,d}.$$

Observe that if $\widetilde{\mathcal{E}}$ is an iterated shtuka of rank 2 over k in the closed stratum, it has exactly one degenerator.

5.2. The goal of this section is to show how to decompose an iterated shtuka into shtukas. Suppose that $\widetilde{\mathcal{E}} = (\mathcal{E}, \mathcal{L}, l, \mathcal{E} \hookrightarrow \mathcal{E}' \hookleftarrow \mathcal{E}'' \Leftarrow \mathcal{E}^\sigma)$ is an iterated shtuka of rank 2 over a scheme S with the pole ∞ and the zero 0.

a) If $\widetilde{\mathcal{E}}$ lies in the open stratum Sht^2, then $\widetilde{\mathcal{E}}$ is in fact a Drinfeld shtuka.

b) If $\widetilde{\mathcal{E}}$ lies in the closed stratum $\overline{\mathrm{Sht}}^{2,\mathrm{closed}}$, we have constructed different maximal line bundles $\mathcal{E}_1$, $\mathcal{E}'_1$, $\mathcal{E}''_1$, $\overline{\mathcal{E}}_1$ of $\mathcal{E}$, $\mathcal{E}'$, $\mathcal{E}''$, $\overline{\mathcal{E}}$, respectively.

We claim that $(\mathcal{E}_1 \hookrightarrow \mathcal{E}'_1 \hookleftarrow \mathcal{E}^\sigma_1)$ is a shtuka of rank 1 over S with the pole ∞. For the injection $\mathcal{E}_1 \hookrightarrow \mathcal{E}'_1$, one takes the natural one. Since $\mathcal{E}^\sigma_1 \cap \overline{\mathcal{E}}_1 = 0$, the composition

$$\mathcal{E}^\sigma_1 \hookrightarrow \overline{\mathcal{E}}/\overline{\mathcal{E}}_1 \xrightarrow{\sim} \mathcal{E}''_1 = \mathcal{E}'_1$$

is in fact an injection. As $\mathcal{E}'_1/\mathcal{E}_1 \xrightarrow{\sim} \mathcal{E}'/\mathcal{E}$, this shtuka of rank 1 has the same pole ∞ as the iterated shtuka $\widetilde{\mathcal{E}}$.

Next, we claim that $(\mathcal{E}/\mathcal{E}_1 \otimes \mathcal{L} \hookleftarrow \overline{\mathcal{E}}_1 \otimes \mathcal{L}^q \hookrightarrow \overline{\mathcal{E}}/\mathcal{E}^\sigma_1 \otimes \mathcal{L}^q)$ is a shtuka of rank 1 with the zero 0. In fact, the injection $\overline{\mathcal{E}}_1 \otimes \mathcal{L}^q \hookrightarrow \mathcal{E}/\mathcal{E}_1 \otimes \mathcal{L}$ is the composition

$$\overline{\mathcal{E}}_1 \otimes \mathcal{L}^q \xrightarrow{\sim} \mathcal{E}''/\mathcal{E}''_1 \otimes \mathcal{L} \hookrightarrow \mathcal{E}'/\mathcal{E}'_1 \otimes \mathcal{L} \xrightarrow{\sim} \mathcal{E}/\mathcal{E}_1 \otimes \mathcal{L}.$$

The injection $\overline{\mathcal{E}}_1 \otimes \mathcal{L}^q \hookrightarrow \overline{\mathcal{E}}/\mathcal{E}^\sigma_1 \otimes \mathcal{L}^q$ follows from the fact that $\mathcal{E}^\sigma_1 \cap \overline{\mathcal{E}}_1 = 0$ in $\overline{\mathcal{E}}$. It is easy to see that this shtuka has the same zero 0 as that of the iterated shtuka $\widetilde{\mathcal{E}}$.

5.3. We observe that to give a polygon $p : [0, 2] \longrightarrow \mathbb{R}$ is equivalent to give the real number $p(1)$. Suppose that d is an integer and $p : [0, 2] \longrightarrow \mathbb{R}$ a convex polygon which is integral with respect to d. We will define a substack $\overline{\mathrm{Sht}}^{2,d,p}$ of $\overline{\mathrm{Sht}}^{2,d}$ containing $\mathrm{Sht}^{2,d,p}$ as an open dense substack as follows:

Suppose that $\widetilde{\mathcal{E}} = (\mathcal{E}, \mathcal{L}, l, \mathcal{E} \hookrightarrow \mathcal{E}' \hookleftarrow \mathcal{E}'' \Leftarrow \mathcal{E}^{\sigma})$ is an iterated shtuka of rank 2 and of degree d over k. It lies in $\overline{\mathrm{Sht}}^{2,d,p}(k)$ if and only if the following condition is verified:

- If $\widetilde{\mathcal{E}}$ lies in the open stratum $\mathrm{Sht}^{2,d}$, one requires that the canonical Harder-Narasimhan polygon of $\mathcal{E}$ is bounded by p.
- If $\widetilde{\mathcal{E}}$ lies in the closed stratum $\overline{\mathrm{Sht}}^{2,d,\mathrm{closed}}$, one denotes by $\mathcal{E}_1$ (resp. $\overline{\mathcal{E}}_1$) be the maximal line bundle of $\mathcal{E}$ (resp. $\overline{\mathcal{E}}$) defined as before, then one requires

$$\deg \mathcal{E}_1 = \frac{d}{2} + p(1),$$

$$\deg \overline{\mathcal{E}}_1 = \frac{d}{2} - p(1) - 1.$$

Drinfeld proved that there exists a unique substack $\overline{\mathrm{Sht}}^{2,d,p}$ of $\overline{\mathrm{Sht}}^{2,d}$ which associates to each scheme S the data of iterated shtukas $\widetilde{\mathcal{E}}$ of rank 2 and of degree d over S such that for every geometric point s of S, the induced iterated shtuka is in $\overline{\mathrm{Sht}}^{2,d,p}(k)$. It contains $\mathrm{Sht}^{2,d,p}$ as an open dense substack.

Here is the desired theorem, *cf.* [Dri89]:

Theorem 5.4 (Drinfeld). *Suppose that d is an integer and $p : [0, 2] \longrightarrow \mathbb{R}$ a polygon which is big enough with respect to X and integral with respect to d. Then the natural morphism*

$$\overline{\mathrm{Sht}}^{2,d,p} \longrightarrow X \times X$$

is proper. In particular, $\overline{\mathrm{Sht}}^{2,d,p}$ is proper.

6. Valuative criterion of properness

In this section, we will try to sketch Drinfeld's proof of Theorem 5.4 using the semistable reduction à la Langton, *cf.* [Lan75].

6.1. Suppose that A is a discrete valuation ring over $\mathbb{F}_q$. We denote by K its fraction field, κ its residual field, π an uniformizing element and $\mathrm{val} : K \longrightarrow \mathbb{Z} \cup \{\infty\}$ the valuation map.

We denote by X_A (resp. X_K, X_κ) the fiber product $X \times \mathrm{Spec}\, A$ (resp. $X \times \mathrm{Spec}\, K$, $X \times \mathrm{Spec}\, \kappa$). Let A_X be the local ring of the scheme X_A at the generic point of the special fiber X_κ. It is a discrete valuation ring containing π as a uniformizing element. The fraction field K_X of A_X can be identified with the fraction field of $F \otimes K$ and the residual field of A_X can be identified with the fraction field of $F \otimes \kappa$. Finally, A_X (resp. K_X) is equipped with a Frobenius endomorphism Frob induced by $\mathrm{Id}_F \otimes \mathrm{Frob}_A$ (resp. $\mathrm{Id}_F \otimes \mathrm{Frob}_K$).

Since X_A is a regular surface, it is well known that the category of vector bundles of rank r over X_A is equivalent to the category of vector bundles of rank r over the generic fiber X_K equipped with a lattice in its generic fiber. Suppose that E is a vector bundle of rank r over X_A, it corresponds to the couple $(\mathcal{E}, M)$ where $\mathcal{E}$ is the restriction of E to the generic fiber X_K, and M is the local ring of E at the generic point of the special fiber X_κ. Let us denote by

$$(\mathcal{E}, M) \mapsto \mathcal{E}(M)$$

the quasi-inverse functor.

In order to prove the above theorem, we use the valuative criterion of properness, i.e., given a commutative diagram

$$\begin{array}{ccc} \operatorname{Spec} K & \longrightarrow & \overline{\operatorname{Sht}}^{2,d,p} \\ \downarrow & & \downarrow \\ \operatorname{Spec} A & \longrightarrow & X \times X, \end{array}$$

one needs to show that up to a finite extension of A, there exists a unique morphism $\operatorname{Spec} A \longrightarrow \overline{\operatorname{Sht}}^{2,d,p}$ such that after putting it in the above diagram, the two triangles are still commutative. In the rest of this section, we will explain how to construct this map. Its uniqueness which is easier to prove will not be discussed

For simplicity, suppose that the morphism $\operatorname{Spec} K \longrightarrow \overline{\operatorname{Sht}}^{2,d,p}$ factors through the open dense substack $\operatorname{Sht}^{2,d,p}$. Then we write $\widetilde{\mathcal{E}} = (\mathcal{E} \hookrightarrow \mathcal{E}' \hookleftarrow \mathcal{E}'' \xleftarrow{\sim} \mathcal{E}^\sigma)$ for the corresponding shtuka. One needs to extend it to a degenerated shtuka in $\overline{\operatorname{Sht}}^{2,d,p}(A)$.

6.2. Suppose that V is the generic fiber of $\mathcal{E}$. Then it is a vector space of dimension r over K_X equipped with an isomorphism $\varphi : V^\sigma \xrightarrow{\sim} V$ or equivalently an injective semi-linear map

$$\varphi : V \longrightarrow V.$$

This means

- For u and v in V, $\varphi(u + v) = \varphi(u) + \varphi(v)$.
- For $t \in K$ and $v \in V$, $\varphi(tv) = \operatorname{Frob}(t) \cdot \varphi(v)$.

Such a couple (V, φ) is called a φ-*space*.

Suppose that M is a lattice in V. We have seen that it induces vector bundles $\mathcal{E}(M)$, $\mathcal{E}'(M)$ and $\mathcal{E}''(M)$ of rank 2 over X_A extending $\mathcal{E}$, $\mathcal{E}'$ and $\mathcal{E}''$, respectively. Furthermore, one has an induced modification $\mathcal{E}(M) \hookrightarrow \mathcal{E}'(M) \hookleftarrow \mathcal{E}''(M)$ whose quotients are automatically supported by the graphs of the pole and the zero $\infty, 0 : \operatorname{Spec} A \longrightarrow X$.

However, the isomorphism $\mathcal{E}^\sigma \xrightarrow{\sim} \mathcal{E}''$ does not always extend to a complete pseudo-homomorphism $(\mathcal{E}(M))^\sigma \Rightarrow \mathcal{E}''(M)$. This can be done if and only if M is a so-called degenerated lattice. By definition, a lattice M of V is called *degenerated* if the following conditions are satisfied:

- $\varphi(M) \subseteq M$.
- The reduction $\overline{\varphi} : \overline{M} \longrightarrow \overline{M}$ is not nilpotent, where $\overline{M} = M/\pi M$.

If $\varphi(M) = M$, M is called a φ-*lattice*. Otherwise, the degenerated lattice M is called an *iterated lattice*.

Suppose that M is a degenerated lattice. Then it induces a degenerated shtuka

$$\mathcal{E}(M) \hookrightarrow \mathcal{E}'(M) \hookleftarrow \mathcal{E}''(M) \Rightarrow (\mathcal{E}(M))^{\sigma}$$

over X_A which extends the shtuka $\mathcal{E} \hookrightarrow \mathcal{E}' \hookleftarrow \mathcal{E}'' \overset{\sim}{\longleftarrow} \mathcal{E}^{\sigma}$ over the generic fiber X_K.

The restriction to the special fiber X_κ gives a degenerated shtuka

$$\mathcal{E}^M \hookrightarrow \mathcal{E}'^M \hookleftarrow \mathcal{E}''^M \Leftarrow (\mathcal{E}^M)^{\sigma} = \overline{\mathcal{E}}^M$$

whose generic fiber is the quotient $M/\pi M$. If M is a φ-lattice, it is a Drinfeld shtuka. If M is an iterated lattice, it is a degenerated shtuka but not a Drinfeld shtuka. One has then maximal line bundle $\mathcal{E}_1^M$, $\overline{\mathcal{E}}_1^M$ of $\mathcal{E}^M$, $\overline{\mathcal{E}}^M$, respectively with $(\mathcal{E}_1^M)^{\sigma} \cap \overline{\mathcal{E}}_1^M = 0$. One remarks that it is an iterated shtuka if and only if $\deg \mathcal{E}_1^M + \deg \overline{\mathcal{E}}_1^M = \deg \mathcal{E}^M - 1$.

Proposition 6.3.

a) *After a finite extension of A, there exists a degenerated lattice in V. Among these lattices, there is a maximal one noted M_0, i.e., every degenerated lattice is contained in M_0.*

b) *Suppose that M_0 is iterated, hence the image $\overline{\varphi}(M_0/\pi M_0)$ is the unique $\overline{\varphi}$-invariant line of $M_0/\pi M_0$, says l. Then the degenerated lattices in V form a descending chain*

$$M_0 \supsetneq M_1 \supsetneq \cdots \supsetneq M_i \supsetneq M_{i+1} \cdots$$

such that for any $i \in \mathbb{N}$, $M_0/M_i \simeq A_X/\pi^i A_X$ and the image of the natural map $M_i/\pi M_i \longrightarrow M_0/\pi M_0$ is the line l. In particular, every degenerated lattice is iterated.

c) *Suppose that M_0 is a φ-lattice. Then for each $\overline{\varphi}$-invariant line l in $M_0/\pi M_0$, there exists a descending chain*

$$M_0 \supsetneq M_{1,l} \supsetneq \cdots \supsetneq M_{i,l} \supsetneq M_{i+1,l} \cdots$$

such that for any $i \in \mathbb{N}$, $M_0/M_{i,l} \simeq A_X/\pi^i A_X$ and the image of the natural map $M_{i,l}/\pi M_{i,l} \longrightarrow M_0/\pi M_0$ is the line l.

These families cover all degenerated lattices when l runs through the set of invariant lines of $M_0/\pi M_0$. In particular, every degenerated lattice except M_0 is iterated.

6.4. We are now ready to prove the valuative criterion of properness. First, suppose that after a finite extension of A, the φ-space V admits a φ-space M_0. If the shtuka

$$\mathcal{E}^{M_0} \hookrightarrow \mathcal{E}'^{M_0} \hookleftarrow \mathcal{E}''^{M_0} \xleftarrow{\sim} (\mathcal{E}^{M_0})^{\sigma} = \overline{\mathcal{E}}^{M_0}$$

associated with M_0 verifies: for every line bundle $\mathcal{L}$ of $\mathcal{E}^{M_0}$, one has the inequality:

$$\deg \mathcal{L} \leq \frac{\deg \mathcal{E}^{M_0}}{2} + p(1),$$

then we are done. Otherwise, there exists a maximal line bundle $\mathcal{L}$ of $\mathcal{E}^{M_0}$ such that

$$\deg \mathcal{L} > \frac{\deg \mathcal{E}^{M_0}}{2} + p(1).$$

For the polygon p is convex enough, i.e., $p(1)$ is large enough, one can show that the generic fiber l of $\mathcal{L}$ is a $\overline{\varphi}$-invariant line in $M_0/\pi M_0$. One considers the descending chain corresponding to this invariant line l:

$$M_0 \supsetneq M_1 \supsetneq \cdots \supsetneq M_i \supsetneq M_{i+1} \cdots$$

Since M_i is always iterated for $i \geq 1$, one has maximal line bundles $\mathcal{E}_1^{M_i}$ of $\mathcal{E}^{M_i}$. One shows that they form a descending chain:

$$\mathcal{E}_1^{M_0} := \mathcal{L} \supseteq \mathcal{E}_1^{M_1} \supseteq \mathcal{E}_1^{M_2} \supseteq \cdots \supseteq \mathcal{E}_1^{M_i} \supseteq \mathcal{E}_1^{M_{i+1}} \supseteq \cdots$$

with $\deg \mathcal{E}_1^{M_{i+1}} = \deg \mathcal{E}_1^{M_i}$ or $\deg \mathcal{E}_1^{M_{i+1}} = \deg \mathcal{E}_1^{M_i} - 1$. One can show that in the latter case, the degenerated shtuka associated to M_i is iterated.

Using the fact that the canonical Harder-Narasimhan polygon of the generic vector bundle $\mathcal{E}$ is bounded by p, one proves that there exists a positive integer i such that $\deg \mathcal{E}_1^{M_i} = \frac{\deg \mathcal{E}^{M_0}}{2} + p(1)$ and $\deg \mathcal{E}_1^{M_{i+1}} = \frac{\deg \mathcal{E}^{M_0}}{2} + p(1) - 1$. The lattice M_i is the desired one.

We can then suppose that for every finite extension of A, the maximal degenerated lattice in V is always iterated. The key observation is that under this hypothesis, we can suppose that the maximal line bundle $\mathcal{E}_1^{M_0}$ satisfies:

$$\deg \mathcal{E}_1^{M_0} > \frac{\deg \mathcal{E}^{M_0}}{2} + p(1).$$

This claim is in fact the hardest part of the proof. Admitting this result, one can then repeat the above arguments to find the desired lattice. We are done.

7. Another proof using the Geometric Invariant Theory

The previous proof is known as the semistable reduction à la Langton. In fact, Langton [Lan75] used a similar strategy to prove the properness of the moduli space of semistable vector bundles over a smooth projective variety.

Another well-known proof of the above result in the case of smooth projective curves is due to Seshadri: he used the Geometric Invariant Theory. We refer a unfamiliar reader to the excellent book [Mum-For-Kir94] for more details about this theory. Roughly speaking, given a quasi-projective scheme Y equipped with an action of a reductive group G, the Geometric Invariant Theory (or GIT for short) gives G-invariant open subschemes U of Y such that the quotient $U//G$ exists. The fundamental theorem of this theory is as follows: suppose that the action of G can be lifted to an ample line bundle of Y which will be called a *polarization*, one can define the open subsets Y^s and Y^{ss} of stable and semistable points of Y with respect to this polarization: $Y^s \subseteq Y^{ss} \subseteq Y$. Then the quotient $Y^{ss}//G$ exists and it is quasi-projective. The geometric points of the quotient $Y^s//G$ are in bijection with the G-orbits of Y^s. Moreover, if Y is projective, the quotient $Y^{ss}//G$ is also projective.

The definitions of stable and semistable points are quite abstract. Fortunately, in practice, one has a powerful tool to determine these points called *the Hilbert-Mumford numerical criterion.*

When the polarization varies, one gets different GIT quotients. However, one can show that given a couple (Y, G) as above, there is only a finite number of GIT quotients, *cf.* [Dol-Hu98, Tha96].

The strategy suggested by Sheshadri to prove the properness of the moduli space of semistable vector bundles of fixed rank and fixed degree over a smooth projective curve is to realize this moduli space as a GIT quotient of a projective scheme by a reductive group. As an immediate corollary, it is projective, hence proper.

Following a suggestion of L. Lafforgue, the author has successfully applied the GIT method to rediscover the Drinfeld compactifications $\overline{\mathrm{Sht}}^{2,d,p}$ and proved that these compactifications are proper over the product $X \times X$.

To do that, we fix a sufficiently convex polygon p_0. Suppose that N is a finite closed subscheme of the curve X and d is an integer. One first defines *the stack* $\mathrm{DegSht}_N^{2,d}$ *classifying degenerated shtukas of rank 2 and degree d with level structure N.* Next, one introduces the open quasi-projective substack $\mathrm{DegSht}_N^{2,d,p_0}$ by the same truncation process presented in the previous sections. It is equipped with different polarizations of a reductive group, says G, indexed by convex polygons $p \leq p_0$. Under several mild conditions on N and d, the different quotients $\mathrm{DegSht}_N^{2,d,p_0}//G$ are exactly the fiber products $\overline{\mathrm{Sht}}^{2,d,p} \times_{X^2} (X - N)^2$ $(p \leq p_0)$. Moreover, as a by-product of the Geometric Invariant Theory, one can show that these quotients are proper over $(X - N)^2$. One varies the level N to obtain the desired result. For more details, see [NgoDac04].

8. Discussion

Both of the previous approaches can be extended to the higher rank case: the semistable reduction à la Langton is done by Lafforgue [Laf98] and the GIT approach is done in [NgoDac04]. The latter one requires a technical condition that the cardinal of the finite field $\mathbb{F}_q$ is big enough with respect to the rank r.

Recently, for any split reductive group G over $\mathbb{F}_q$ and any sequence of dominant coweights $\underline{\mu} = (\mu_1, \mu_2, \ldots, \mu_n)$, Ngo [Ngo03] and Varshavsky [Var04] have introduced the stack $\mathrm{Sht}_{G,\underline{\mu}}$ of G-shtukas associated to $\underline{\mu}$: it classifies the following data:

- a G-torsor $\mathcal{E}_0$ over X,
- n points $x_1, x_2, \ldots, x_n \in X$,
- modifications $\mathcal{E}_0 \rightsquigarrow \mathcal{E}_1 \rightsquigarrow \cdots \rightsquigarrow \mathcal{E}_n$ such that, for each integer $1 \leq i \leq n$, the modification $\mathcal{E}_i \rightsquigarrow \mathcal{E}_{i+1}$ is of type μ_i at x_i,
- an isomorphism $\mathcal{E}_0^\sigma \xrightarrow{\sim} \mathcal{E}_n$.

One can raise the question of compactifying certain interesting open substacks of finite type of these stacks $\mathrm{Sht}_{G,\underline{\mu}}$. The semistable reduction à la Langton seems to be difficult to generalize. However, the GIT method can be adapted without difficulties to other groups, *cf.* [NgoDac04, NgoDac07].

References

[Dol-Hu98] I. Dolgachev, Y. Hu, *Variation of geometric invariant theory quotients*, Inst. Hautes Études Sci. Publ. Math., **87** (1998), 5–56.

[Dri87] V. Drinfeld, *Varieties of modules of F-sheaves*, Functional Analysis and its Applications, **21** (1987), 107–122.

[Dri89] V. Drinfeld, *Cohomology of compactified manifolds of modules of F-sheaves of rank* 2, Journal of Soviet Mathematics, **46** (1989), 1789–1821.

[Gie77] D. Gieseker, *On the moduli of vector bundles on an algebraic surface*, Annals of Math., **106** (1977), 45–60.

[Har-Nar75] G. Harder, M.S. Narasimhan, *On the cohomology groups of moduli spaces of vector bundles on curves*, Math. Ann., **212** (1975), 215–248.

[Har77] R. Hartshorne, *Algebraic geometry*, Graduate Texts in Mathematics, **52** (1977), Springer-Verlag, New York-Heidelberg.

[Laf97] L. Lafforgue, *Chtoucas de Drinfeld et conjecture de Ramanujan-Petersson*, Astérisque, **243** (1997).

[Laf98] L. Lafforgue, *Une compactification des champs classifiant les chtoucas de Drinfeld*, J. Amer. Math. Soc., **11**(4) (1998), 1001–1036.

[Laf02a] L. Lafforgue, *Cours à l'Institut Tata sur les chtoucas de Drinfeld et la correspondance de Langlands*, prépublication de l'IHES, M/02/45.

[Laf02] L. Lafforgue, *Chtoucas de Drinfeld et correspondance de Langlands*, Invent. Math., **147** (2002), 1–241.

[Lan75] S.G. Langton, *Valuative criteria for families of vector bundles on algebraic varieties*, Annals of Math., **101** (1975), 88–110.

[Lau-MB99] G. Laumon, L. Moret-Bailly, *Champs algébriques*, Ergebnisse der Mathematik **39**, Springer, 1999.

[Mum-For-Kir94] D. Mumford, J. Fogarty, F. Kirwan, *Geometric Invariant Theory*, 3rd enlarged edition, Ergebnisse der Mathematik und ihrer Grenzgebiete **34**, Springer, 1994.

[Ngo03] Ngo Bao Chau, *D-chtoucas de Drinfeld à modifications symétriques et identité de changement de base*, Annales Scientifiques de l'ENS, **39** (2006), 197–243.

[NgoDac04] T. NGO DAC, *Compactification des champs de chtoucas et théorie géométrique des invariants*, to appear in Astérisque.

[NgoDac07] T. NGO DAC, *Compactification des champs de chtoucas: le cas des groupes reductifs*, preprint.

[Tha96]   M. Thaddeus, *Geometric invariant theory and flips*, J. Amer. Math. Soc., **9**(3) (1996), 691–723.

[Var04]   Y. Varshavsky, *Moduli spaces of principal F-bundles*, Selecta Math., **10** (2004), 131–166.

Ngo Dac Tuan
CNRS – Université de Paris Nord
LAGA – Département de Mathématiques
99 avenue Jean-Baptiste Clément
F-93430 Villetaneuse, France
e-mail: ngodac@math.univ-paris13.fr